Plant-Associated Medicinal Nano-Chemistry

Plant-Associated Medicinal Nano-Chemistry is a collection of 15 meticulously curated chapters that interconnects the transformative synergy between plant-derived bioactives and advanced nanotechnological platforms. This book reveals the pathways of green nanoparticle synthesis from plant sources, charting a dense trajectory through their complex applications—from targeted drug release and biosensing to innovative food preservation methods. A necessity for nanomedicine and pharmacognosy visionaries, and environmentally aware chemists, this handbook blends innovation and sustainability.

Salient features:

- The book delivers holistic and in-depth coverage of plant-mediated nanoparticle synthesis with detailed descriptions of biomedical, therapeutic, and diagnostic applications.
- Interdisciplinary integration of various themes related to nanotech, phytochemistry, molecular biology, toxicology, biosensing, food science, and regulatory affairs offers a truly rational approach.
- Inclusion of topics such as nanocellulose production, noscapine-potentiated cancer therapies, and real-time detection of molecular hazards highlights the pioneering innovations in the field of drug development.
- Each chapter reflects the latest experimental findings and theoretical frameworks enriched by the insights of domain experts and promises to deliver scholar-driven excellence.
- Chapters on eco-friendly synthesis routes, emphasizing biodegradable technologies, minimal environmental impact, and ethical standards in nano-biomedicine, support the idea of sustainability.

Plant-Associated Medicinal Nano-Chemistry

Edited by Pankaj Kumar Chaurasia,
Shashi Lata Bharati, Sunita Singh, and
Ashutosh Mani

CRC Press
Taylor & Francis Group
Boca Raton London New York

CRC Press is an imprint of the
Taylor & Francis Group, an informa business

Designed cover image: Pankaj Kumar Chaurasia, Shashi Lata Bharati, Sunita Singh, and Ashutosh Mani

First edition published 2026
by CRC Press
2385 NW Executive Center Drive, Suite 320, Boca Raton FL 33431

and by CRC Press
4 Park Square, Milton Park, Abingdon, Oxon, OX14 4RN

CRC Press is an imprint of Taylor & Francis Group, LLC

ISBN: 978-1-032-90808-3 (hbk)
ISBN: 978-1-032-90809-0 (pbk)
ISBN: 978-1-003-55990-0 (ebk)

DOI: 10.1201/9781003559900

Typeset in Times
by Apex CoVantage, LLC

Contents

Preface

Plant-Associated Medicinal Nano-Chemistry is a collection of 15 chapters contributed by the experts of their field. This book explores the significance of plant-based medicinal nano-chemistry. The chapters present useful and promising medicinal information on the synthesis and bio-medical applications of various types of nanoparticles synthesized using plant sources. Several other bio-medically significant topics viz. the role of plant-associated nanotechnology in biosensoring and bio-medical applications, herb-based medicines' bioavailability, plant-based carbon dots and their biomedical applications, toxicity profiles of nanoparticles synthesized using plants, plant-derived nanocellulose, regulatory and ethical considerations, bio-based nano-emulsions in food preservation, cancer treatment, and herbal nanotechnology in detection of hazardous constituents are also covered comprehensively in this book.

Chapter 1 deals with the comprehensive approach towards the role of plants in the field of nanotechnology. **Chapter 2** explores the biosynthesis of gold nanoparticles using plant sources and their detailed biomedical applications. **Chapter 3** describes the plant-assisted biogenic synthesis of silver nanoparticles having valuable pharmaceutical significances. **Chapter 4** involves the detailed discussions on the plant-based synthesis of non-noble metal nanoparticles with their notable medicinal applications. The role of plant-based nanotechnology in the biosensoring and medicinal applications has been explored in **Chapter 5.** The significance of the nanotechnology in enhancing herb-based medicines' bioavailability and activity has been explored in **Chapter 6**. Toxicity profiles of nanoparticles prepared using plant sources have been discussed in **Chapter 7**. Plant-based carbon dots and their bio-medical applications have been explored in **Chapter 8**. The plant-based biogenic synthesis of nanoparticles of the d^8 (nickel, palladium, and platinum) system along with their medicinal roles have been described in **Chapter 9**. **Chapter 10** deals with the plant-assisted biogenic synthesis of medicinally significant zinc oxide nanoparticles. **Chapter 11** deals with the extraction techniques and therapeutic applications of plant-derived nanocellulose. Regulatory and ethical considerations in plant-based nanomedicine are explored in **Chapter 12**. The role of bio-based nanoemulsions in food preservation is discussed in **Chapter 13**. A unique topic, 'Revolutionizing Cancer Treatment: Nanoparticle-Enhanced Noscapine for Management of Cancer', is covered as **Chapter 14**. Lastly, **Chapter 15** describes the role of herbal nanotechnology in the detection of health hazardous molecules.

BOOK EDITORS
Dr. Pankaj Kumar Chaurasia
Dr. Shashi Lata Bharati
Dr. Sunita Singh
Dr. Ashutosh Mani

About the Editors

Pankaj Kumar Chaurasia, Ph.D. is Assistant Professor in the PG Department of Chemistry, L.S. College, B.R. Ambedkar Bihar University, Muzaffarpur. He has more than seven years of experience in teaching at postgraduate (PG) and undergraduate (UG) levels. He has a good academic as well as research career. He qualified 'National Eligibility Test' in 2009 as CSIR-JRF (NET) and has been awarded with SRF-NET in 2012. He received Ph.D. in chemistry. He also worked as guest faculty (2016–2017) in Department of Chemistry, University of Allahabad, Prayagraj (A Central University of India). He was awarded 'Research Associateship' by CSIR-New Delhi (India) in 2017 and worked as CSIR-RA in Motilal Nehru NIT, Allahabad/Prayagraaj (India). In his research career, he has more than 80 publications out of which more than 50 journal publications, 25 book chapters, and 9 books, and many of the publications are in the pipeline. He has guided one Ph.D. student and is also guiding three Ph.D. research scholars. He has worked in different areas of chemistry and has research expertise in biological chemistry, biotechnology, microbiology, enzymology, organometallic chemistry, and environmental chemistry.

Shashi Lata Bharati (she/her), Ph.D. is Assistant Professor in the Department of Chemistry, Northeastern Regional Institute of Science and Technology, Nirjuli, Arunachal Pradesh (India). She has a good academic as well as research career. She was awarded 'UGC-DSA fellowship' for meritorious students during her Ph.D. programme. She obtained her Ph.D. degree in chemistry. She was awarded with UGC-post doctoral fellowship for women in 2013 by UGC New Delhi and worked as a post-doctoral fellow in the Department of Chemistry of DDU Gorakhpur University, Gorakhpur (India).

She has published more than 55 publications and 8 edited books with national and international journals/publishers of repute. She has guided many M.Sc. project students and one Ph.D. research scholar. She has expertise in the field of inorganic chemistry, organometallic chemistry, biological chemistry, and enzymology.

Sunita Singh is Assistant Professor of Chemistry in the Navyug Kanya Mahavidyalaya, University of Lucknow since 2019. She pursued her Ph.D. in the year 2015 on the topic, 'Chemistry, antioxidant and antimicrobial activities of essential oils and oleoresins of spices' from DDU Gorakhpur University, Gorakhpur. She was awarded junior research and senior research fellowships from the University Grant Commission. She also worked as Research Assistant (2010–2011) on the project sponsored by CST 'Chemistry, Antioxidant and Antimicrobial Activities of Oleoresins Extracted from Cardamom, Black Pepper and Caraway'. Her work was mainly focused on quantitative and qualitative analyses of essential oils and oleoresins of spices namely *Piper nigrum*, *Nigella sativa*, *Mentha longifolia*, *Anethum graveolans*, *Brassica juncea*, and *Sinapis alba*. Chemistry, antioxidant and antimicrobial efficacies of essential oil and oleoresins were investigated using different techniques.

She has published 38 research articles in journals of national and international repute. She has authored and coauthored nine book chapters with national and international publishing houses. She has edited books with reputed publishers. She is also on the review boards of journals namely *Advances in Clinical Toxicology*, *MOJ Food Processing & Technology*, and *Pharmaceutical Drug Regulatory Affairs Journal*. She has attended in total 30 conferences and webinars of national and international repute and delivered more than 15 invited talks and oral presentations. She has presented her work at various global events. She is also an active member of the Association of Chemistry Teachers (India), the International Clinical Aromatherapy Network, and the Global Harmonization Initiative.

Ashutosh Mani earned his M.Sc. and a Ph.D. in bioinformatics from the University of Allahabad, India. He worked as Visiting Fellow at the School of Life Sciences, Jawaharlal Nehru University New Delhi, India to complete part of his Ph.D. work. He joined as a MIUR fellow at the University of Cagliari, Italy and is currently Associate Professor in the Department of Biotechnology, Motilal Nehru National Institute of Technology Allahabad, India. He has more than 12 years' teaching and research experience in the field of molecular and structural biology, genomics, proteomics, and computational biology. His areas of interest include herbal drug discovery against cancer, Alzheimer's disease and diabetes, and plant stress tolerance. Apart from editing two books, he has published more than 50 articles in national and international journals of repute. He has supervised eight Ph.D. students and several undergraduate and postgraduate students. He has completed six sponsored research projects and has received research grants from funding agencies like DST, SERB, CSTUP, and DBT. He has organized several international conferences and workshops in the area of biotechnology. He is guest editor and reviewer of various international journals in the area of life sciences. He has reviewed research project proposals for leading funding agencies.

Contributors

Gholamreza Abdi
Department of Biotechnology, Persian Gulf Research Institute
Persian Gulf University
Bushehr, Iran

Surendra Agrawal
Datta Meghe College of Pharmacy
Datta Meghe Institute of Higher Education, Sawangi (M)
Wardha, India

Cristóbal Noe Aguilar
Bioprocesses and Bioproducts Research Group, Food Research Department, School of Chemistry
Universidad Autónoma de Coahuila
Saltillo, México

Yogita Ale
Department of Pharmaceutics, Uttaranchal Institute of Pharmaceutical Sciences
Uttaranchal University, Dehradun
Uttarakhand, India

Ayman Younes Allam
Department of Food Science and Technology, Faculty of Agriculture, Menoufia University,
Shibin El Kom, Egypt

Muyideen Olaitan Bamidele
Bioprocesses and Bioproducts Research Group, Food Research Department, School of Chemistry
Universidad Autónoma de Coahuila
Saltillo, México

Nilendu Basak
Department of Microbiology, University of Kalyani, Kalyani, Nadia, West Bengal, India
Department of Civil and Environmental Engineering, Indian Institute of Technology Delhi,
New Delhi, India

Laxmi Kant Bhardwaj
Amity Institute of Environmental Toxicology, Safety and Management,
Amity University, Noida,
Uttar Pradesh, India

Ambar Bongirwar
MedPharm Ltd., Durham, NC, USA

Mansi Butola
Department of Pharmaceutics, Uttaranchal Institute of Pharmaceutical Sciences, Uttaranchal University,
Dehradun, Uttarakhand, India

Mahadev Chambhare
Department of Botany, Amruteshwar Arts, Commerce & Science College, Vinzar,
Pune, India

Chetan Singh Chauhan
Bhupal Noble's University
Udaipur, Rajasthan, India

Pankaj Kumar Chaurasia
P.G. Department of Chemistry, L.S. College, B.R. Ambedkar Bihar University, Muzaffarpur
Bihar, India

Monica Lizeth Chavez
Bioprocesses and Bioproducts Research Group, Food Research Department, School of Chemistry
Universidad Autónoma de Coahuila
Saltillo, México

Atif Aziz Chowdhury
Faculty of Agricultural, Environmental and Food Sciences, Free University of Bozen-Bolzano
Piazza Università, 1
Bolzano, Italy

Sandoval Jose Cortes
Bioprocesses and Bioproducts Research Group, Food Research Department, School of Chemistry
Universidad Autónoma de Coahuila
Saltillo, México

Urvee Dave
Department of Pharmacy, Bhupal Noble's University, Udaipur
Rajasthan, India

Sumit Durgapal
Uttaranchal Institute of Pharmaceutical Sciences, Uttaranchal University, Dehradun
Uttarakhand, India

María Liliana Flores-López
Centro de Investigacion e Innovacion Científica y Tecnológica, Universidad Autónoma de Coahuila
Saltillo, México

K. Gayathri
Department of Microbiology, Sacred Heart College (Autonomous), Tirupattur
Tamil Nadu, India

Arun Ghuge
Department of Biology, Government Institute of Forensic Science, Rabindranath Tagore Road, Civil Lines
Nagpur, India

Himanko Gogoi
Department of Pharmaceutics, Uttaranchal Institute of Pharmaceutical Sciences, Uttaranchal University, Dehradun
Uttarakhand, India

Indarchand Gupta
Department of Biotechnology, Government Institute of Science, Nipatniranjan Nagar, Caves Road, Chhatrapati Sambhajinagar
India

Neeraj Mohan Gupta
Department of Chemistry, Govt. Postgraduate College,
Guna, India

Harshita Jain
Amity Institute of Environmental Sciences, Amity University, Noida, Uttar
Pradesh, India

Jyotirmayee
School of Biotechnology, Institute of Science, Banaras Hindu University, Varanasi
Uttar Pradesh, India

Rahul Kamble
Department of Chemistry, Amruteshwar Arts, Commerce & Science College
Pune, India

Harpreet Kaur
University Institute of Pharmaceutical Sciences, Panjab University
Chandigarh, India

K. Kesavardhini
Department of Microbiology, Sacred Heart College (Autonomous), Tirupattur
Tamil Nadu, India

Shweta Khanna
Department of Physics and Nanotechnology, SRM University, Chennai, Kattankulathur
Tamil Nadu, India

Hrishikesh Khodade
PG Research Centre in Botany, PDEA's Prof. Ramkrishna More ACS College, Pune, India
Affiliated to Savitribai Phule Pune University, Pune
Department of Botany, S.M. Joshi College, Hadapsar, Pune, India

Rakesh K Kore
Department of Pharmacy, Guru Ghasidas University
Bilaspur, India

Gauree Kukreti
School of Pharmaceutical Sciences and Technology, Sardar Bhagwan Singh University, Balawala, Dehradun,
Uttarakhand, India

B. Lokeshwari
Department of Microbiology, Sacred Heart College (Autonomous), Tirupattur
Tamil Nadu, India

Ayush Madan
Department of Biotechnology, School of Research & Technology, People's University, Bhanpur, Bhopal
Madhya Pradesh, India
Institute of Tropical Aquaculture and Fisheries (AKUATROP), Universiti Malaysia Terengganu, Kuala Nerus, Terengganu, Malaysia

M. Manigandan
Department of Microbiology, Sacred Heart College (Autonomous), Tirupattur
Tamil Nadu, India

Abhishek Kumar Mishra
School of Pharmacy, The Assam Kaziranga University, Jorhat
Assam, India

Shachi Mishra
Department of Chemistry, Jai Prakash University, Chapra
Bihar, India

Pradeep K Naik
Research Centre of Excellence in Natural Products and Therapeutics, Department of Biotechnology and Bioinformatics, Sambalpur University, Jyoti Vihar, Burla, Sambalpur
Odisha, India

Shashi Pandey
Department of Chemistry, University of Lucknow, Lucknow
Uttar Pradesh, India

Neetu Panwar
School of Applied Sciences, Shri Venkateshwara University
Gajraula, India

Jitendra Patel
Datta Meghe College of Pharmacy, Datta Meghe Institute of Higher Education, Sawangi (M)
Wardha, India

Vijay Kumar Patel
Ram Sharan Roy College of Pharmacy, Panapur, Makkanpur
Vaishali, Bihar

Sayantani Paul
Department of Chemistry, University of Kalyani, Kalyani, Nadia
West Bengal, India—

Olga Berenice Álvarez Pérez
Greencorp Biorganiks
México

Pratyush Pragyandipta
Research Centre of Excellence in Natural Products and Therapeutics, Department of Biotechnology and Bioinformatics, Sambalpur University, Jyoti Vihar, Burla, Sambalpur
Odisha, India

Rashmi Priya
Department of Chemistry, Jai Prakash University
Chapra, Bihar, 841302, India

Harish Rajak
Department of Pharmacy, Guru Ghasidas University
Bilaspur, India

Shilpa Rana
Department of Pharmacology, Uttaranchal Institute of Pharmaceutical Sciences, Uttaranchal University, Dehradun
Uttarakhand, India

Sonali Rastogi
Chitkara College of Pharmacy, Chitkara University, Rajpura, Punjab—140401, India; Department of Pharmacy, Venkateshwara Group of Institutions, Jatoli, Roorkee—bypass road
Meerut, India

Prangya Rath
Amity Institute of Environmental Sciences, Amity University, Noida,
Uttar Pradesh, India

Kundan Singh Rawat
Prof. Rajendra Singh Nanoscience and Nanotechnology Centre, Department of Chemistry D.S.B. Campus, Kumaun University Nainital
Uttarakhand, India

Sandeep Sangle
Department of Forensic Biology, Government Institute of Forensic Science, Nipatniranjan Nagar, Caves Road
Chhatrapati Sambhajinagar, India

Megha Sapkota
Department of Forensic Science, JSPM University Pune, Wagholi
Pune, India

P. Saranraj
Department of Microbiology, Sacred Heart College (Autonomous), Tirupattur
Tamil Nadu, India

Vanshika Sautha
Department of Pharmaceutics, Uttaranchal Institute of Pharmaceutical Sciences, Uttaranchal University, Dehradun
Uttarakhand, India

Pranshul Sethi
Department of Pharmacology, College of Pharmacy, Shri Venkateshwara University, Gajraula, India; Chitkara College of Pharmacy, Chitkara University, Rajpura
Punjab, India

Avanish Chandra Sharma
Department of Chemistry, Navyug Kanya Mahavidyalaya, University of Lucknow, Lucknow
Uttar Pradesh, India

Divya Sharma
Department of Biotechnology, St. John's College, Dr. B.R. Ambedkar University, Agra
Uttar Pradesh, India

Deepak Shelke
Department of Botany, Amruteshwar Arts, Commerce & Science College, Vinzar
Pune, India

Ekta Shirbhate
Department of Pharmacy, Guru Ghasidas University
Bilaspur, India

Alka Singh
School of Pharmaceutical Sciences and Technology, Sardar Bhagwan Singh University, Balawala, Dehradun,
Uttarakhand, India

Paramjeet Singh
Aurobindo Pharma, Durham, NC, USA

Sunita Singh
Department of Chemistry, Navyug Kanya Mahavidyalaya, University of Lucknow, Lucknow
Uttar Pradesh, India

Vaibhav Singh
Department of Pharmacy, Guru Ghasidas University
Bilaspur, India

Vartika Singh
Amity Institute of Global Warming & Ecological Study, Amity University Noida
Uttar Pradesh, India

Hiralal Sonawane
PG Research Centre in Botany, PDEA's Prof. Ramkrishna More ACS College, Pune, MS, India, Affiliated to Savitribai Phule Pune University
Pune, India

Ahmed A. Tayeld
Department of Fish Processing and Biotechnology, Faculty of Aquatic and Fisheries Sciences, Kafrelsheikh University
Kafrelsheikh, Egypt

Akanksha Upadhyay
Department of Chemistry, Anugrah Narayan College, Patliputra University
Patna, India

Abhinav Verma
Department of Pharmaceutics, School of Pharmacy, Shri Venkateshwara University
Gajraula, India

Malkhey Verma
School of Biotechnology, Institute of Science, Banaras Hindu University, Varanasi
Uttar Pradesh, India

Shalu Verma
Uttaranchal Institute of Pharmaceutical Sciences, Uttaranchal University, Dehradun
Uttarakhand, India

Subham K. Vishwakarma
Department of Pharmacy, Guru Ghasidas University
Bilaspur, India

1 Introduction on Plants in Nano-Biotechnology

Ayman Younes Allam, Gholamreza Abdi, Sunita Singh, Avanish Chandra Sharma, and Ahmed A. Tayel

1.1 INTRODUCTION TO NANO-BIOTECHNOLOGY

Nano-biotechnology, an interdisciplinary science integrating nanotechnology and biology, has revolutionized various fields, particularly medicine, agriculture, and environmental sciences (Iravani, 2011; Khan et al., 2020a). By manipulating materials at the nanometer scale (1–100 nm), scientists harness unique physicochemical properties such as high surface-to-volume ratios, enhanced reactivity, and quantum confinement effects (Chaudhuri & Paria, 2012; Rai et al., 2018b). These characteristics enable the design of advanced materials with superior performance for biomedical devices, drug delivery systems, biosensors, and eco-friendly agricultural practices (Bhattacharya & Mukherjee, 2008; Nasrollahzadeh et al., 2020b). Traditionally, the synthesis of nanoparticles involved physical and chemical methods that often required toxic solvents, high energy input, and environmentally hazardous reagents (Ahmed et al., 2016b; Barani et al., 2021a). Consequently, there has been a growing shift toward "green synthesis" approaches, which employ biological systems such as plants, microorganisms, and enzymes to produce nanoparticles under mild, eco-friendly conditions (Mittal et al., 2013; Iravani et al., 2014). Among biological systems, plants have emerged as particularly advantageous for nanoparticle synthesis due to their wide availability, rapid production capability, and rich phytochemical composition (Khalil et al., 2012; Narayanan & Sakthivel, 2010). Phytochemicals—including flavonoids, terpenoids, saponins, alkaloids, and phenolic acids—play dual roles as reducing and stabilizing agents during nanoparticle fabrication (Singh et al., 2018; Aziz et al., 2021). These natural compounds reduce metal ions into zero-valent nanoparticles and prevent their aggregation through capping mechanisms (Iravani, 2011; Rai & Ingle, 2012). The plant-mediated synthesis of nanoparticles aligns closely with green chemistry principles, offering several advantages such as simplicity, cost-effectiveness, scalability, and the elimination of hazardous by-products (Thakkar et al., 2010; Awwad & Salem, 2012). Factors such as plant species, extraction method, pH, temperature, and precursor concentration significantly influence the morphology, size, and stability of the synthesized nanoparticles (Fariq et al., 2017; Makarov et al., 2014). Numerous plant species have been successfully employed to produce metal and metal oxide nanoparticles. For instance, *Azadirachta indica* (neem), *Moringa oleifera*, *Aloe vera*, and *Camellia sinensis*

DOI: 10.1201/9781003559900-1

(green tea) have been used to synthesize silver, gold, zinc oxide, and titanium dioxide nanoparticles with potent antibacterial, antioxidant, and anticancer properties (Huang et al., 2007; Elbagory et al., 2016; Kharissova et al., 2013). The applications of plant-derived nanoparticles span several sectors. In medicine, they facilitate targeted drug delivery, photothermal therapy, antimicrobial treatments, and diagnostic imaging (Bhattacharya & Mukherjee, 2008; Zhang et al., 2016). In agriculture, nanoparticles enhance fertilizer efficiency, pest control, and crop yield through smart delivery systems (Kah et al., 2019; Pérez-de-Luque, 2017). Environmental applications include water purification, pollutant degradation, and heavy metal removal from wastewater (Nasrollahzadeh et al., 2019; Mishra et al., 2018). Despite the promising prospects, several challenges hinder the full-scale commercialization of plant-based nanoparticle synthesis. Issues such as variability in nanoparticle size, lack of standardization, scalability concerns, and potential cytotoxicity must be thoroughly addressed (Iravani et al., 2014; Shaikh et al., 2019a). Moreover, comprehensive toxicological assessments and regulatory frameworks are needed to ensure environmental and human safety (Siddiqi et al., 2018; Sharma et al., 2019a). Nano-biotechnology represents the intersection of nanotechnology and biological sciences, focusing on manipulating materials at atomic and molecular scales (Khan et al., 2020b; Rai et al., 2018a). At nanoscale dimensions, materials exhibit unique physicochemical properties, including enhanced surface area, reactivity, and quantum effects (Chaudhuri & Paria, 2012; Singh et al., 2021). These distinctive features enable applications across biomedicine, agriculture, environmental remediation, and food industries (Iravani, 2011; Nasrollahzadeh et al., 2020b). Plants have emerged as critical agents in green nano-biotechnology, offering eco-friendly routes for nanoparticle synthesis (Iravani, 2011; Khalil et al., 2012). Conventional physical and chemical synthesis methods often involve toxic chemicals, high energy inputs, and environmental hazards (Ahmed et al., 2016a; Barani et al., 2021b). In contrast, plant-mediated synthesis utilizes natural phytochemicals such as flavonoids, alkaloids, terpenoids, and phenolic acids as reducing and stabilizing agents (Rai & Yadav, 2013; Mittal et al., 2013), aligning with green chemistry principles.

The green synthesis approach is cost-effective, sustainable, and scalable (Iravani et al., 2014; Thakkar et al., 2010). Plant extracts simplify nanoparticle fabrication by eliminating the need for complex equipment, toxic solvents, or hazardous reaction conditions (Shankar et al., 2004; Aziz et al., 2021). Additionally, the diversity of phytochemicals across plant species allows control over nanoparticle size, shape, crystallinity, and surface chemistry (Narayanan & Sakthivel, 2010; Singh et al., 2018), enhancing their functional versatility. Several mechanisms govern nanoparticle formation through plants. Initially, metal ions interact with biomolecules, leading to reduction, nucleation, growth, and stabilization phases (Mittal et al., 2013; Iravani, 2011). Key factors influencing these processes include pH, temperature, metal salt concentration, and the specific plant species used (Fariq et al., 2017; Makarov et al., 2014). Plant-derived nanoparticles have demonstrated vast applications. In medicine, they are utilized for targeted drug delivery, antimicrobial coatings, cancer therapy, and imaging (Bhattacharya & Mukherjee, 2008; Zhang et al., 2016). In agriculture, nano-formulations improve nutrient delivery, pest control, and crop protection

(Kah et al., 2019; Pérez-de-Luque, 2017). Environmental applications include wastewater treatment and heavy metal remediation (Nasrollahzadeh et al., 2019; Mishra et al., 2018). Moringa oleifera, Azadirachta indica, and Aloe vera are among widely studied plants for nanoparticle synthesis (Kharissova et al., 2013; Awwad & Salem, 2012). Their rich phytochemical content facilitates rapid and efficient nanoparticle production (Rai & Ingle, 2012; Elbagory et al., 2016). The resulting nanoparticles, such as silver, gold, zinc oxide, and titanium dioxide, exhibit potent antibacterial, antioxidant, and catalytic properties (Ahmed et al., 2016a; Huang et al., 2007). Despite significant advances, challenges persist. Standardization of synthesis protocols, nanoparticle reproducibility, scalability, and biosafety evaluations remain critical bottlenecks (Iravani et al., 2014; Singh et al., 2021). Furthermore, thorough toxicological assessments are needed to ensure the safe deployment of plant-synthesized nanoparticles (Siddiqi et al., 2018; Shaikh et al., 2019a). Future research must emphasize mechanistic insights, optimize green synthesis protocols, and expand the exploration of novel plant species for nanoparticle fabrication (Chugh & Kumar, 2018; Zahin et al., 2020). Integrating omics technologies with nano-biotechnology could open new avenues for the rational design of plant-based nanomaterials (Kuppusamy et al., 2016a; Sharma et al., 2019b). In conclusion, plants offer a promising, sustainable platform for advancing nano-biotechnology. By merging the natural biochemical diversity of plants with nanoscale engineering, green nano-biotechnology holds tremendous potential for revolutionizing multiple sectors, from healthcare to environmental conservation (Chaudhuri & Paria, 2012; Iravani, 2011).

Future research directions should focus on elucidating the molecular mechanisms underlying plant-mediated nanoparticle synthesis, exploring a broader diversity of plant species, integrating omics technologies, and optimizing large-scale production protocols. Such efforts will not only enhance the scientific understanding of green nanotechnology but also accelerate its practical applications across multiple industries. In summary, plants represent a sustainable, versatile, and powerful platform for nano-biotechnology, offering promising solutions to global challenges in health, agriculture, and environmental management (Figure. 1.1).

1.1.1 Definition and Scope

Nano-biotechnology is a multidisciplinary field that merges concepts from biology, chemistry, physics, and materials science to develop nanoscale tools and devices for biomedical, agricultural, and environmental applications (Siddiqi et al., 2020). It involves the synthesis, characterization, and application of materials with dimensions typically ranging from 1 to 100 nanometers, where unique physical and chemical properties emerge (Barani et al., 2021a). At this scale, quantum effects, high surface-area-to-volume ratios, and enhanced reactivity allow for novel functionalities that are unattainable in bulk materials (Nasrollahzadeh et al., 2021). Plant-based nano-biotechnology, a subfield often termed "green nanotechnology," focuses on using plant systems—such as leaves, stems, roots, seeds, and flowers—for the biosynthesis of nanoparticles. The phytochemicals present in plants, including alkaloids, phenolics, terpenoids, and flavonoids, act as natural reducing, capping, and

FIGURE 1.1 Nano-biotechnology, an interdisciplinary science integrating nanotechnology and biology.

stabilizing agents during nanoparticle formation (Abdallah et al., 2022; Khan et al., 2023). This method provides an eco-friendly, cost-effective, and scalable alternative to traditional physical and chemical nanoparticle synthesis methods that often involve toxic chemicals and high energy consumption (El-Saadony et al., 2021; Singh et al., 2024). The scope of nano-biotechnology extends beyond particle synthesis. It includes the development of nanocarriers for targeted drug delivery, nanobiosensors for disease diagnostics, nanofertilizers for sustainable agriculture, and nanoremediation technologies for environmental cleanup (Alsharif et al., 2022; Almalki et al., 2023). Recent advances also encompass gene editing tools like CRISPR-Cas systems, where nanoscale delivery platforms are improving gene transfer efficiency (Chaudhary et al., 2022; Wani et al., 2023). Moreover, nano-biotechnology intersects with emerging disciplines such as synthetic biology, precision agriculture, personalized medicine, and food packaging technologies (Mukherjee et al., 2022; Duhan et al., 2022). In this context, plants are increasingly viewed as natural bioreactors capable of producing functionalized nanoparticles with enhanced therapeutic or agricultural potential (Kumar et al., 2023; Ojha et al., 2024). Thus, the definition and scope of nano-biotechnology highlight its transformative potential to address critical challenges in global health, food security, and environmental sustainability using innovative, biologically inspired approaches (Figure 1.2).

1.1.2 Importance and Applications

The integration of plants into nano-biotechnology offers remarkable advantages in multiple domains due to their inherent biochemical diversity, renewable nature, and

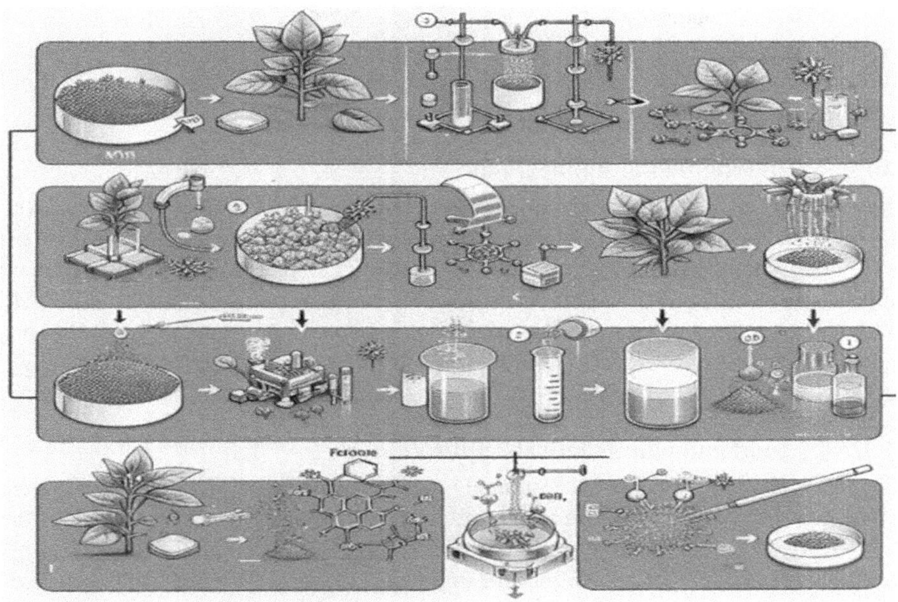

FIGURE 1.2 Nano-biotechnology is a multidisciplinary field that merges concepts from biology, chemistry, physics, and materials science to develop nanoscale tools and devices for biomedical, agricultural, and environmental applications.

compatibility with green chemistry principles (Sivasankarapillai et al., 2020; Singh et al., 2022).

1.1.2.1 Medical Applications

In medicine, plant-derived nanoparticles have opened new frontiers in diagnostics, therapeutics, and regenerative medicine. Nanoparticles synthesized using *Moringa oleifera*, *Azadirachta indica*, and *Curcuma longa* extracts have demonstrated potent anticancer, antibacterial, and anti-inflammatory activities (Rani et al., 2021; Shaik et al., 2022). Targeted drug delivery systems using gold and silver nanoparticles enhance the bioavailability of chemotherapeutic agents while minimizing systemic toxicity (Alam et al., 2021; Balashanmugam et al., 2023). Plant-synthesized nanoparticles are also being utilized in photothermal therapy, bioimaging, wound healing, and antiviral treatments, including innovative therapies against emerging viral infections (Chowdhury et al., 2022; Khalil et al., 2024).

1.1.2.2 Agricultural Applications

In agriculture, nano-formulations based on plant-derived nanoparticles offer sustainable solutions to enhance crop productivity while reducing environmental impact (Subramanian et al., 2021; Vishwakarma et al., 2022). Nano-fertilizers and nano-pesticides improve nutrient use efficiency and pest management, reducing the excessive use of conventional agrochemicals (Awan et al., 2023; Singh et al., 2024). Moreover, plant-mediated nanoparticles are applied for seed priming, soil remediation, and plant disease management, contributing to climate-resilient agricultural practices (Palanikumar et al., 2022; Abdul Samad et al., 2023).

1.1.2.3 Environmental Applications

Environmental applications include water purification, pollutant degradation, and heavy metal detoxification. Green-synthesized iron oxide, titanium dioxide, and silver nanoparticles from plant extracts are effective catalysts in wastewater treatment and environmental remediation strategies (Fatima et al., 2022; Khan et al., 2024). Plant-based nanomaterials are also employed in biosensors for detecting pollutants, enabling early environmental monitoring and mitigation strategies (Bhatia et al., 2022).

1.1.2.4 Food and Nutraceutical Applications

The incorporation of nano-encapsulated bioactives from plants enhances food preservation, shelf life, and nutritional profiles (Ravichandran et al., 2021; Al-Shabib et al., 2023). Nanotechnology applications in edible coatings, intelligent packaging, and nutraceutical delivery are gaining momentum, offering clean-label and functional food innovations (Sathiyabama & Manikandan, 2023).

1.1.2.5 Challenges and Future Perspectives

Despite these promising applications, challenges such as nanoparticle toxicity, standardization of green synthesis protocols, and regulatory issues must be addressed (Hashemi et al., 2022; Shafiq et al., 2023). The translation from laboratory-scale research to industrial applications requires multidisciplinary collaborations and the establishment of safety evaluation frameworks (Khan et al., 2022; Sharma et al., 2024). In conclusion, the importance of plant-based nano-biotechnology lies in its potential to drive sustainable innovations across health, agriculture, food, and environmental sectors. Future research should focus on expanding plant species diversity, mechanistic insights into nanoparticle synthesis, and risk assessment to fully harness the benefits of this emerging technology (Mishra et al., 2022; Iravani et al., 2023).

1.1.3 Plants as Essential Components in Nano-Biotechnology

The integration of plants into nano-biotechnology represents a transformative approach to the green synthesis of nanoparticles (NPs), offering sustainable and eco-friendly alternatives to conventional physical and chemical methods. Plants act as natural bio-factories, harnessing their rich phytochemical profiles to mediate the synthesis, stabilization, and functionalization of nanoparticles in a single-step process (Iravani, 2011). This method not only minimizes environmental hazards but also reduces production costs, making it a promising platform for large-scale applications. Plant extracts derived from various parts such as leaves, roots, stems, flowers, fruits, and seeds are rich in bioactive compounds, including flavonoids, terpenoids, alkaloids, phenolic acids, and proteins, which facilitate the reduction of metal ions and cap the resulting nanoparticles (Kuppusamy et al., 2016a). These phytochemicals simultaneously act as reducing and stabilizing agents, eliminating the need for hazardous reagents typically used in traditional nanoparticle synthesis (Iravani, 2011; Mittal et al., 2013). Several plant species have been successfully employed in nanoparticle synthesis. For example, *Azadirachta indica* (neem) has been utilized for the synthesis of silver and gold nanoparticles, demonstrating strong antimicrobial activities (Ahmed et al., 2016b). Similarly, *Camellia sinensis* (green tea) extracts, rich in catechins and other polyphenols, have been widely used to synthesize gold and silver nanoparticles with potent antioxidant properties (Mittal et al., 2013). Other plants such as *Pelargonium graveolens* (geranium) and various *Citrus* species have also been reported to mediate rapid and efficient nanoparticle formation (Kuppusamy et al., 2016b). The general mechanism of plant-mediated nanoparticle synthesis involves three primary steps: (1) reduction of metal ions by phytochemicals, (2) nucleation and growth of nanoparticles, and (3) stabilization through capping by biomolecules (Ahmed et al., 2016a). The process is influenced by several factors, including the pH of the extract, temperature, reaction time, and the concentration of metal precursors and plant metabolites.

Plant-based nanoparticles have found extensive applications across multiple fields. In biomedicine, they exhibit antimicrobial, anticancer, and anti-inflammatory activities, making them attractive candidates for drug delivery and therapeutic interventions (Iravani, 2011). In agriculture, nano-fertilizers and nano-pesticides synthesized through plant-mediated methods improve crop yields while reducing chemical inputs (Mittal et al., 2013). Moreover, in environmental remediation, these nanoparticles serve as efficient agents for water purification and pollutant degradation. Despite the considerable advantages, challenges remain in standardizing plant-based nanoparticle synthesis. Variability in plant phytochemical composition due to geographical, seasonal, and environmental factors can lead to inconsistencies in nanoparticle size, shape, and stability (Kuppusamy et al., 2016a). Future research should focus on optimizing extraction methods, understanding the molecular mechanisms of biosynthesis, and developing genetically engineered plants tailored for nanoparticle production. In conclusion, plants are indispensable components in nano-biotechnology, offering an environmentally responsible, cost-effective, and scalable method for nanoparticle synthesis. Continued interdisciplinary research integrating botany, nanotechnology, and material science will further enhance the development of plant-based nanomaterials for diverse industrial and biomedical applications.

1.1.4 Unique Features of Plant Systems for Nanotechnology

Plant systems offer several unique advantages that make them highly suitable for applications in nanotechnology. Unlike traditional chemical and physical nanoparticle synthesis, plants utilize naturally occurring biomolecules to mediate the reduction and stabilization of nanoparticles in an environmentally benign manner (Iravani, 2011). Plants are abundant sources of a wide array of secondary metabolites, such as flavonoids, alkaloids, terpenoids, saponins, and phenolic acids, which act as natural reducing and capping agents. These compounds facilitate the conversion of metal ions into stable nanoparticles while simultaneously controlling their size, shape, and dispersity (Kuppusamy et al., 2016a). One of the most significant features of plant-based synthesis is its scalability. Plant materials are readily available, inexpensive, and renewable, allowing large-scale production of nanoparticles without the need for complex infrastructure (Mittal et al., 2013). Additionally, plant-mediated nanoparticle synthesis generally occurs at ambient temperatures and pressures, further reducing energy consumption compared to conventional methods. Another critical advantage is the biocompatibility and functionalization potential of plant-derived nanoparticles. The phytochemicals that coat the nanoparticles often impart additional biological activities, such as antioxidants, antimicrobial, and anticancer properties, enhancing their effectiveness in biomedical and environmental applications (Ahmed et al., 2016b). Furthermore, the eco-friendly nature of plant-based synthesis aligns well with the growing emphasis on green chemistry and sustainable technological development. The inherent diversity among plant species also offers a broad platform for tailoring nanoparticles with specific properties. Variations in phytochemical composition between different plant types or even different parts of the same plant can influence nanoparticle characteristics, enabling customization for targeted applications (Kuppusamy et al., 2016a).

1.1.5 Plant-Based Nanoparticles

Plant-based nanoparticles are metallic, or metal oxide nanoparticles synthesized using plant extracts as reducing and stabilizing agents. These nanoparticles typically include silver (AgNPs), gold (AuNPs), zinc oxide (ZnONPs), iron oxide (Fe3O4NPs), and titanium dioxide (TiO2NPs), among others (Iravani, 2011). Silver nanoparticles synthesized using plant systems are widely recognized for their potent antimicrobial activities. For example, extracts from *Azadirachta indica* have been used to produce AgNPs that exhibit significant antibacterial properties against a range of pathogenic bacteria (Ahmed et al., 2016b). Similarly, gold nanoparticles synthesized using *Camellia sinensis* extracts show excellent antioxidant and potential anticancer activities (Mittal et al., 2013). Zinc oxide nanoparticles, synthesized through plant-mediated approaches, have demonstrated strong photocatalytic and antimicrobial properties, making them valuable for applications in environmental remediation and food packaging (Kuppusamy et al., 2016a). Iron oxide nanoparticles synthesized via plant extracts have been explored for magnetic resonance imaging (MRI) and targeted drug delivery, leveraging their magnetic properties for biomedical applications. The morphology, size, and stability of these nanoparticles can be finely tuned by optimizing reaction parameters such as pH, temperature, metal ion concentration, and extract volume. Plant-based nanoparticles often display superior stability and enhanced biological activities compared to those synthesized via chemical routes due to the presence of biofunctional phytochemicals on their surfaces. In conclusion, plant-mediated nanoparticle synthesis represents a versatile, green, and efficient method for producing nanoparticles with diverse applications, from medicine to environmental technology. Continued research into plant-based nanotechnology holds promise for the development of next-generation nanomaterials with minimal ecological footprint.

1.1.6 Interactions Between Nanoparticles and Plants

The interactions between NPs and plants have become a focal point in agricultural, environmental, and nanotechnology research. Nanoparticles, typically ranging between 1 and 100 nm, possess unique properties such as high surface area-to-volume ratio, enhanced reactivity, and tunable surface functionalities, which significantly influence their behavior in biological systems (Rico et al., 2011). Their interactions with plants depend on numerous factors, including the particle size, chemical composition, surface charge, concentration, exposure route, and plant species. Plants can uptake nanoparticles primarily through their roots from soil or hydroponic solutions and through foliar exposure via stomatal openings (Ma et al., 2010). Following uptake, nanoparticles may be translocated through the xylem or phloem, reaching distant tissues, including stems, leaves, and fruits (Hong et al., 2014a). The efficiency of uptake and translocation is governed by nanoparticle characteristics such as size (smaller particles generally translocate more easily), hydrophobicity, and surface functionalization (Wang et al., 2016). At optimal, low concentrations, certain nanoparticles can promote plant growth and productivity. For example, zinc oxide (ZnO) nanoparticles enhance seed germination, chlorophyll production, and nutrient uptake (Dimkpa et al., 2012a). Titanium dioxide (TiO_2) nanoparticles have been reported to improve photosynthetic rates and stress tolerance by facilitating better light absorption and antioxidant activity (Gao et al., 2008). Similarly, iron oxide (Fe_3O_4) nanoparticles can enhance iron availability in iron-deficient soils, promoting plant development (Rui et al., 2016b). However, excessive or prolonged exposure to nanoparticles may cause phytotoxic effects. Nanoparticle-induced toxicity commonly results from the overproduction of reactive oxygen species (ROS), leading to oxidative stress, lipid peroxidation, protein oxidation, DNA damage, and ultimately impaired physiological functions (Singh et al., 2017a). Plants counteract oxidative stress by activating enzymatic antioxidants such as superoxide dismutase (SOD), catalase (CAT), and peroxidase (POD) (Zuverza-Mena et al., 2016). The method of nanoparticle synthesis also significantly influences their interaction with plants. Green-synthesized nanoparticles, produced using biological extracts, tend to be more biocompatible and less toxic than those synthesized chemically (Iravani, 2011). Surface modifications, including polymeric coatings or functionalization with biomolecules, can enhance nanoparticle stability and reduce potential toxicity (Tripathi et al., 2017a). Environmental factors play a critical role in modulating nanoparticle behavior. Soil properties such as pH, texture, organic matter content, and microbial activity affect nanoparticle aggregation, dissolution, and bioavailability (Servin et al., 2015). For instance, acidic soils can enhance the solubility and mobility of metal-based nanoparticles, thereby increasing their phytotoxic potential (McLaughlin et al., 2010b). Despite promising applications in precision agriculture, such as nano-fertilizers, nano-pesticides, and controlled-release systems, concerns remain regarding the long-term accumulation of nanoparticles in edible plant parts and subsequent entry into the food chain (Wang et al., 2017). Thus, rigorous risk assessments and development of environmentally sustainable nanotechnology applications are crucial (Khodakovskaya et al., 2012b). In conclusion, the interactions between nanoparticles and plants are multifaceted, involving both beneficial and adverse outcomes depending on a variety of intrinsic and extrinsic factors. Future research should prioritize understanding nanoparticle behavior under realistic field conditions, optimizing safe dosages, and developing guidelines for responsible nanoparticle usage in agriculture.

1.1.7 Uptake Mechanisms

Plants can absorb NPs primarily through root and foliar pathways. Root uptake occurs via two main routes: apoplastic and symplastic transport (Ma et al., 2010; Wang et al., 2016). The apoplastic route allows NPs to move between cell walls, while the symplastic route involves endocytosis and passage through plasmodesmata (Hong et al., 2014a; Tripathi et al., 2017b). Particle size, surface charge, and coating significantly influence uptake efficiency, with nanoparticles smaller than 20 nm generally exhibiting greater mobility and absorption (Khodakovskaya et al., 2009; Sun et al., 2014b). Foliar uptake occurs through stomata, trichomes, cuticular cracks, and hydrophilic pores (Rico et al., 2011; Kurepa et al., 2010). After foliar entry, NPs can translocate via the phloem to different parts of the plant, affecting both local and systemic physiological responses (Servin et al., 2015; Raliya et al., 2015). The chemical nature of the NPs also determines their fate inside the plant. For instance, metallic nanoparticles such as ZnO and AgNPs may dissolve into ions, while carbon-based NPs like carbon nanotubes maintain structural integrity and accumulate in specific tissues (Rui et al., 2016a; Larue et al., 2012). Surface functionalization, such as polymer or organic molecule coating, improves nanoparticle stability, reduces aggregation, and enhances plant uptake (Nair et al., 2010; Tripathi et al., 2017a). Endocytosis appears to be a major cellular uptake mechanism for several nanoparticle types, particularly in the case of small, functionalized particles (Ma et al., 2010; Wang et al., 2017). Once internalized, nanoparticles can be transported through vascular systems, sequestered in vacuoles, metabolized, or transformed into secondary products, depending on their physicochemical properties (Hong et al., 2014b; Schwab et al., 2016).

1.1.8 Effects on Plant Physiology

Nanoparticles can significantly influence plant physiological processes, exerting both beneficial and detrimental effects depending on dose, exposure duration, and plant species (Dimkpa et al., 2012b; Servin et al., 2015). At low concentrations, certain nanoparticles stimulate plant growth by enhancing seed germination, root elongation, and biomass accumulation (Gao et al., 2008; Raliya et al., 2015). For instance, TiO_2 nanoparticles improve photosynthetic efficiency by increasing chlorophyll content and promoting light absorption (Zheng et al., 2005; Hong et al., 2014b). Nanoparticles can modulate antioxidant defense systems. Exposure to metal-based NPs like ZnO or CuO can upregulate antioxidant enzymes, including superoxide dismutase (SOD), catalase (CAT), and peroxidase (POD), helping plants manage reactive oxygen species (ROS) (Singh et al., 2017b; Zuverza-Mena et al., 2016). Moderate ROS levels can act as signaling molecules, triggering adaptive stress responses (Rico et al., 2011; Wang et al., 2016). However, high concentrations of nanoparticles often induce toxicity. Excessive ROS production may overwhelm antioxidant defenses, causing oxidative damage to membranes, proteins, and DNA (Khodakovskaya et al., 2012a; Rui et al., 2016b). AgNPs and CuO-NPs, for example, have been reported to inhibit photosynthesis, reduce chlorophyll content, disrupt nutrient uptake, and decrease overall plant biomass (Ma et al., 2010; Sun et al., 2014a). Additionally, nanoparticle exposure can disturb hormonal balances, affecting auxin and gibberellin signaling pathways critical for growth and development (Tripathi et al., 2017b; Nair et al., 2010). Changes in soil microbial communities due to nanoparticle release can further impact plant health indirectly by altering nutrient cycling and plant-microbe symbioses (McLaughlin et al., 2010a; Schwab et al., 2016). Furthermore, the accumulation of NPs in edible plant parts raises concerns regarding food safety and human health risks (Rico et al., 2011; Servin et al., 2015). In conclusion, while nanoparticles hold promise for enhancing agricultural productivity by improving plant growth and stress resilience, careful management is essential to avoid phytotoxic effects and environmental risks.

1.1.9 Current Research and Future Directions

The interactions between NPs and plants have emerged as a pivotal research area due to the increasing application of nanomaterials in agriculture, food technology, and environmental remediation. Current studies focus extensively on understanding the mechanisms of nanoparticle uptake, translocation, and accumulation in plants, as well as their physiological and molecular effects (Ma et al., 2010; Rico et al., 2011). Significant advancements have been made in developing nanofertilizers, nanopesticides, and nanosensors that improve plant productivity and stress tolerance (DeRosa et al., 2010; Khot et al., 2012). Nanofertilizers, characterized by slow-release and targeted delivery properties, enhance nutrient use efficiency and minimize environmental pollution compared to conventional fertilizers (Naderi & Danesh-Shahraki, 2013; Liu & Lal, 2015). For example, nano-encapsulated nitrogen fertilizers have demonstrated improved nitrogen assimilation in maize and wheat (Subramanian et al., 2015; Mukhopadhyay et al., 2016). Similarly, nanopesticides offer controlled release mechanisms, reducing the overall quantity of chemical usage and mitigating environmental hazards (Kah et al., 2013; Chen & Yada, 2011). Plant responses to nanoparticle exposure are complex and depend on several factors, including nanoparticle type, size, concentration, surface modification, plant species, and developmental stage (Tripathi et al., 2017b; Wang et al., 2017). For instance, silver nanoparticles (AgNPs) at low concentrations can stimulate seed germination and root growth, whereas high concentrations often result in phytotoxic effects such as chlorosis, inhibited growth, and oxidative stress (Zuverza-Mena et al., 2016; Rico et al., 2011). Current research also explores the role of nanoparticles in modulating plant defense mechanisms. Studies have shown that nanoparticles can upregulate antioxidant enzyme activities such as superoxide dismutase (SOD) and catalase (CAT), thus enhancing plant resilience to abiotic stresses like drought and salinity (Dimkpa et al., 2012b; Raliya et al., 2015). Furthermore, nanoparticles such as cerium oxide (CeO_2) have been demonstrated to scavenge reactive oxygen species (ROS) directly, thereby mitigating oxidative damage in plants (Heckman et al., 2013; Ma et al., 2015). Despite these promising outcomes, significant knowledge gaps remain. One major concern is nanoparticle phytotoxicity, particularly at high concentrations or with prolonged exposure (Wang et al., 2016; Singh et al., 2017a). Research indicates that nanoparticles can disrupt cellular homeostasis, impair photosynthesis, interfere with hormonal signaling, and cause genotoxic effects (Khodakovskaya et al., 2009; Sun et al., 2014b). In addition, the long-term fate of nanoparticles in plant-soil ecosystems remains poorly understood (McLaughlin et al., 2010b; Schwab et al., 2016). Another challenge lies in the difficulty of accurately tracing and quantifying nanoparticles within plant tissues due to limitations in imaging and analytical techniques (Larue et al., 2012; Hong et al., 2014b). Novel approaches such as synchrotron-based X-ray fluorescence microscopy, hyperspectral imaging, and isotope labeling are being developed to address this limitation (Hernandez-Viezcas et al., 2013; Spielman-Sun et al., 2017). Furthermore, the impact of engineered nanoparticles on plant-associated microbial communities remains an area of concern. Disruptions in beneficial plant-microbe interactions, such as those involving nitrogen-fixing bacteria or mycorrhizal fungi, could have detrimental consequences for plant health and ecosystem sustainability (Ge et al., 2011; Priester et al., 2012).

1.1.9.1 Mechanistic Understanding: Elucidating the Molecular Pathways and Gene Networks Modulated by Nanoparticle Exposure

The mechanistic understanding of NP toxicity or therapeutic effects requires a deep dive into the molecular pathways and gene networks that are modulated upon exposure. Nanoparticles, due to their unique physicochemical properties (e.g., size, shape, surface charge, and composition), can interact with cellular components in ways that bulk materials cannot. This interaction initiates complex signaling cascades that can influence various cellular processes, including oxidative stress, inflammation, apoptosis, cell proliferation, and gene expression.

1.1.9.2 Omics Approaches to Nanotoxicology

Omics technologies, including transcriptomics, proteomics, and metabolomics, offer powerful tools to uncover these molecular mechanisms. Transcriptomics involves the study of gene expression, helping to identify alterations in the mRNA levels of specific genes following nanoparticle exposure. This provides insight into which genes are upregulated or downregulated, potentially revealing key regulatory pathways or biomarkers of exposure and toxicity (Andersson et al., 2020). Proteomics, on the other hand, analyzes the complete set of proteins expressed in a cell, providing information on how nanoparticle exposure affects protein synthesis, degradation, and function. By examining changes in protein expression, modifications, and interactions, proteomics can pinpoint critical signaling pathways or proteins involved in nanoparticle-induced cellular responses (Bouwmeester et al., 2011). For instance, nanoparticles have been shown to affect proteins involved in immune responses, apoptosis, and cellular repair mechanisms (Jiang et al., 2021). Metabolomics provides a snapshot of the global metabolic changes in response to nanoparticle exposure. Since nanoparticles can interact with metabolic pathways, metabolomic profiling helps in identifying alterations in metabolites like lipids, amino acids, and carbohydrates, which can provide further insights into the cellular stress, energy production, and oxidative damage induced by nanoparticles (Xia et al., 2019). These data are crucial for understanding the broader physiological impacts of NP exposure.

1.1.9.3 Gene Regulatory Networks and Pathways

In addition to identifying individual molecular targets, omics approaches allow for the construction of gene regulatory networks. These networks reveal how genes and proteins are interconnected and how nanoparticle exposure may disrupt normal cellular functions. For example, studies have shown that nanoparticles can modulate key pathways such as the MAPK/ERK, NF-κB, PI3K/Akt, and Nrf2 signaling pathways, all of which are involved in inflammation, oxidative stress, and apoptosis (Liu et al., 2020b; Zhang et al., 2020).

1.1.9.4 High-Throughput Screening and Systems Biology

High-throughput screening of gene expression and protein data, combined with advanced bioinformatics tools, facilitates the identification of crucial molecular targets and mechanisms. Systems biology approaches integrate omics data to build comprehensive models that predict the behavior of biological systems in response to nanoparticle exposure (Ding et al., 2018). These models enable researchers to simulate various cellular responses to nanoparticles, aiding in the identification of biomarkers and therapeutic targets.

1.1.9.5 Challenges and Future Directions

Although omics approaches have provided valuable insights into the molecular effects of nanoparticles, several challenges remain. One of the primary obstacles is the complexity of nanoparticle behavior in biological systems. Factors such as nanoparticle size, shape, surface charge, and coating can significantly influence the interaction with cells and tissues, complicating the interpretation of results (Nel et al., 2006). Additionally, the diversity of nanoparticle types and their varying toxicity profiles necessitate the development of more standardized experimental models and data analysis techniques. Future research should focus on integrating **multi-omics approaches** to capture the full range of cellular responses. The combination of transcriptomics, proteomics, and metabolomics, when integrated with advanced computational models, holds the potential to unravel the intricate molecular mechanisms and gene networks modulated by nanoparticle exposure. Moreover, understanding the **dose-response relationships** and **long-term effects** of nanoparticles on human health and the environment is essential to ensuring safe and effective use. In conclusion, omics-based approaches are transforming the field of nanotoxicology by offering a deeper mechanistic understanding of how nanoparticles interact with biological systems at the molecular level. As the field progresses, it will be essential to incorporate these findings into regulatory frameworks to ensure the safety of nanoparticles in consumer products and medical applications.

1.1.10 Nanomaterial Design: Developing Biocompatible, Biodegradable, and Targeted Nanomaterials

The design of nanomaterials plays a crucial role in their safety and efficacy for biomedical and environmental applications. For nanoparticles to be used effectively in medical treatments or industrial applications, they must possess specific properties that not only minimize toxicity but also maximize their therapeutic or functional potential. Three key design considerations—biocompatibility, biodegradability, and targeted functionality—are essential in optimizing nanomaterials.

1.1.10.1 Biocompatibility

Biocompatibility refers to the ability of a nanomaterial to interact safely with biological systems without causing harmful immune responses, toxicity, or long-term adverse effects. This characteristic is critical for ensuring that nanoparticles can be safely used in drug delivery, imaging, and tissue engineering applications. Nanomaterials should be designed to minimize cytotoxicity and promote cell survival, proliferation, and function. Factors such as surface chemistry, charge, and size influence biocompatibility, as they determine how nanoparticles interact with cellular membranes, proteins, and other biomolecules (Zhang et al., 2018). Modifications like PEGylation (attaching polyethylene glycol) are commonly used to improve the biocompatibility of nanoparticles by enhancing their solubility and reducing protein adsorption, thus preventing immune recognition.

1.1.10.2 Biodegradability

Biodegradable nanomaterials are particularly important in medical applications, where nanoparticles must eventually break down and be safely eliminated from the body without causing harm. The design of biodegradable nanoparticles ensures that they do not accumulate in tissues over time, which could lead to toxicity or chronic inflammation. Natural polymers such as chitosan, alginate, and poly(lactic-co-glycolic acid) (PLGA) are commonly used as bases for biodegradable nanomaterials due to their ability to degrade into non-toxic byproducts, such as lactic acid or carbon dioxide, when exposed to physiological conditions (Sahoo et al., 2017). The degradation rate can be controlled by modifying the chemical structure, allowing for sustained drug release in controlled applications.

1.1.10.3 Targeted Functionality

Targeted drug delivery is a significant area where nanomaterials are designed to improve the efficacy of treatments by delivering therapeutic agents specifically to the desired tissues or cells. Functionalizing nanoparticles with targeting ligands, such as antibodies, peptides, or small molecules, allows them to bind selectively to receptors overexpressed on the target cells, thus reducing off-target effects and enhancing therapeutic outcomes (Kim et al., 2020). For example, cancer therapy benefits from targeted nanoparticles that selectively deliver chemotherapy drugs to tumor cells, sparing healthy tissues. In conclusion, the development of biocompatible, biodegradable, and targeted nanomaterials is essential for optimizing their safety and effectiveness in biomedical and industrial applications. This approach helps minimize adverse effects while maximizing the desired outcomes, particularly in drug delivery, diagnostic imaging, and tissue engineering.

1.1.10.4 Standardization of Methodologies: Harmonizing Experimental Protocols

In the rapidly growing field of nanotoxicology and nanobiotechnology, standardization of experimental methodologies is essential for ensuring the reliability, reproducibility, and comparability of results across different studies. NPs exhibit unique physicochemical properties that can influence their interactions with biological systems, including plants, animals, and humans. However, the lack of standardized protocols for nanoparticle characterization, plant exposure methods, and toxicity assessments has hindered the ability to compare data and draw consistent conclusions from different studies. Therefore, the harmonization of experimental procedures is necessary for advancing the understanding of nanoparticle behavior in biological systems.

1.1.10.5 Nanoparticle Characterization

A key component in the study of nanoparticles is their characterization. The size, shape, surface charge, surface chemistry, and aggregation state of nanoparticles significantly influence their biological activity and toxicity. Despite the widespread use of various techniques, such as transmission electron microscopy (TEM), dynamic light scattering (DLS), and atomic force microscopy (AFM), there is no universally accepted standard for nanoparticle characterization (Rocco et al., 2020). Differences in instrument calibration, measurement protocols, and sample preparation can lead to discrepancies in the reported properties of nanoparticles, which in turn affect their observed biological effects. Establishing unified guidelines for nanoparticle characterization would enable more accurate comparisons between studies and provide a clearer understanding of the relationship between nanoparticle properties and their biological effects.

1.1.10.6 Plant Exposure Methods

The methods by which plants are exposed to nanoparticles are also diverse and inconsistent across studies. Nanoparticles can be applied to plants through soil, foliar spray, or root uptake, and each exposure route can lead to different results due to variations in particle uptake, distribution, and interaction with plant tissues. Standardizing exposure techniques, such as determining optimal nanoparticle concentrations, exposure durations, and application methods, would help reduce experimental variability and improve the reproducibility of findings (Georgieva et al., 2019). Furthermore, it is important to consider factors such as plant species, developmental stage, and environmental conditions, all of which can influence nanoparticle uptake and toxicity. By developing consensus guidelines for plant exposure protocols, researchers can more effectively compare the impact of nanoparticles across different plant species and experimental conditions.

1.1.10.7 Toxicity Assessments

Toxicity assessment methodologies are another area where standardization is crucial. Nanoparticle toxicity is evaluated using a variety of assays, such as cell viability tests, oxidative stress assays, and gene expression analyses. However, the lack of standardized protocols for assessing toxicity in plants and other organisms results in variability in the reported outcomes. For example, different assays may measure distinct aspects of toxicity, such as cell membrane integrity, antioxidant activity, or gene expression, but they are not always comparable due to differences in methodology and endpoints. Standardized toxicity testing protocols, including the use of control groups, appropriate nanoparticle concentrations, and consistent endpoints, would ensure that results from different studies are comparable and can be used to build a more robust understanding of nanoparticle safety (Liu et al., 2020a). Harmonizing experimental protocols across nanoparticle characterization, plant exposure methods, and toxicity assessments is essential for advancing the field of nanotoxicology and nanobiotechnology. Standardization will improve the reproducibility of findings, enhance the comparability of data across studies, and provide a clearer understanding of the potential risks and benefits of nanoparticles in agricultural and environmental applications. Establishing global consensus on these methodologies will pave the way for more reliable and robust research outcomes, ultimately facilitating the safe use of nanomaterials in various industries.

1.1.10.8 Long-term Environmental Studies: Understanding Chronic Effects of Nanoparticle Exposure

Long-term environmental studies focusing on the chronic effects of NP exposure are essential for understanding the broader ecological risks posed by nanoparticles. While short-term studies have provided valuable insights into the immediate toxicological effects of nanoparticles on plants, animals, and microorganisms, there is a significant gap in our understanding of their long-term environmental impacts. To better inform risk assessments, it is crucial to examine nanoparticle exposure under realistic environmental conditions, including interactions with other environmental stressors such as pollutants, climate change, and land use changes. These studies are key for evaluating the persistence, bioaccumulation, and ecological consequences of nanoparticle exposure over extended periods.

1.1.10.9 Chronic Effects and Persistence in the Environment

One of the primary concerns regarding nanoparticles is their potential persistence in the environment. Unlike conventional chemicals, nanoparticles have unique properties such as high surface area, reactivity, and stability, which can influence their degradation rates and interactions with environmental matrices (air, water, soil). Long-term studies are required to assess how nanoparticles degrade over time and whether their persistence could lead to chronic exposure in ecosystems. Research is needed to investigate how different nanoparticle types—such as metal-based, carbon-based, or polymeric nanoparticles—differ in terms of their longevity in the environment and potential for accumulation in various environmental compartments, such as soil, sediment, and water bodies (Rochman et al., 2019).

1.1.11 Interaction with Environmental Stressors

Environmental stressors, such as pollutants, temperature fluctuations, and UV radiation, can significantly influence the behavior and toxicity of nanoparticles in the environment. For example, the presence of heavy metals or organic pollutants in water bodies may enhance the toxicity of nanoparticles, as these materials can interact synergistically or alter the bioavailability of toxic substances. Long-term studies that simulate real-world conditions of environmental stress, such as the effects of elevated temperatures, drought, or pollution, are necessary to understand how nanoparticles behave under these combined stressors (Bouwmeester et al., 2018). These studies can help predict whether nanoparticles could exacerbate the effects of other pollutants, increase the bioavailability of harmful substances, or alter ecosystem processes over time.

1.1.12 Bioaccumulation and Ecotoxicity

One of the most significant concerns regarding the long-term environmental presence of nanoparticles is their potential for bioaccumulation in organisms at various trophic levels. Chronic exposure to nanoparticles in the environment could lead to their accumulation in plants, animals, and microorganisms, potentially disrupting food webs and altering biodiversity. Research on bioaccumulation patterns of nanoparticles in both aquatic and terrestrial organisms is critical for understanding how nanoparticles enter and move through ecosystems. Long-term exposure studies can reveal whether nanoparticles accumulate in tissues over time, leading to sublethal effects such as reproductive failure, impaired growth, or behavioral changes in wildlife (Pérez et al., 2020). Furthermore, the impacts of such bioaccumulation on human health, particularly through the consumption of contaminated food, must also be assessed.

1.1.13 Regulatory Implications and Risk Assessments

The findings from long-term environmental studies are crucial for developing robust risk assessment frameworks for nanoparticles. Understanding the chronic effects of nanoparticle exposure will allow regulatory agencies to set safety thresholds, recommend safe handling practices, and ensure that environmental policies are based on comprehensive and accurate data. Risk assessments should not only consider the direct toxic effects of nanoparticles but also account for their persistence, bioaccumulation potential, and interactions with other environmental stressors. This holistic approach will provide a more complete picture of the potential risks posed by nanoparticles to ecosystems and human health. Long-term environmental studies are fundamental for understanding the chronic effects of nanoparticle exposure, especially under realistic environmental conditions. These studies will provide critical information on the persistence, bioaccumulation, and ecological consequences of nanoparticles in the environment, enabling better risk assessments and more effective regulations. By considering interactions with multiple stressors, these studies can help predict how nanoparticles might behave in complex, dynamic ecosystems and inform strategies to mitigate potential environmental harm.

1.1.14 Food Safety and Human Health: Investigating Nanoparticle Accumulation in Edible Plant Parts

The increasing use of NPs in agriculture, including as fertilizers, pesticides, and in food packaging materials, raises significant concerns regarding food safety and human health. One of the primary concerns is the potential for nanoparticles to accumulate in edible plant parts, such as fruits, vegetables, and grains. Understanding the pathways by which nanoparticles enter plants and accumulate in their edible tissues is essential for assessing the risks associated with their consumption. Moreover, evaluating the potential human health risks posed by the ingestion of nanoparticle-contaminated food is crucial for the development of regulatory frameworks that ensure food safety.

1.1.15 Mechanisms of Nanoparticle Uptake and Accumulation in Plants

Plants can take up nanoparticles through various routes, including root absorption from the soil, foliar uptake through leaves, and, in some cases, via uptake from the air (Sharma et al., 2019c). The extent of nanoparticle accumulation in plant tissues depends on several factors, including the physicochemical properties of the nanoparticles, such as size, surface charge, and composition, as well as the type of plant species and growth conditions. Smaller nanoparticles tend to have higher uptake efficiency due to their ability to pass through plant cell walls and membranes more easily (Lin & Xing, 2007).

Once nanoparticles enter plant cells, they can be transported through the plant's vascular system to various tissues, including roots, stems, leaves, and edible parts such as fruits or seeds. The accumulation of nanoparticles in edible plant parts depends on factors such as particle size, surface coating, and the presence of specific transporters or channels in plant cells that facilitate nanoparticle movement. For example, studies have shown that nanoparticles can accumulate in plant tissues like the epidermis, mesophyll, and vascular tissues, leading to potential risks of contamination of edible plant parts (Fouad et al., 2018).

1.1.16 Potential Human Health Risks from Nanoparticle Consumption

The ingestion of nanoparticles through contaminated food is a potential health risk for humans, and the long-term exposure to certain nanoparticles could have adverse health effects. Studies have demonstrated that nanoparticles can cross biological barriers such as the intestinal epithelium, enter the bloodstream, and accumulate in various organs (Gurunathan et al., 2018). The unique properties of nanoparticles, including their high surface area and reactivity, may cause them to interact with cellular components in ways that bulk materials do not. This can lead to oxidative stress, inflammation, and toxicity in human cells. Nanoparticles, such as silver, titanium dioxide, and zinc oxide, are commonly used in agricultural applications and food products. When consumed, these nanoparticles may cause toxicity through several mechanisms, including the generation of reactive oxygen species (ROS), disruption of cellular signaling pathways, or interference with cellular metabolism (Kuo et al., 2020).

These effects could manifest as gastrointestinal disturbances, immune system alterations, or even more severe conditions, such as cancer or organ damage, depending on the properties of the nanoparticles and the level of exposure. In addition to direct toxicological effects, the bioaccumulation of nanoparticles in human tissues over time could pose cumulative risks. For instance, prolonged exposure to certain nanoparticles could lead to the accumulation of these particles in vital organs, such as the liver, kidneys, or brain, potentially causing long-term health complications (Buzea et al., 2007). The potential for chronic exposure underscores the need for thorough investigation into the bioaccumulation and toxicity of nanoparticles in food products.

1.1.17 Regulatory Frameworks and Food Safety Standards

Given the potential risks posed by nanoparticle contamination in food, it is essential to develop regulatory frameworks and food safety standards that address these concerns. Regulatory agencies, such as the U.S. Food and Drug Administration (FDA) and the European Food Safety Authority (EFSA), are working to establish guidelines for the safe use of nanomaterials in food production. However, the lack of standardized testing protocols for nanoparticle toxicity and bioaccumulation presents challenges in assessing the risks accurately (European Commission, 2011). To ensure food safety, comprehensive studies are required to evaluate the levels of nanoparticle accumulation in edible plant parts, the effects of chronic exposure through diet, and the potential health risks associated with consuming nanoparticle-contaminated food. Additionally, guidelines for nanoparticle concentrations in food products must be established based on scientific evidence regarding the safe thresholds for human health.

The accumulation of nanoparticles in edible plant parts poses significant concerns for food safety and human health. As nanoparticles continue to be integrated into agricultural practices and food products, it is crucial to understand their uptake mechanisms in plants and evaluate the potential health risks associated with their consumption. This knowledge will help inform the development of regulatory frameworks that ensure food safety, protect public health, and guide the responsible use of nanomaterials in the food industry.

1.1.18 Integration with Sustainable Agriculture: Utilizing Nanoparticles to Promote Sustainable Practices

The integration of NPs into sustainable agriculture presents a promising approach to addressing the challenges of food security, environmental protection, and the efficient use of resources. Nanotechnology, particularly the use of NPs in precision farming, offers numerous advantages for reducing agrochemical inputs, enhancing crop resilience, and promoting environmentally friendly agricultural practices. As global populations increase and the demand for food rises, integrating nanoparticles into agricultural systems can help meet these challenges while minimizing the environmental footprint of farming.

1.1.19 Precision Farming and Resource Efficiency

One of the key benefits of integrating nanoparticles into sustainable agriculture is their potential to improve precision farming techniques. Precision farming involves the use of advanced technologies to monitor and manage crop growth, soil health, and environmental conditions at a highly localized level. Nanoparticles can be employed in sensors and delivery systems that allow for the precise application of water, nutrients, and pesticides, reducing waste and improving overall resource efficiency (Kolluru et al., 2020). For example, nanosensors can detect soil moisture levels or nutrient deficiencies in real-time, enabling farmers to apply water and fertilizers only when and where they are needed, thereby reducing overuse and minimizing environmental impacts. Similarly, nanoparticle-based fertilizers, such as slow-release formulations, can provide nutrients to plants in a controlled manner, reducing nutrient runoff into water systems and enhancing fertilizer use efficiency (Jin et al., 2020). This precision application of agrochemicals not only helps in reducing the quantity of chemicals used but also minimizes the environmental risks associated with excessive fertilization, such as soil degradation, water pollution, and greenhouse gas emissions.

1.1.20 Reduced Agrochemical Inputs

The use of nanoparticles in agriculture has the potential to significantly reduce the reliance on traditional chemical pesticides and herbicides. Nanoparticles can be engineered to deliver active ingredients more effectively and specifically to target pests or pathogens, reducing the need for broad-spectrum chemical applications. For instance, nanoparticle-based pesticide formulations can be designed to release their active ingredients slowly, ensuring a sustained effect and reducing the frequency of pesticide applications (Soni et al., 2018). This targeted approach decreases the overall volume of chemicals needed, thus lowering the environmental load of agrochemicals and reducing the risk of resistance development in pests. Moreover, nanoparticles can also enhance the effectiveness of biocontrol agents, which are natural alternatives to chemical pesticides. For example, nanoparticles can be used to encapsulate and protect beneficial microorganisms or biocontrol agents, ensuring that they remain viable and active for longer periods in the field, even under harsh environmental conditions (González-Morales et al., 2020).

1.1.21 Enhanced Crop Resilience and Stress Tolerance

Another important area where nanoparticles can contribute to sustainable agriculture is in enhancing crop resilience to environmental stressors. Nanotechnology can be used to improve plant tolerance to abiotic stresses such as drought, salinity, and extreme temperatures. For example, nanoparticle-based formulations have been shown to improve water retention in plants, boost antioxidant defense mechanisms, and enhance nutrient uptake under stress conditions (Sastry et al., 2020). By improving the stress tolerance of crops, nanoparticles can help farmers maintain productivity even in the face of changing climate conditions, thus contributing to more resilient agricultural systems.

Furthermore, nanoparticles can be used to stimulate plant growth and development by enhancing the efficiency of photosynthesis or promoting the synthesis of protective molecules like proteins and enzymes. This could lead to higher crop yields with fewer inputs, contributing to sustainable intensification in agriculture, where more food is produced without additional environmental harm (Zhang et al., 2021). Integrating nanoparticles into sustainable agricultural practices offers a range of opportunities for improving efficiency, reducing environmental impact, and enhancing food security. Through precision farming, reduced agrochemical inputs, and enhanced crop resilience, nanoparticles can play a crucial role in advancing sustainable agriculture. However, as with any emerging technology, careful consideration must be given to the potential risks and challenges associated with their use, including environmental persistence and toxicity. Ongoing research and development are needed to optimize nanoparticle formulations and ensure their safe and effective application in agricultural systems. Nanoparticles hold considerable promise in revolutionizing agriculture by offering innovative solutions to enhance crop productivity, promote sustainability, and reduce the environmental footprint of farming. With their unique properties, such as high surface area, reactivity, and the ability to be engineered for specific functions, nanoparticles can optimize nutrient delivery, increase stress tolerance, reduce pesticide use, and enhance overall crop health. However, despite these exciting prospects, the integration of nanoparticles into agricultural practices requires careful consideration of both the potential benefits and risks. To realize the full potential of nanoparticles in agriculture, interdisciplinary research is crucial. Plant scientists, nanotechnologists, toxicologists, and policymakers must work together to address the many challenges associated with nanoparticle use in agricultural systems. For instance, the safety and environmental impact of nanoparticles need to be rigorously assessed, ensuring that their widespread use does not inadvertently harm ecosystems, soil health, or human health through the accumulation of nanoparticles in the food chain. Studies on the biodegradability, bioaccumulation, and long-term environmental effects of nanoparticles are essential for understanding their ecological footprint and ensuring their responsible application. In addition to environmental concerns, it is important to optimize the design and functionality of nanoparticles for agricultural use. This involves developing biocompatible and biodegradable nanoparticles that can deliver agrochemicals, fertilizers, or biocontrol agents effectively while minimizing adverse impacts on non-target organisms. Nanomaterials must be tailored for specific crops, environments, and stress conditions to maximize their efficacy and reduce waste. Advances in nanotechnology should focus on enhancing the sustainability of agricultural systems by reducing the need for harmful chemicals, improving resource use efficiency, and promoting the resilience of crops to changing climate conditions. Collaboration between scientific disciplines is equally vital for the development of standardized methodologies and regulatory frameworks that govern nanoparticle use in agriculture. Uniform experimental protocols for nanoparticle characterization, plant exposure methods, and toxicity assessments will help create a more cohesive body of knowledge, enabling researchers to compare results across studies and establish safe usage limits. Policymakers also play a crucial role in creating guidelines that balance innovation with caution, ensuring that the benefits of nanoparticles are maximized while minimizing potential risks. Ultimately, while nanoparticles offer immense promise for the future of sustainable agriculture, their responsible and thoughtful integration into farming practices will require a concerted, collaborative effort. Through continued research, dialogue, and careful policy development, nanotechnology can be harnessed to support more efficient, resilient, and environmentally friendly agricultural systems that can help feed the world while preserving the planet for future generations.

1.1.22 Emerging Trends in Nanotechnology for Agriculture

Nanotechnology has been evolving rapidly, offering promising advancements in agricultural practices. As the potential applications of NPs in agriculture expand, several emerging trends are shaping the future of this field. These trends focus on improving crop yields, enhancing sustainability, and reducing the environmental impacts of conventional farming methods. The following are some key emerging trends in nanotechnology for agriculture:

1.1.22.1 Smart Nano-Enabled Fertilizers

One of the most promising applications of nanoparticles in agriculture is the development of smart, nano-enabled fertilizers. These fertilizers are designed to release nutrients slowly and efficiently, enhancing nutrient uptake by plants while minimizing nutrient runoff. This trend is particularly crucial in addressing the environmental concerns associated with traditional fertilizer use, which can lead to soil and water pollution. Nanoparticle-based fertilizers offer more targeted nutrient delivery and are tailored to meet specific plant needs, thus improving crop productivity with fewer resources (Jin et al., 2020).

1.1.22.2 Nanoparticles in Precision Agriculture

Precision agriculture involves the use of advanced technologies to optimize field-level management of crops, based on varying environmental conditions. Nanotechnology is playing a significant role in this trend by enhancing the precision and efficiency of farming practices. For instance, nanoparticle-based sensors are being used to monitor soil moisture, temperature, and nutrient levels in real-time. These sensors provide valuable data, enabling farmers to apply water, fertilizers, and pesticides more efficiently, reducing waste and environmental impacts (Kolluru et al., 2020). As precision agriculture continues to grow, the integration of nanoparticles will likely become more sophisticated, offering further enhancements in crop management and resource use efficiency.

1.1.22.3 Enhanced Crop Protection and Biocontrol

Another emerging trend is the use of nanoparticles in crop protection, particularly in reducing the dependence on chemical pesticides. Nanoparticles can encapsulate and deliver biocontrol agents, such as beneficial bacteria, fungi, or viruses, directly to the target pests. This approach ensures more controlled and targeted pest management, reducing the need for broad-spectrum chemical treatments that can harm non-target organisms and ecosystems. Additionally, nanoparticle formulations can be designed to release active ingredients gradually, providing long-term protection against pests and pathogens (Soni et al., 2018). This trend supports sustainable farming practices by reducing pesticide use while promoting natural pest control methods.

1.1.22.4 Nanomaterial-Based Crop Enhancers

Nanoparticles are increasingly being used to enhance plant growth and stress tolerance. Nanomaterials can improve nutrient uptake, boost photosynthetic efficiency, and increase plant resilience to abiotic stressors such as drought, salinity, and temperature extremes. For example, nanomaterials like silica and titanium dioxide have been shown to improve root development and increase water retention in plants, helping crops thrive under adverse environmental conditions (Sastry et al., 2020). As climate change continues to exacerbate stress conditions, the use of nanomaterials to improve crop resilience is likely to become a key trend in sustainable agriculture.

1.1.22.5 Nano-Biosensors for Disease Detection

Nanotechnology is also being applied to detect plant diseases early through nano-biosensors, which offer rapid, cost-effective, and highly sensitive diagnostics. These biosensors can identify pathogens or disease-related biomarkers at very low concentrations, allowing for early intervention and better management of crop diseases. By enabling precise and timely disease detection, these sensors can reduce the need for chemical pesticides and prevent the spread of diseases, thus improving crop health and yield while minimizing chemical inputs (González-Morales et al., 2020).

1.1.22.6 Nano-Enhanced Packaging for Agricultural Products

The development of nano-enhanced packaging materials is another emerging trend. Nanoparticles, particularly those with antimicrobial and antioxidant properties, can be incorporated into packaging materials to extend the shelf life of agricultural products. These materials help reduce food waste by maintaining the freshness and quality of perishable items such as fruits, vegetables, and meats. Nano-enabled packaging also has the potential to monitor the quality of stored goods through embedded sensors that can detect spoilage or contamination in real-time (González-Morales et al., 2020).

1.1.22.7 Regulatory and Safety Frameworks

As the use of nanoparticles in agriculture grows, the establishment of regulatory and safety frameworks becomes increasingly important. The application of nanotechnology in food and agriculture raises concerns regarding the potential risks to human health, ecosystems, and biodiversity. There is a growing emphasis on developing standardized protocols for the safe use of nanoparticles, including thorough testing of their environmental impact, bioaccumulation, and toxicity. Additionally, regulatory bodies are working to establish guidelines that balance innovation with caution, ensuring that nanoparticle applications are safe for both consumers and the environment (Kolluru et al., 2020).

Nanotechnology is revolutionizing agriculture by enabling smarter, more efficient, and environmentally sustainable farming practices. The emerging trends highlighted earlier underscore the potential of nanoparticles to enhance crop yields, improve resource use efficiency, reduce chemical inputs, and promote sustainable farming practices. However, the widespread adoption of nanotechnology in agriculture will require careful consideration of potential risks and the development of regulatory frameworks to ensure safety and sustainability. With continued interdisciplinary research and innovation, the future of nanotechnology in agriculture holds great promise for addressing the challenges of global food production in a sustainable and environmentally responsible manner.

1.1.23 Potential Applications of Nanotechnology in Agriculture and Medicine

Nanotechnology, the science of manipulating materials at the nanoscale, has emerged as a transformative technology with vast potential applications in both agriculture and medicine. By leveraging the unique properties of NPs, such as high surface area, reactivity, and the ability to be engineered for specific tasks, nanotechnology offers innovative solutions that can revolutionize these sectors. The following are some of the most promising applications of nanotechnology in agriculture and medicine.

1.1.23.1 Applications in Agriculture

- **Nanoparticles for Precision Farming** Nanotechnology is reshaping the way farming is practiced, particularly through precision agriculture. Nanoparticles can be used to develop smart fertilizers and pesticides that deliver nutrients or chemicals more efficiently to plants, reducing wastage and minimizing environmental impact. For example, nano-enabled fertilizers can provide a slow release of nutrients directly to plant roots, optimizing nutrient uptake and reducing nutrient runoff into the environment. This precision allows farmers to use fewer chemicals, which is beneficial for both cost savings and environmental sustainability (Jin et al., 2020).
- **Nano-Biosensors for Early Disease Detection** Another significant application is the use of nano-biosensors to detect plant diseases early. These sensors can identify pathogens or disease markers at very low concentrations, enabling early intervention. This helps reduce the need for broad-spectrum pesticides and enables targeted treatment. With their high sensitivity and ability to detect specific pathogens, these sensors can provide real-time monitoring of crop health and help farmers manage diseases more effectively, leading to healthier crops and higher yields (González-Morales et al., 2020).
- **Nano-Encapsulation for Pest Control** Nanoparticles can be used to encapsulate biological control agents, such as beneficial bacteria or fungi, and release them gradually over time. This ensures that the active agents are delivered directly to the target pests, reducing the need for chemical pesticides. This technology also offers long-term pest control, improving crop protection without harming the environment or non-target organisms. Nano-encapsulation can improve the efficiency of biocontrol agents, providing sustainable alternatives to traditional pest management methods (Soni et al., 2018).
- **Enhanced Crop Growth and Stress Resistance** Nanotechnology has also shown promise in enhancing crop growth and improving plant resistance to environmental stressors such as drought, salinity, and extreme temperatures.

Nanoparticles like silica and titanium dioxide have been found to promote root development and improve water retention in plants. Additionally, certain nanoparticles can stimulate the production of antioxidants in plants, which help mitigate oxidative stress caused by adverse environmental conditions. This makes crops more resilient and improves productivity, especially in areas affected by climate change (Sastry et al., 2020).

1.1.23.2 Applications in Medicine

- **Drug Delivery Systems** One of the most significant applications of nanotechnology in medicine is the development of targeted drug delivery systems. Nanoparticles can be engineered to carry drugs directly to specific cells or tissues, improving the efficacy of the treatment and minimizing side effects. For instance, nanoparticles can be designed to bind to cancer cells and deliver chemotherapy drugs directly to the tumor, sparing healthy tissues and reducing the toxicity associated with traditional chemotherapy. This targeted approach has the potential to revolutionize cancer treatment and other diseases requiring precise drug delivery (Kumar et al., 2020).
- **Nanodiagnostics** Nanotechnology has also advanced the field of diagnostics by enabling the development of highly sensitive and rapid diagnostic tools. Nano-biosensors can detect disease markers at the molecular level, allowing for early detection of diseases such as cancer, diabetes, and infectious diseases. These sensors can provide real-time data, enabling quicker and more accurate diagnoses. Additionally, nanotechnology can improve the sensitivity of imaging techniques, such as MRI and CT scans, by enhancing the contrast between tissues, which can lead to earlier detection of abnormalities and improved patient outcomes (Yin et al., 2020).
- **Nanomaterials for Regenerative Medicine** Nanomaterials are increasingly being used in regenerative medicine to promote tissue repair and regeneration. Nanostructured scaffolds can be used to support the growth of new tissue or organs by mimicking the natural extracellular matrix. These scaffolds provide a supportive environment for cells to grow and proliferate, aiding in the regeneration of damaged tissues such as bone, skin, or nerve tissue. Nanoparticles can also be used to deliver growth factors or other bioactive molecules that promote healing and tissue regeneration (Gao et al., 2020).
- **Antimicrobial Nanomaterials** In the medical field, nanoparticles with antimicrobial properties are being used to combat infections, particularly those caused by antibiotic-resistant bacteria. Nanomaterials like silver nanoparticles and copper nanoparticles have been shown to exhibit strong antibacterial activity. These nanoparticles can be incorporated into medical devices, wound dressings, and coatings for implants to prevent infections. The ability of nanoparticles to disrupt bacterial cell walls and interfere with cellular functions makes them a promising solution to address the growing issue of antibiotic resistance (Kolluru et al., 2020).

Nanotechnology offers a wide range of applications in both agriculture and medicine, with the potential to revolutionize these industries. In agriculture, nanoparticles are enhancing crop productivity, pest control, and disease management while promoting sustainability. In medicine, nanotechnology is advancing drug delivery, diagnostics, regenerative medicine, and infection control, offering more effective and targeted treatments. However, as the use of nanotechnology expands, it is essential to address concerns regarding the safety, environmental impact, and ethical considerations associated with these technologies. Continued research and development will be crucial in unlocking the full potential of nanotechnology in these fields.

1.1.24 Conclusion and Implications

Nanotechnology has rapidly evolved as a powerful tool with significant potential to revolutionize both agriculture and medicine. In agriculture, its applications in precision farming, disease detection, pest control, and crop enhancement promise substantial benefits in terms of productivity, sustainability, and environmental protection. In medicine, the ability of nanomaterials to improve drug delivery, diagnostic accuracy, and tissue regeneration is paving the way for more effective treatments and advanced therapeutic options. However, these advancements come with several challenges and considerations that must be carefully addressed to maximize the benefits of nanotechnology while minimizing potential risks.

1.1.24.1 Agricultural Implications

Nanotechnology in agriculture holds the promise of improving food security and promoting sustainable farming practices. The development of nano-based fertilizers, pesticides, and herbicides that are more efficient and less harmful to the environment is a significant breakthrough. For instance, nano-formulations allow for precise delivery of nutrients to plants, reducing wastage and optimizing plant growth. The use of nano-enabled biosensors provides farmers with real-time monitoring of crop health, enabling timely interventions and more efficient disease management. Additionally, nanomaterials can enhance plant resilience to environmental stresses, such as drought, salinity, and temperature extremes, which are becoming increasingly important as climate change continues to impact global agricultural production. Despite these advantages, there are several concerns associated with the use of nanomaterials in agriculture. First, the potential for nanoparticle accumulation in the soil and water could have long-term ecological consequences, particularly with regards to soil health and aquatic ecosystems. While nanomaterials are designed to be biodegradable and less toxic, the long-term environmental impact of their widespread use is still not fully understood. Furthermore, there is a need for standardized testing protocols to assess the safety and efficacy of nano-based agricultural products. Regulatory frameworks that balance innovation with caution are critical to ensuring that nanotechnology is applied responsibly in agriculture.

1.1.24.2 Medical Implications

In the medical field, nanotechnology has the potential to significantly enhance patient care by enabling targeted therapies, more accurate diagnostics, and regenerative treatments. One of the most promising applications is the development of targeted drug delivery systems, which can minimize side effects and improve the efficacy of drugs, especially in cancer treatment. Nanoparticles can be engineered to deliver drugs directly to specific cells or tissues, bypassing healthy cells and thereby reducing the toxicity typically associated with conventional treatments. This targeted approach is especially important in chemotherapy, where the need to minimize damage to healthy cells is paramount. Nanotechnology also holds promise in the realm of diagnostics. The development of highly sensitive nano-biosensors can enable the early detection of diseases at the molecular level, facilitating timely interventions and improving patient outcomes. Moreover, nanomaterials are being used to enhance medical imaging techniques, making it possible to detect abnormalities with greater accuracy and at earlier stages of disease progression. In regenerative medicine, nanomaterials offer the potential for creating scaffolds that can support tissue growth and repair, making significant strides toward tissue engineering and organ regeneration. However, the use of nanotechnology in medicine also raises several concerns that must be addressed. The long-term health implications of nanoparticles, especially in terms of toxicity and accumulation in human tissues, are not yet fully understood. Furthermore, the regulatory and ethical considerations surrounding the use of nanomaterials in medical applications must be carefully evaluated to ensure patient safety and equity in access to treatments. Standardized testing methods and regulatory guidelines are essential to ensure that nanotechnology is used safely and effectively in the medical field.

1.1.25 Future Directions and Challenges

While nanotechnology offers immense potential, the future success of its applications in agriculture and medicine depends on overcoming several challenges. First, more research is needed to better understand the interactions between nanoparticles and biological systems, including their potential toxicity, accumulation, and biodegradation pathways. This research will inform the development of safer, more effective nanomaterials. Additionally, the scalability of nanotechnology-based solutions remains a challenge. While laboratory studies have shown promising results, the translation of these technologies into large-scale, practical applications in both agriculture and medicine requires significant investment and development. The integration of nanotechnology with existing agricultural and medical practices must also be carefully managed. In agriculture, this means ensuring that nano-based products are accessible and affordable for farmers, especially those in developing countries. In medicine, the potential for personalized treatments using nanotechnology is vast, but this requires the development of individualized nanomaterial-based therapies that are both effective and affordable for a broad range of patients.

1.2 CONCLUSION

In conclusion, nanotechnology has the potential to significantly transform both agriculture and medicine, offering solutions that could address some of the most pressing challenges facing these sectors today. From increasing agricultural productivity and sustainability to advancing medical treatments and diagnostics, the applications of nanotechnology are vast and varied. However, to fully realize these benefits, it is crucial to address the associated risks and uncertainties through continued research, responsible innovation, and the development of clear regulatory frameworks. By doing so, we can ensure that nanotechnology is used in a way that maximizes its positive impact on society, the environment, and human health. Collaborative efforts among scientists, policymakers, and industry leaders will be essential to unlocking the full potential of nanotechnology in agriculture and medicine while safeguarding against its potential risks.

REFERENCES

Abdallah, Y., Ogunyemi, S. O., Sharma, A., & Al-Khattaf, F. S. (2022). Green synthesis of metallic nanoparticles via plant extracts: Recent advances and future prospects. *Materials Science and Engineering: C, 134*, 112636. https://doi.org/10.1016/j.msec.2021.112636

Abdul Samad, A., Tahir, M., & Javaid, S. (2023). Nanotechnology-based seed priming: Prospects for sustainable agriculture. *Environmental Nanotechnology, Monitoring & Management, 20*, 100754. https://doi.org/10.1016/j.enmm.2023.100754

Ahmed, S., Ahmad, M., Swami, B. L., & Ikram, S. (2016a). A review on plants extract mediated synthesis of silver nanoparticles for antimicrobial applications: A green expertise. *Journal of Advanced Research, 7*(1), 17–28. https://doi.org/10.1016/j.jare.2015.02.007

Ahmed, S., Saifullah, Ahmad, M., Swami, B. L., & Ikram, S. (2016b). Green synthesis of silver nanoparticles using Azadirachta indica aqueous leaf extract. *Journal of Radiation Research and Applied Sciences, 9*(1), 1–7.

Alam, T., Ghosh, I., & Ghosh, S. (2021). Advances in plant-mediated silver nanoparticles and their applications: A review. *Journal of Nanostructure in Chemistry, 11*, 1–25. https://doi.org/10.1007/s40097-020-00344-1

Almalki, W. H., Alzahrani, S. M., & Alghamdi, H. A. (2023). Role of nanotechnology in agriculture and environmental sustainability. *Environmental Research, 222*, 115187. https://doi.org/10.1016/j.envres.2023.115187

Al-Shabib, N. A., Husain, F. M., & Khan, J. M. (2023). Plant-based edible coatings and nanocoatings for food preservation. *Food Chemistry Advances, 2*, 100142. https://doi.org/10.1016/j.focha.2022.100142

Alsharif, S. M., El-Keblawy, A., & Soliman, S. S. (2022). Applications of plant-mediated nanoparticles in sustainable agriculture. *Plants, 11*(8), 1087. https://doi.org/10.3390/plants11081087

Andersson, M., et al. (2020). Transcriptomic responses to nanoparticle exposure. *Toxics, 8*(3), 75.

Awan, H. A., Abbas, A., & Rafique, M. (2023). Nano-pesticides for sustainable agriculture: A systematic review. *Journal of Environmental Chemical Engineering, 11*(3), 109266. https://doi.org/10.1016/j.jece.2023.109266

Awwad, A. M., & Salem, N. M. (2012). Green synthesis of silver nanoparticles by mulberry leaves extract. *Nanoscience and Nanotechnology, 2*(4), 125–128.

Aziz, S. B., et al. (2021). Green nanotechnology: A review on green synthesis of silver nanoparticles. *Environmental Nanotechnology, Monitoring & Management, 15*, 100428.

Balashanmugam, P., Kaliyan, B. K., & Krishnaveni, M. (2023). Recent trends in phytosynthesized nanoparticles for biomedical applications. *Biomedicine & Pharmacotherapy, 161*, 114427. https://doi.org/10.1016/j.biopha.2023.114427

Barani, M., Bilal, M., Rahdar, A., & Kyzas, G. Z. (2021a). Nanotechnology as a tool for revolutionizing modern agriculture. *Nanomaterials, 11*(1), 140. https://doi.org/10.3390/nano11010140

Barani, M., et al. (2021b). Nanotechnology applications for food safety and quality management. *Food and Chemical Toxicology, 147*, 111864.

Bhatia, S., Verma, A., & Verma, R. (2022). Application of biosynthesized nanoparticles in environmental remediation. *Chemosphere*, *286*, 131744. https://doi.org/10.1016/j.chemosphere.2021.131744

Bhattacharya, R., & Mukherjee, P. (2008). Biological properties of "naked" metal nanoparticles. *Advanced Drug Delivery Reviews*, *60*(11), 1289–1306.

Bouwmeester, H., et al. (2011). Nanomaterials in food and feed. *Toxicology*, *7*(1), 51–58.

Bouwmeester, H., et al. (2018). Environmental exposure to engineered nanoparticles and their effects on ecosystems. *Environmental Science & Technology*, *52*(3), 1099–1107.

Buzea, C., et al. (2007). Nanomaterials and nanoparticles: Sources and toxicity. *Biointerfaces*, *52*(1), 1–14.

Chaudhary, S., Sharma, P., & Kumar, A. (2022). CRISPR-Cas and nanotechnology synergy: A breakthrough for sustainable crop improvement. *Plant Cell Reports*, *41*(4), 861–877. https://doi.org/10.1007/s00299-022-02785-3

Chaudhuri, R. G., & Paria, S. (2012). Core/shell nanoparticles: Classes, properties, synthesis mechanisms, characterization, and applications. *Chemical Reviews*, *112*(4), 2373–2433.

Chen, H., & Yada, R. (2011). Nanotechnologies in agriculture: New tools for sustainable development. *Trends in Food Science & Technology*, *22*(11), 585–594.

Chowdhury, D., Kumar, V., & Kumari, R. (2022). Advances in nanotechnology-based antimicrobial therapies for viral infections. *Current Nanomaterials*, *7*(1), 47–59. https://doi.org/10.2174/2405461506666210906095122

Chugh, S., & Kumar, S. (2018). Green synthesis of nanoparticles using plant extracts: A review. *Environmental Chemistry Letters*, *16*(1), 113–127.

DeRosa, M. C., Monreal, C., Schnitzer, M., Walsh, R., & Sultan, Y. (2010). Nanotechnology in fertilizers. *Nature Nanotechnology*, *5*(2), 91.

Dimkpa, C. O., McLean, J. E., Britt, D. W., Johnson, W. P., Boyanov, M. I., & Anderson, A. J. (2012a). Fate of ZnO and Zn ions derived from ZnO nanoparticles in the plant environment. *Environmental Science & Technology*, *46*(4), 2161–2168. https://doi.org/10.1021/es202624t

Dimkpa, C. O., McLean, J. E., Latta, D. E., Manangón, E., Britt, D. W., Johnson, W. P., Boyanov, M. I., & Anderson, A. J. (2012b). CuO and ZnO nanoparticles: Phytotoxicity, metal speciation, and induction of oxidative stress in sand-grown wheat. *Journal of Nanoparticle Research*, *14*(9), 1125. https://doi.org/10.1007/s11051-012-1125-9

Ding, L., et al. (2018). Systems biology approaches in nanotoxicology. *Environmental Toxicology and Pharmacology*, *62*, 118–126.

Duhan, J. S., Kumar, R., & Kumar, N. (2022). Nanotechnology: The new perspective in precision agriculture. *Biocatalysis and Agricultural Biotechnology*, *41*, 102015. https://doi.org/10.1016/j.bcab.2022.102015

Elbagory, A. M., et al. (2016). A review on green synthesis of silver nanoparticles using plant extracts. *Nanomaterials*, *6*(11), 215.

El-Saadony, M. T., Shafi, M. E., & Alagawany, M. (2021). Green synthesis of metal nanoparticles using agricultural wastes and their applications: A review. *Environmental Nanotechnology, Monitoring & Management*, *15*, 100431. https://doi.org/10.1016/j.enmm.2021.100431

European Commission. (2011). Commission recommendation of 18 October 2011 on the definition of nanomaterial (2011/696/EU). *Official Journal of the European Union L*, *275*, 38–40.

Fariq, A., Khan, M. S., & Danish, M. (2017). Green synthesis of nanoparticles using plant extracts. *Biomaterials Research*, *21*(1), 13.

Fatima, S., Panda, P. K., & Jha, E. (2022). Green synthesized nanoparticles for wastewater remediation: A review. *Environmental Research*, *207*, 112235. https://doi.org/10.1016/j.envres.2021.112235

Fouad, H. A., et al. (2018). Uptake, translocation, and accumulation of nanoparticles in plants: Implications for food safety. *Nanotoxicology*, *12*(6), 834–847.

Gao, C., et al. (2020). Nanomaterials in regenerative medicine. *Materials Today*, *23*(3), 45–59.

Gao, F., Hong, F., Liu, C., Zheng, L., Su, M., & Wu, X. (2008). Mechanism of nano-anatase TiO_2 on promoting photosynthetic carbon reaction of spinach. *Biological Trace Element Research*, *119*(1), 68–76. https://doi.org/10.1007/s12011-007-0063-6

Ge, Y., Schimel, J. P., & Holden, P. A. (2011). Evidence for negative effects of TiO_2 and ZnO nanoparticles on soil bacterial communities. *Environmental Science & Technology*, *45*(4), 1659–1664.

Georgieva, V., et al. (2019). Nanoparticle exposure in plants: Methods and implications. *Science of the Total Environment*, *654*, 22–36.

González-Morales, S. I., et al. (2020). Nanoparticles as biocontrol agents in agriculture: Current status and future directions. *Nanomaterials*, 10(11), 2211.

Gurunathan, S., et al. (2018). The role of nanoparticles in food safety and human health. *Science of the Total Environment*, *639*, 1–9.

Hashemi, S. R., Mousavi, S. M., & Farhoodi, M. (2022). Plant-based nanoparticles: Toxicity, characterization, and sustainable applications. *Environmental Research*, *210*, 112917. https://doi.org/10.1016/j.envres.2022.112917

Heckman, K. L., DeCoteau, W., Estevez, A. Y., Reed, K. J., & Lipinski, W. J. (2013). Custom cerium oxide nanoparticles protect against a free radical mediated autoimmune degenerative disease in the brain. *ACS Nano*, *7*(12), 10582–10596.

Hernandez-Viezcas, J. A., Castillo-Michel, H., Servin, A., Peralta-Videa, J. R., & Gardea-Torresdey, J. L. (2013). Spectroscopic techniques for the analysis of engineered nanomaterials in plants: A review. *Environmental Science: Processes & Impacts*, *15*(1), 113–122.

Hong, J., Wang, C., Wagner, D. C., Gardea-Torresdey, J. L., & He, F. (2014a). Metal-based nanoparticle interactions with plants: Physiological effects and mechanisms. *Environmental Science: Nano*, *1*(6), 508–522. https://doi.org/10.1039/C4EN00071G

Hong, J., Wang, C., Zhang, C., & Wang, Z. (2014b). Translocation and physiological effects of foliar-applied ZnO nanoparticles in maize (Zea mays L.). *Plant Physiology and Biochemistry*, *83*, 163–172. https://doi.org/10.1016/j.plaphy.2014.08.004

Huang, J., et al. (2007). Biosynthesis of silver and gold nanoparticles by novel sundried Cinnamomum camphora leaf. *Nanotechnology*, *18*(10), 105104.

Iravani, S. (2011). Green synthesis of metal nanoparticles using plants. *Green Chemistry*, *13*(10), 2638–2650. https://doi.org/10.1039/C1GC15386B

Iravani, S., Korbekandi, H., Mirmohammadi, S. V., & Zolfaghari, B. (2014). Synthesis of silver nanoparticles: Chemical, physical and biological methods. *Research in Pharmaceutical Sciences*, *9*(6), 385–406.

Iravani, S., Varma, R. S., & Ahmad, B. (2023). Green synthesis of functional nanomaterials from plants: Mechanisms, characterization, and applications. *Green Chemistry Letters and Reviews*, *16*(1), 1–23. https://doi.org/10.1080/17518253.2023.2164781

Jiang, Y., et al. (2021). Proteomic responses to nanoparticles in cells and tissues. *Nature Nanotechnology*, *16*(8), 899–910.

Jin, X., et al. (2020). Nanotechnology in fertilizers and plant growth regulators: A review. *Agronomy*, *10*(9), 1356.

Kah, M., Beulke, S., Tiede, K., & Hofmann, T. (2013). Nanopesticides: State of knowledge, environmental fate, and exposure modeling. *Critical Reviews in Environmental Science and Technology*, *43*(16), 1823–1867.

Kah, M., et al. (2019). Nanotechnology for agricultural productivity: An update. *Critical Reviews in Plant Sciences*, *38*(4), 272–291.

Khalil, A. T., Ovais, M., & Khamlich, S. (2024). Phytofabricated nanomaterials as antiviral agents: Progress and perspectives. *Current Pharmaceutical Design*, *30*(3), 333–348. https://doi.org/10.2174/1381612829666230206141055

Khalil, M. M. H., et al. (2012). Green synthesis of silver nanoparticles using olive leaf extract and its antibacterial activity. *Arabian Journal of Chemistry*, *7*(6), 1131–1139.

Khan, I., Saeed, K., & Khan, I. (2020a). Nanoparticles: Properties, applications and toxicities. *Arabian Journal of Chemistry*, *12*(7), 908–931.

Khan, I., Saeed, K., & Khan, I. (2020b). Nanoparticles: Properties, applications and toxicities. *Arabian Journal of Chemistry*, *13*(7), 908–931.

Khan, I., Saeed, K., & Khan, I. (2023). Nanotechnology applications in agriculture: Plant-based nanoparticle synthesis and their roles. *Environmental Chemistry Letters*, *21*, 451–478. https://doi.org/10.1007/s10311-022-01422-7

Khan, S., Nazir, S., & Anjum, A. A. (2024). Plant-based nanomaterials for environmental applications: Current status and future prospects. *Environmental Advances*, *10*, 100362. https://doi.org/10.1016/j.envadv.2022.100362

Khan, Z., Alsharif, S., & El-Sheikh, M. A. (2022). Green nanoparticle synthesis and their toxicological effects: A systematic review. *Journal of Hazardous Materials*, *430*, 128430. https://doi.org/10.1016/j.jhazmat.2022.128430

Kharissova, O. V., Dias, H. R., Kharisov, B. I., Pérez, B. O., & Pérez, V. M. J. (2013). The greener synthesis of nanoparticles. *TrAC Trends in Analytical Chemistry*, *30*(10), 1351–1360.

Khodakovskaya, M., Dervishi, E., Mahmood, M., Xu, Y., Li, Z., Watanabe, F., & Biris, A. S. (2009). Carbon nanotubes are able to penetrate plant seed coat and dramatically affect seed germination and plant growth. *ACS Nano*, *3*(10), 3221–3227. https://doi.org/10.1021/nn900887m

Khodakovskaya, M. V., de Silva, K., Biris, A. S., Dervishi, E., & Villagarcia, H. (2012a). Carbon nanotubes induce growth enhancement of tobacco cells. *ACS Nano*, *6*(3), 2128–2135. https://doi.org/10.1021/nn204643g

Khodakovskaya, M. V., Kim, B. S., Kim, J. N., Alimohammadi, M., Dervishi, E., Mustafa, T., & Cernigla, C. E. (2012b). Carbon nanotubes as plant growth regulators: Effects on tomato growth, reproductive development, and gene expression. *Small*, *8*(6), 911–918. https://doi.org/10.1002/smll.201102654

Khot, L. R., Sankaran, S., Maja, J. M., Ehsani, R., & Schuster, E. W. (2012). Applications of nanomaterials in agricultural production and crop protection. *Crop Protection*, *35*, 64–70.

Kim, J., Park, S., & Lee, J. (2020). Recent advances in plant-mediated synthesis of nanoparticles and their applications: A review. *Environmental Technology & Innovation*, *20*, 101046. https://doi.org/10.1016/j.eti.2020.101046

Kolluru, S. S., et al. (2020). Application of nanoparticles in precision farming: A review. *Environmental Science and Pollution Research*, *27*, 30797–30811.

Kumar, S., et al. (2020). Targeted drug delivery using nanoparticles: A review. *International Journal of Pharmaceutics*, *572*(1), 1–16.

Kumar, V., Prasad, R., & Prasad, S. (2023). Phytosynthesized nanoparticles: A review of recent advances and future perspectives. *Journal of Nanostructure in Chemistry*, *13*, 1–25. https://doi.org/10.1007/s40097-022-00430-7

Kuo, C. H., et al. (2020). Nanoparticle toxicity in humans: Current concerns and research challenges. *Frontiers in Toxicology*, *9*, 45.

Kuppusamy, P., Yusoff, M. M., Maniam, G. P., & Govindan, N. (2016a). Biosynthesis of metallic nanoparticles using plant derivatives and their new avenues in pharmacological applications. An overview. *Nanomaterials*, *6*(12), 203.

Kuppusamy, P., Yusoff, M. M., Maniam, G. P., & Govindan, N. (2016b). Biosynthesis of metallic nanoparticles using plant derivatives and their new avenues in pharmacological applications—An updated report. *Saudi Pharmaceutical Journal*, *24*(4), 473–484. https://doi.org/10.1016/j.jsps.2014.11.013

Kurepa, J., Smalle, J. A., Van Montagu, M., & Inzé, D. (2010). Nanoparticles and plants: Good, bad, and unknown. *Plant Physiology*, *154*(2), 478–481. https://doi.org/10.1104/pp.110.162735

Larue, C., Castillo-Michel, H., Sobanska, S., Cécillon, L., Bureau, S., Barthès, V., . . . Sarret, G. (2012). Fate of pristine TiO_2 nanoparticles in the wheat plant: Cytotoxicity and elemental distribution analysis. *Environmental Science & Technology*, *46*(15), 7637–7643. https://doi.org/10.1021/es301308x

Lin, D., & Xing, B. (2007). Phytotoxicity of nanoparticles: Inhibition of seed germination and root growth. *Environmental Pollution*, *150*(2), 243–250. https://doi.org/10.1016/j.envpol.2007.01.016

Liu, J., et al. (2020a). Nanoparticle-induced inflammation and gene regulation pathways. *Journal of Nanoscience and Nanotechnology*, *20*(10), 6085–6097.

Liu, R., & Lal, R. (2015). Potentials of engineered nanoparticles as fertilizers for increasing agronomic productions. *Science of the Total Environment*, *514*, 131–139.

Liu, X., et al. (2020b). Toxicity assessment of nanoparticles: Harmonization of experimental protocols. *Environmental Toxicology and Chemistry*, *39*(9), 2422–2432.

Ma, X., Geiser-Lee, J., Deng, Y., & Kolmakov, A. (2010). Interactions between engineered nanoparticles (ENPs) and plants: Phytotoxicity, uptake and accumulation. *Science of the Total Environment*, *408*(16), 3053–3061. https://doi.org/10.1016/j.scitotenv.2010.03.031

Ma, Y., et al. (2015). Phytotoxicity and uptake of cerium oxide nanoparticles by wheat plants. *Environmental Science: Nano*, *2*(2), 144–153.

Makarov, V. V., et al. (2014). Green nanotechnologies: Synthesis of metal nanoparticles using plants. *Acta Naturae*, *6*(1), 35–44.

McLaughlin, M. J., Hamon, R. E., McLaren, R. G., Speir, T. W., & Rogers, S. L. (2010a). Review: A bioavailability-based rationale for controlling metal and metalloid contamination of agricultural land in Australia and New Zealand. *Australian Journal of Soil Research*, *38*(6), 1037–1086. https://doi.org/10.1071/SR99112

McLaughlin, M. J., Tiller, K. G., Naidu, R., & Stevens, D. P. (2010b). The behaviour and environmental impact of contaminants in fertilizers. *Australian Journal of Soil Research*, *34*(1), 1–54. https://doi.org/10.1071/SR9960001

Mishra, M., Kumari, P., & Bharti, R. (2022). Sustainable nanotechnology using plant-based approaches: Recent progress. *Materials Science for Energy Technologies*, *5*, 239–251. https://doi.org/10.1016/j.mset.2022.05.001

Mishra, S., et al. (2018). Application of nanotechnology in water and wastewater treatment. *Environmental Nanotechnology, Monitoring & Management*, *9*, 100260.

Mittal, A. K., Chisti, Y., & Banerjee, U. C. (2013). Synthesis of metallic nanoparticles using plant extracts. *Biotechnology Advances*, *31*(2), 346–356. https://doi.org/10.1016/j.biotechadv.2013.01.003

Mukherjee, A., Goswami, M., & Das, S. (2022). Applications of nanotechnology in sustainable agriculture and food systems. *Nanotechnology Reviews*, *11*(1), 1566–1582. https://doi.org/10.1515/ntrev-2021-0108

Mukhopadhyay, S. S., et al. (2016). Nanotechnology in agriculture: Prospects and constraints. *Nanotechnology, Science and Applications*, *9*, 63–71.

Naderi, M. R., & Danesh-Shahraki, A. (2013). Nanofertilizers and their roles in sustainable agriculture. *International Journal of Agriculture and Crop Sciences*, *5*(19), 2229–2232.

Nair, R., Varghese, S. H., Nair, B. G., Maekawa, T., Yoshida, Y., & Kumar, D. S. (2010). Nanoparticulate material delivery to plants. *Plant Science*, *179*(3), 154–163. https://doi.org/10.1016/j.plantsci.2010.04.012

Narayanan, K. B., & Sakthivel, N. (2010). Biological synthesis of metal nanoparticles by microbes. *Advances in Colloid and Interface Science*, *156*(1–2), 1–13.

Nasrollahzadeh, M., et al. (2019). Plant-mediated green synthesis of nanoparticles: A review. *Inorganic Chemistry Communications*, *107*, 107495.

Nasrollahzadeh, M., et al. (2020a). Recent advances in green synthesis of nanoparticles using plant extracts and their applications. *Green Chemistry Letters and Reviews*, *13*(2), 109–134.

Nasrollahzadeh, M., et al. (2020b). Recent advances in green synthesis of nanoparticles using plant extracts. *Advances in Colloid and Interface Science*, *277*, 102125.

Nasrollahzadeh, M., Sajjadi, M., & Iravani, S. (2021). Green synthesis of metal nanoparticles using plants: Mechanisms and applications. *Advances in Colloid and Interface Science*, *289*, 102316. https://doi.org/10.1016/j.cis.2020.102316

Nel, A., et al. (2006). Toxic potential of materials at the nanolevel. *Science*, *311*(5761), 622–627.

Ojha, S., Shukla, S., & Kumar, P. (2024). Role of plant extracts in green nanoparticle synthesis for biomedical and agricultural applications. *Plant Physiology Reports*, *29*, 1–18. https://doi.org/10.1007/s40502-023-00699-1

Palanikumar, L., Subbaiya, R., & Karmegam, N. (2022). Bio-inspired nanomaterials for sustainable agriculture: A review. *Environmental Nanotechnology, Monitoring & Management*, *18*, 100732. https://doi.org/10.1016/j.enmm.2022.100732

Parisi, C., Vigani, M., & Rodríguez-Cerezo, E. (2015). Agricultural nanotechnologies: What are the current possibilities? *Nano Today*, *10*(2), 124–127.

Pérez, S., et al. (2020). Bioaccumulation and ecotoxicity of engineered nanoparticles in aquatic and terrestrial organisms. *Science of the Total Environment*, *731*, 138955.

Pérez-de-Luque, A. (2017). Interaction of nanomaterials with plants: What do we need for real applications in agriculture? *Frontiers in Environmental Science*, *5*, 12.

Priester, J. H., et al. (2012). Soybean susceptibility to manufactured nanomaterials with evidence for food quality and soil fertility interruption. *Proceedings of the National Academy of Sciences*, *109*(37), E2451–E2456.

Rai, M., & Ingle, A. (2012). Role of nanotechnology in agriculture with special reference to management of insect pests. *Applied Microbiology and Biotechnology*, *94*(2), 287–293.

Rai, M., & Yadav, A. (2013). Plants as potential synthesizers of precious metal nanoparticles: Progress and prospects. *International Journal of Nanotechnology*, *10*(9–10), 878—902.

Rai, M., Ingle, A. P., & Gupta, I. (2018a). Green synthesis of metal nanoparticles using microorganisms and plants. *Encyclopedia of Biomedical Engineering*, *3*, 32–44.

Rai, M., Ingle, A. P., & Pandit, R. (2018b). Nanotechnology for agriculture and food science. *Nanotechnology*, *29*(43), 432001.

Raliya, R., Nair, R., Chavalmane, S., Wang, W. N., & Biswas, P. (2015). Mechanistic evaluation of translocation and physiological impact of titanium dioxide and zinc oxide nanoparticles on the tomato (Solanum lycopersicum L.) plant. *Metallomics*, *7*(12), 1584–1594. https://doi.org/10.1039/C5MT00168D

Rani, P., Kumari, S., & Saini, P. (2021). Phyto-mediated synthesis of metal nanoparticles: Biomedical applications and future prospects. *Biotechnology Reports*, *30*, e00600. https://doi.org/10.1016/j.btre.2021.e00600

Ravichandran, R., Palanisamy, T., & Harikrishnan, R. (2021). Nanotechnology in food science: Applications and challenges. *Current Opinion in Food Science*, *39*, 60–65. https://doi.org/10.1016/j.cofs.2020.12.002

Rico, C. M., Majumdar, S., Duarte-Gardea, M., Peralta-Videa, J. R., & Gardea-Torresdey, J. L. (2011). Interaction of nanoparticles with edible plants and their possible implications in the food chain. *Journal of Agricultural and Food Chemistry*, *59*(8), 3485–3498. https://doi.org/10.1021/jf104517j

Rocco, M., et al. (2020). Standardizing nanoparticle characterization for environmental and biological studies. *Nature Nanotechnology*, *15*(9), 697–709.

Rochman, C. M., et al. (2019). Chronic exposure to nanomaterials in the environment: Implications for ecosystems and risk assessment. *Environmental Toxicology and Chemistry*, *38*(10), 2071–2080.

Rui, M., Ma, C., Zhang, C., Zhao, Q., & Guo, J. (2016a). Transformation of ceria nanoparticles in cucumber plants is influenced by phosphate. *Environmental Pollution*, *219*, 579–586. https://doi.org/10.1016/j.envpol.2016.06.077

Rui, M., Ma, C., Zhao, J., Shen, Z., Zhou, X., & Zhang, Z. (2016b). Iron oxide nanoparticles as a potential iron fertilizer for peanut (Arachis hypogaea). *Frontiers in Plant Science*, *7*, 815. https://doi.org/10.3389/fpls.2016.00815

Sahoo, C., Sahu, S., & Tripathy, S. (2017). Green synthesis of metallic nanoparticles using plant extracts and their applications: A review. *Materials Today: Proceedings*, *4*(9), 4111–4120. https://doi.org/10.1016/j.matpr.2017.06.010

Sastry, N. S. R., et al. (2020). Enhancing plant growth under stress conditions using nanoparticle-based formulations. *Agricultural Sciences*, *11*(3), 378–394.

Sathiyabama, M., & Manikandan, A. (2023). Green synthesis of nanoparticles and their applications in food and agriculture: Recent trends. *Sustainable Chemistry and Pharmacy*, *32*, 100887. https://doi.org/10.1016/j.scp.2022.100887

Schwab, F., Zhai, G., Kern, M., Turner, A., Schnoor, J. L., & Wiesner, M. R. (2016). Barriers, pathways and processes for uptake, translocation and accumulation of nanomaterials in plants–Critical review. *Nanotoxicology*, *10*(3), 257–278. https://doi.org/10.3109/17435390.2015.1048326

Servin, A., Elmer, W., Mukherjee, A., De la Torre-Roche, R., Hamdi, H., White, J. C., . . . Gardea-Torresdey, J. (2015). A review of the use of engineered nanomaterials to suppress plant disease and enhance crop yield. *Journal of Nanoparticle Research*, *17*(2), 92. https://doi.org/10.1007/s11051-015-2907-1

Shafiq, M., Hussain, T., & Raza, Z. A. (2023). Risk assessment of plant-derived nanoparticles: Challenges and regulatory perspectives. *Environmental Advances*, *11*, 100339. https://doi.org/10.1016/j.envadv.2022.100339

Shaik, M. R., Khan, M., & Adil, S. F. (2022). Green synthesis of nanoparticles using plant extracts and their biomedical applications: A review. *Saudi Journal of Biological Sciences*, *29*(4), 2297–2311. https://doi.org/10.1016/j.sjbs.2021.12.065

Shaikh, S., et al. (2019a). Plant-based green synthesis of nanoparticles. *Nanotechnology Reviews*, *8*(1), 116–127.

Shaikh, S., Nazam, N., Rizvi, S. M. D., Ahmad, K., Baig, M. H., Lee, E. J., & Choi, I. (2019b). Mechanistic insights into the biological activities of chitosan: A review. *International Journal of Biological Macromolecules*, *121*, 1256–1268.

Shankar, S. S., Rai, A., Ahmad, A., & Sastry, M. (2004). Rapid synthesis of Au, Ag, and bimetallic Au core–Ag shell nanoparticles using Neem (Azadirachta indica) leaf broth. *Journal of Colloid and Interface Science*, *275*(2), 496–502.

Sharma, D., et al. (2019a). Plants mediated biosynthesis of silver nanoparticles and their applications: A review. *Environmental Nanotechnology, Monitoring & Management*, *12*, 100234.

Sharma, G., et al. (2019b). Green synthesis of nanoparticles and their potential applications. *Environmental Nanotechnology, Monitoring & Management*, *12*, 100255.

Sharma, V., et al. (2019c). Nanoparticles in agriculture: A review of their potential applications and risks. *Environmental Science and Pollution Research*, *26*, 453–463.

Sharma, D., Gupta, A., Singh, P., & Kumar, R. (2024). Advances in plant-mediated synthesis of nanoparticles: Mechanistic insights, applications, and future perspectives. *Journal of Environmental Chemical Engineering*, *12*(3), 113457. https://doi.org/10.1016/j.jece.2024.113457

Siddiqi, K. S., Husen, A., & Rao, R. A. K. (2018). A review on biosynthesis of silver nanoparticles and their biocidal properties. *Journal of Nanobiotechnology*, *16*(1), 14.

Siddiqi, K. S., Husen, A., & Rao, R. A. K. (2020). A review on biosynthesis of silver nanoparticles and their biocidal properties. *Journal of Nanobiotechnology*, *18*(1), 1–29. https://doi.org/10.1186/s12951-020-00650-4

Singh, D., et al. (2017a). Impact of engineered nanoparticles on the growth and development of plants: A review. *Environmental Nanotechnology, Monitoring & Management*, *7*, 1–21.

Singh, D., Kumar, A., & Singh, K. (2017b). Engineered nanoparticles: Plant growth promoting and phytotoxicity. In *Handbook of nanomaterials for industrial applications* (pp. 543–556). Elsevier.

Singh, J., Dutta, T., & Kim, K. H. (2022). The green synthesis of nanoparticles: A review on their current status and future prospects. *Environmental Research*, *206*, 112589. https://doi.org/10.1016/j.envres.2021.112589

Singh, J., Lee, B. K., & Mahajan, M. (2017c). Influence of nanoparticles on the germination of seeds: present status and prospects. *Advances in Natural Sciences: Nanoscience and Nanotechnology*, *8*(1), 013001. https://doi.org/10.1088/2043-6254/aa5657

Singh, J., et al. (2018). Green synthesis of metallic nanoparticles: Applications and limitations. *Materials Letters*, *210*, 75–78.

Singh, N., Bhatia, S., & Singh, P. (2024). Green nanotechnology for sustainable agriculture: Challenges and opportunities. *Sustainable Energy Technologies and Assessments*, *59*, 102854. https://doi.org/10.1016/j.seta.2023.102854

Singh, P., Kim, Y. J., Zhang, D., & Yang, D. C. (2021). Biological synthesis of nanoparticles from plants and microorganisms. *Trends in Biotechnology*, *39*(4), 347–360.

Sivasankarapillai, V. S., Shukla, S. S., & Thakur, M. S. (2020). Recent advances in plant-mediated nanoparticle synthesis: Mechanisms, characterization, and applications. *Journal of Nanostructure in Chemistry*, *10*(3), 255–291. https://doi.org/10.1007/s40097-020-00343-2

Soni, S., et al. (2018). Nanotechnology in pest control: A review. *International Journal of Environmental Science and Technology*, *15*, 2455–2470.

Spielman-Sun, E., et al. (2017). Fate and transformation of silver nanoparticles in soils: A review of the science. *Environmental Science: Nano*, *4*(5), 1250–1267.

Subramanian, S. B., Manikandan, M., & Ramesh, V. (2021). Green synthesis, characterization, and applications of plant-based metallic nanoparticles: A review. *Environmental Nanotechnology, Monitoring & Management*, *16*, 100526. https://doi.org/10.1016/j.enmm.2021.100526

Subramanian, S. B., Murugan, K., Panneerselvam, C., Madhiyazhagan, P., & Hwang, J. S. (2015). Green synthesis of silver nanoparticles using plant extracts and their potential biomedical applications. *Advances in Natural Sciences: Nanoscience and Nanotechnology*, *6*(3), 035008. https://doi.org/10.1088/2043-6262/6/3/035008

Sun, H., Yang, X., Wang, W., Ge, C., & Wang, X. (2014a). Phytotoxicity, translocation, and biotransformation of $NaYF_4$ upconversion nanoparticles in a rice plant. *Small*, *10*(23), 4379–4388. https://doi.org/10.1002/smll.201400859

Sun, L., et al. (2014b). Physiological responses of plants to engineered nanoparticles: A review. *Environmental Reviews*, *22*(3), 298–308.

Thakkar, K. N., Mhatre, S. S., & Parikh, R. Y. (2010). Biological synthesis of metallic nanoparticles. *Nanomedicine: Nanotechnology, Biology and Medicine*, *6*(2), 257–262. https://doi.org/10.1016/j.nano.2009.07.002

Tripathi, D. K., Singh, S., Singh, S., Srivastava, P. K., Singh, V. P., Singh, S., . . . Chauhan, D. K. (2017a). Uptake, accumulation and toxicity of silver nanoparticle in autotrophic plants, and heterotrophic microbes: A concentric review. *Frontiers in Microbiology*, *8*, 7. https://doi.org/10.3389/fmicb.2017.00007

Tripathi, D. K., Singh, V. P., & Sharma, S. (2017b). Role of nanoparticles in improving plant performance under environmental stress. In *Nanoscience and plant–soil systems* (pp. 313–330). Springer.

Vishwakarma, K., Singh, J., & Yadav, S. K. (2022). Role of nanotechnology in sustainable agriculture: A review. *Frontiers in Nanotechnology*, *3*, 622179. https://doi.org/10.3389/fnano.2021.622179

Wang, P., Lombi, E., Zhao, F. J., & Kopittke, P. M. (2016). Nanotechnology: A new opportunity in plant sciences. *Trends in Plant Science*, *21*(8), 699–712. https://doi.org/10.1016/j.tplants.2016.04.005

Wang, Z., Xie, X., Zhao, J., Liu, X., Feng, W., & White, J. C. (2017). Xylem- and phloem-based transport of CuO nanoparticles in maize (*Zea mays* L.). *Environmental Science & Technology*, *51*(9), 5154–5162. https://doi.org/10.1021/acs.est.7b00616

Wani, A. B., Jaiswal, P., & Kumar, P. (2023). Nanoscale delivery systems in plant biotechnology: Recent developments. *Plant Cell, Tissue and Organ Culture (PCTOC), 155*, 309–325. https://doi.org/10.1007/s11240-023-02499-3

Xia, T., et al. (2019). Metabolomics and nanoparticle toxicity. *Science Advances*, *5*(1), eaav9981.

Yin, Z., et al. (2020). Nanotechnology in medical diagnostics: Progress and challenges. *Nanomedicine: Nanotechnology, Biology and Medicine*, *24*, 123–132.

Zahin, N., et al. (2020). Plant-mediated green synthesis of nanoparticles. *Nanomaterials*, *10*(12), 2352.

Zhang, X., Li, Q., Xu, D., Zhang, Y., & Wang, J. (2018). Green synthesis of metal nanoparticles using plant extracts and their applications in environmental and biomedical fields. *Journal of Cleaner Production*, *172*, 114–124. https://doi.org/10.1016/j.jclepro.2017.10.194

Zhang, X. F., Liu, Z. G., Shen, W., & Gurunathan, S. (2016). Silver nanoparticles: Synthesis, characterization, properties, applications, and therapeutic approaches. *International Journal of Molecular Sciences*, *17*(9), 1534.

Zhang, Y., et al. (2020). Nanoparticles and inflammation: The role of NF-κB signaling. *International Journal of Nanomedicine*, *15*, 4795–4810.

Zhang, Y., et al. (2021). The role of nanoparticles in enhancing crop resilience and productivity. *Science of the Total Environment*, *765*, 142694.

Zheng, L., Hong, F., Lu, S., & Liu, C. (2005). Effect of nano-TiO_2 on strength of naturally aged seeds and growth of spinach. *Biological Trace Element Research*, *104*(1), 83–91. https://doi.org/10.1385/BTER:104:1:083

Zuverza-Mena, N., Armendariz, R., Peralta-Videa, J. R., & Gardea-Torresdey, J. L. (2016). Effects of silver nanoparticles on radish sprouts: Root growth reduction and modifications in the nutritional value. *Frontiers in Plant Science*, *7*, 90. https://doi.org/10.3389/fpls.2016.00090

2 Gold Nanoparticles
Green Synthesis, Characterization, and Biomedical Applications

P. Saranraj, Sunita Singh, M. Manigandan, K. Kesavardhini, B. Lokeshwari, and K. Gayathri

2.1 INTRODUCTION

Biomedical and technical uses of gold nanoparticles (AuNPs) encompass medication and gene delivery, protein and pathogen detection, DNA labeling, the field of tissue engineering, and also imaging techniques such as MRI. Their highly impressive properties of optics, synthesis, biocompatibility, and control over size and shape have been extensively utilized. AuNPs are usually functionalized with amines, oligonucleotides, peptides, antibodies, lipids, or stabilizers like sodium citrate, transferrin, and acetyl tri-methyl ammonium bromide to provide stabilized AuNPs against aggregation (Giljohann *et al.*, 2010). The numerous recognized benefits of AuNPs have been undermined by potential toxicity. The whole point of trying to limit agents to non-toxic categories is to result in some improved biocompatibility with the nanoparticles (Bennur *et al.*, 2016). Gold from HAuCl4 solutions transforms into nanoparticles during the creation of AuNPs.

Salomi *et al.* (2023) assert that the dimensions and morphology of nanoparticles are significantly affected by the pH level, temperature, rate, and, crucially, the choice of decreasing and stabilizing agents. Electrochemical techniques using amino-terminated ionic liquids have successfully synthesized nanoparticles as small as 1.7 nm. It also showed the efficacy of plant extracts like tea leaves, *Citrifolia* roots, and bayberry tannin in synthesizing gold nanoparticles. These green synthesis techniques are based on the principle that traditional chemical synthesis methods produce hazardous byproducts that can cause biological and environmental risks (Das *et al.*, 2009).

Nanomaterials are produced through eco-friendly methods for diverse applications in biotechnology, including pharmacology, tissue engineering, medicine, drug delivery, and agriculture (Luo *et al.*, 2014). The morphology of AuNPs determines their physical, chemical-based, electrical, and optical properties in areas such as bioimaging, electronic nanotechnology, and optoelectronics (Mosquera-Romero *et al.*, 2022). AuNPs are used in the medical domain for their sensitivity and selectivity in detecting hazardous heavy metal ions. They have been employed to coat

DOI: 10.1201/9781003559900-2

medical apparatus, effectively inhibiting infections caused by *Candida albicans* and drug-resistant bacteria. Plant-derived AuNPs exhibit significant free radical scavenging activity, enhancing their applicability in antioxidative and therapeutic applications.

The diminished cytotoxicity of AuNPs compared to other metal nanoparticles like silver, platinum, and copper makes them more suitable for biomedical applications. The distinctive size-dependent surface plasmon resonance of gold nanoparticles enhances their efficacy in biological applications (Raveendran *et al.,* 2006). The formation of AuNPs involves physical processes such as thermolysis or irradiation and chemical processes such as the reduction by sodium citrate or ascorbic acid of gold (III) ions; there are many other chemical processes, such as through the use of surfactants or metal oxides to inhibit nanoparticle aggregation. Natural reduction and stabilizing agents have recently gained significance because to their environmentally benign and biocompatible. The use of "green" synthesis aligns with the growing emphasis on sustainable nanotechnology (Vedha *et al.,* 2023).

The application of AuNPs such as biomedicine, catalysis, remediation of the environment and technologies for conversion is not comprehensive. Their exotic plasmonic properties enable them to function as excellent sensors and catalysts in the surface chemistry that underpins drug delivery targeted for use in precision medicine applications. It has enhanced scalability in the application of AuNPs for industrial use in biocompatible green synthesis techniques (Mishra *et al.,* 2022). The effective large-scale synthesis of AuNPs of these qualities, using renewable bioresources, and optimizing the synthesis to the particular requirements for different applications. Gold nanoparticles in nanotechnology have distinctive physical and chemical attributes (Timoszyk, 2018). Synthesis through greener methods reduces environmental and health issues related to previous chemical synthesis and shows that AuNPs are truly feasible solutions to sustainable technology problems. It is eeded for future improvement in medicine and biotechnology applications, for this becomes significant in next-generation materials (Naveena and Prakash, 2013).

2.2 GREEN SYNTHESIS OF GOLD NANOPARTICLES

2.2.1 Importance of Green Synthesis of Gold Nanoparticles

The utilization of AuNPs requires creative, multidisciplinary approaches in their preparation to enhance stability and safety. Traditional chemical synthesis methods frequently have toxicity and environmental drawbacks. Green synthesis has emerged as a viable environmentally friendly process (Saranraj *et al.,* 2024). This process creates biocompatible nanoparticles by using natural bioresources like microorganisms, plant extracts, and biomolecules, rather than dangerous chemicals (Abhijith and Thakur, 2012). Numerous nanoparticles, including cobalt ferrite nanoparticles made from grape peel and pulp extracts and cobalt-zinc ferrite nanoparticles mediated by honey, have been successfully produced using green synthesis. In green synthesis, biological organisms are usually used to bio-reduce a metal salt precursor, like chloroauric acid (Basavegowda *et al.,* 2013).

TABLE 2.1
Bioagent-Based Green Synthesis of Gold Nanoparticles and Their Applications

Bioagents	Type of Bioagents	Role in Synthesis	Applications	References
Plant extracts	Phytochemicals	Reducing and stabilization of gold ions	Catalysis, drug delivery, and antimicrobial activities	Nangia *et al.* (2009)
Bacteria	Microorganisms	Enzymatic reduction of gold ions	Biosensors and biomedicine	Rajan *et al.* (2015)
Fungi	Microorganisms	Enzyme secretion for nanoparticle synthesis	Cancer therapy and nanomedicine	Vinay Gopal *et al.* (2013)
Algae	Photosynthetic organisms	Reduction via metabolites	Environmental applications and biomedical uses	Zhang *et al.* (2011)
Proteins	Biomolecules	Capping and stabilizing agents	Health monitoring and medicine delivery	Salam *et al.* (2012)
Polysaccharides	Biopolymers	Stabilization and reduction	Precise medicine delivery and tissue repair	Yu *et al.* (2016)

The molecules and metabolites included in these biological substances, including proteins, phenols, as fatty acids, and the presence of sugars and enzymes are crucial for reducing and stabilizing agents during nanoparticle synthesis (Wu *et al.*, 2022). Three elements of the process depend on environmental sustainability, natural reducing agents from biological resources, environmentally friendly solvents like water as reaction medium, and non-toxic capping agents like polyphenols and other secondary metabolites that improve stability and prevent particle aggregation (Aljabali *et al.*, 2018).

Green synthesis is termed as a new possible area that brings advantages like reduction of toxicity, low cost, and scalabilities in hammering out nanoparticles for drug delivery, environmental remediation, diagnostics, and catalysis in a very wide variety of sizes and functions (Tatarchuk *et al.*, 2021). This is the way of developing sustainable nanotechnology by integrating with materials science, biology, and chemistry through its green synthesis. It has great potential for bringing a much-needed breakthrough to nanoparticle synthesis that would facilitate better analysis and optimization of industrial processes to develop future technology needs with efficient and eco-friendly alternatives (Table 2.1; Geethalakshmi and Sarada, 2013).

2.3 BIOSYNTHESIS OF PLANT-BASED GOLD NANOPARTICLES

AuNPs can be produced using the one-pot method in a straightforward and eco-friendly manner. The material can be finely chopped or powdered after being rinsed with distilled water (Kuppusamy *et al.*, 2016). The bioactive ingredients must be extracted by heating the previously mentioned ingredients to particular temperatures above purified water (Wang *et al.,* 2012). A colour shift from the transparent gold solution is indicative of the conversion of gold ions to AuNPs when the plant extract is mixed with varying concentrations of salt solution.

The production of nanoparticles using natural plant components like tannins, flavonoids, alkaloid compounds, steroids, and saponins as stabilizing and reducing agents is more environmentally friendly and economically viable. Among the metals that can be transformed into green nanoparticles utilizing plant resources, copper, cobalt, silver, gold, magnetite, and zinc oxide call particular attention (Drummer *et al.,* 2021). For their maximum metabolite concentration, leaves have always been the most often selected plant components. They are usually produced from *Salix alba* L., a stable AuNP shows anti-inflammatory and antipyretic properties. By a reduction of chemical components in the extract of the leaves, the interaction of gold ions with this aqueous solution produced colloidally stabilized AuNPs (Driskell *et al.,* 2011). It is useful for many applications, as these nanoparticles stay stable at different pH and salinity levels (Velmurugan *et al.,* 2014). Indicating its possible efficiency in biological and medical uses, functionalized *Salix alba* AuNPs show antifungal, analgesic, and muscle relaxant characteristics. Applied utilizing a mixture of extracts from leaves with aqueous gold ions, the synthesis is adjustable and conducted in a single pot, with all dimensions of nanoparticle shape controlled by varying quantities of plant-derived metabolites (Begum *et al.,* 2022).

2.4 PHYSICAL AND CHEMICAL CHARACTERIZATION OF GOLD NANOPARTICLES

2.4.1 Ultraviolet–Visible Spectroscopy (UV-VIS)

The Localized Surface Plasmon Resonance (LSPR) band refers to the resonance effects created by surface electrons moving on a metal that oscillate in harmony with incoming electromagnetic waves. Islam *et al.* (2019) assert that it happens for each particle whose size is equal or less than activating light. An absorption spectra peak at 520 nm indicates the plasmon wavelength of resonance. LSPR spectrum is done to check the alterations of size, allotment, aggregate, and surface alteration of the nanoparticles (Wang *et al.,* 2008). Ultraviolet-visible spectroscopy can identify AuNP aggregation. Aggregate structure forms the Surface Plasmon Resonance (SPR) band and indicates near about 600 nm. Combined with other complementary analysis techniques, UV-Vis spectroscopy may enhance the dimension characterization (Vijayakumar *et al.,* 2017). With Beer-Lambert law that relates UV absorbance with sample concentration and path length, AuNP solution concentration has been calculated. The authors Mukherjee *et al.* (2001) would recommend the absorbance band in the 500–600 nm range where the LSPR for AuNPs occurs.

2.4.2 Fourier Transform Infrared Spectroscopy (FTIR)

FTIR spectroscopy is a non-destructive technique for identification of the chemical components of nanoparticles, the functional chemicals connected to the nanoparticle being absorbed or transmitted. Among the functional groups seen in AuNPs' FTIR spectra are amines, carboxylic acid phosphates, and cysteine residues (Yamini *et al.*, 2019). Phenolic and alcoholic groups have been shown to cause peak changes in absorption because of their potential to interact with biomolecules and alter the C=O peak. Flavonoids, tannins, protein molecules, alkaloids, limonene pectin, and carboxylic acid all aid in the reduction of gold ions to promote functionalizing and therapeutic delivery (Yang *et al.*, 2014).

2.4.3 Dynamic Light Scattering (DLS)

The aggregation state and size distributions of AuNP can be accurately assessed by dynamic light scattering. Described are changes in size with respect to interactions, stability, ligand binding, and aggregation of conjugated AUNPs (Ghosh *et al.*, 2008).

2.4.4 Scanning Electron Microscope (SEM)

Morphological size, shape, and distribution patterns of particles utilize data created from the SEM, which observes how sample morphology changes with treatment (Jeyaraj *et al.*, 2017). It measures pore sizes and differences in cell morphology to ascertain antibacterial activity of the nanoparticles. AuNPs have a wide variety of morphologies and signals created under SEM study, which are important to comprehend their nature (Poojary *et al.*, 2016).

2.4.5 Transmission Electron Microscopy (TEM)

In nanoscale levels, transmission electron microscopy, in some cases referred to as TEM, possesses maximum resolving power when imaging gold nanoparticles, and so it is an essential technique used in nanotechnology for particle shape and size measurement. Low-angle rotating metal-shadowing TEM and cryogenic transmission-electron microscopy (cryo-TEM) are TEM techniques used in the modification of instruments to achieve surface modifications of concern. In addition to providing qualitative, quantitative, and structural details, high-resolution TEM can potentially display the individual atom on a particular nanoparticle. Limitations of TEM include constrained image magnification and inability to analyze materials thinner than 300 nm. Samples larger than 100 nm or 10–20 nm were therefore examined. Eight round AuNPs between 14 and 33 nm in size were examined with the use of TEM.

2.4.6 Atomic Force Microscope (AFM)

This is an analytic measurement tool for nanoparticle hardness, surface topography, size, and shape. The three-dimensional high-resolution imaging it can also provide will enable one to view single nanoparticles or their agglomerate. Through its volume and surface area distribution, the AFM assists in assessing the size and physical features of nanoparticles (Pal *et al.*, 2011). This will mainly help with the investigation of rough surfaces due to surface modification or nanoparticle clustering. AFM possesses immense applications in nanoscience and nanotechnology at the nanoscale and immense flexibility with a plethora of nanoparticles (Sublemontier *et al.*, 2014).

2.4.7 Thermogravimetric Analysis (TGA)

TGA is an accurate and efficient method to analyze the chemical structure and purity of nanomaterials. Thermal values are illustrated in terms of the TGA decomposition curve (Mahoney *et al.*, 2020). Usually, the remaining mass can present signs of the presence of contaminants, metal catalysts from synthesis processes, or residual inorganics, becoming extremely crucial for nanomaterials as it can measure organics, assess melting characteristics on the surface, and oxidize resistance to high temperatures (Alaallah *et al.*, 2023).

2.4.8 X-ray Photoelectron Spectroscopy (XPS)

XPS is a high-precision spectroscopic technique conducted in an ultra-high vacuum, focusing on the atomic surface properties of various substances, including coatings, the catalysts, and nanomaterials, at nanometre-thick atomic layers. This non-destructive analysis may allow quantitative and qualitative measurement of a sample without changing its original state; hence, it applies to various other samples, solids, and liquids to gases (Shankar *et al.*, 2004). The exterior chemistry of a nanoparticle primarily examines their binding and ligand-coordination interactions (Flieger *et al.*, 2021). Therefore, XPS is an important technique for surface investigations establishing oxidation states, chemical interactions, and surface functionalities during the growth and characterization of complex materials and nanostructures.

2.4.9 X-ray Diffraction (XRD)

XRD is the most suitable technique for evaluating the atomic arrangement and crystallization of AuNPs. Diffraction patterns and constructive interference produced by X-ray interactions with matter offer details about the crystalline structure of that substance. In this regard, XRD offers important information on atomic structure and crystallinity, therefore facilitating investigation and characterization. Fundamental tools for qualitative and quantitative evaluations of materials are diffraction peaks exposing the Face-Centered Cubic (FCC) crystalline structure (Khalil *et al.*, 2012).

2.4.10 Thermogravimetric Analysis (TGA) and Differential Scanning Calorimetry (DSC)

Thermal properties and purity of nanomaterials are evaluated using TGA and DSC. *Aspergillus trinidadensis* improves the thermally stable characteristics of gold nanoparticles, as TGA analysis shows. DSC gives the thermal degradation, adsorption, and melting or crystallization properties of nanomaterials, claims Kanwar *et al.* (2019). Fundamental approaches of thermal characterization are TGA and DSC procedures. As capping agents improve the stiffness and thermal properties of nanoparticles, therefore making them ideal for many uses, the surface changes (Sarkar *et al.*, 2012).

2.5 ADVANTAGES OF GREEN SYNTHESIS OF GOLD NANOPARTICLES

Figure 2.1 summarizes the core advantages of the green synthesis pathway for gold nanoparticles. These benefits include superior safety, cost-effectiveness, and the use of renewable resources. The resulting AuNPs are biocompatible and possess valuable biomedical properties, making the process ideal for large-scale, sustainable production.

2.6 BIOMEDICAL APPLICATIONS OF GOLD NANOPARTICLES

2.6.1 Antimicrobial Activity

Gold nanoparticles possess a radius of 25 nm and demonstrate significant antibacterial efficacy against *Staphylococcus aureus* and *Escherichia coli*, with inhibition zones measuring 18 mm for *Staphylococcus aureus* and 25–32 mm for *Escherichia*

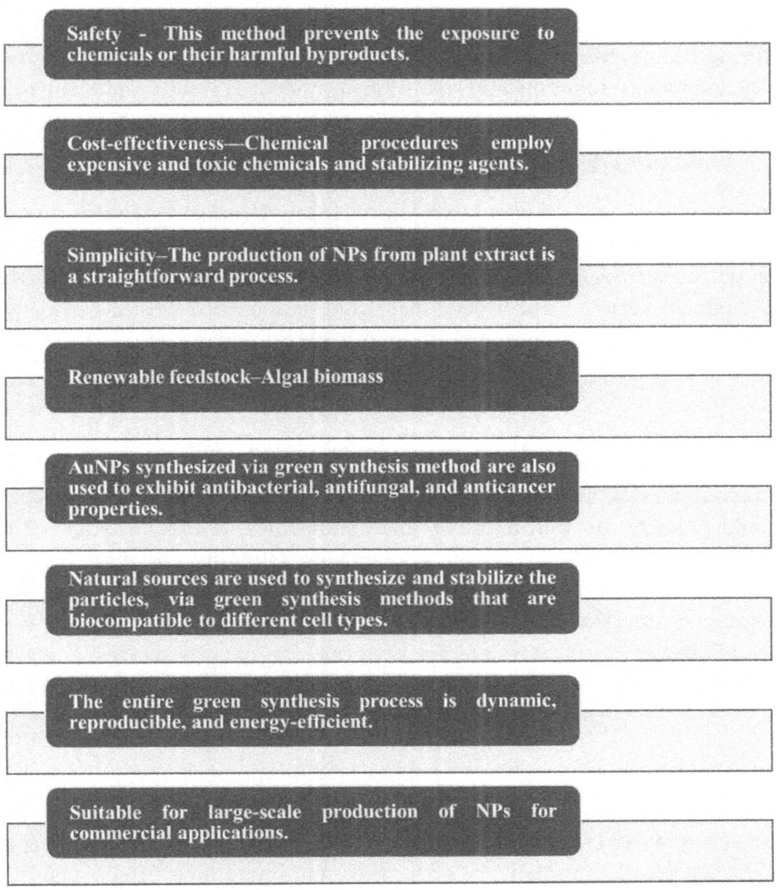

FIGURE 2.1 Advantages of green synthesis of gold nanoparticles.

coli. They are ineffectual against Gram-negative bacteria, including *Klebsiella pneumoniae* and *Proteus mirabilis*. Gold nanoparticles synthesized from natural sources including garlic, basil, and rosemary exhibit significant antibacterial effects (Li *et al.,* 2021): Gold nanoparticle interactions with sulfur and phosphate groups destroy bacterial cell membranes, resulting in DNA damage and reactive oxygen species generation (Philip, 2009).

2.6.2 Anti-Tuberculosis Effect

Gold nanoparticles synthesized through green chemistry from flora, e.g. *Nigella arvensis* and *Chenopodium formosanum*, have been found to possess important pharmacological efficacy. These have been established to be effective for treating tuberculosis, and studies conducted on ethanolic and hydroalcoholic extracts have revealed anti-tuberculosis activity (Lowry *et al.,* 2016). AgNPs have exhibited catalytic activity in reducing the amount of methylene blue, which is a dye frequently utilized in laboratory environments. AgNPs that are derived from plants can induce cytotoxicity in cancer cell lines, indicating their potential uses for cancer therapies. Crucially in the treatment of cancer and infectious diseases, AgNPs show biocompatibility and natural reducing and stabilizing properties (Sujitha and Kannan, 2013).

2.6.3 Anticancer Effect

Gold nanoparticles, which are synthesized from *Mentha longifolia* and Carica papaya, hold potential applications for cancer cell treatment (Suman *et al.,* 2014). Silver nanoparticles from *Mentha longifolia* have demonstrated anti-proliferative activity and death in various cancer cell lines; gold nanoparticles from *Carica papaya* have demonstrated anti-tumor activity against breast and liver cancer cells (Yusefi *et al.,* 2021).

2.6.4 Antioxidant Effect

The antioxidant effect is the protection against free radical damage to nucleic acids, lipids, and proteins, as well as other macromolecules. It can be either of natural or synthetic origin, synthesized from polyphenols, vitamins, carotenoids, as well as intrinsic systems (Kumar *et al.,* 2008). Synthetic antioxidants, such as phenolics and nano-oxidants, are also vital. Antioxidant activity depends on their active groups and the number of times they interact with reactive oxygen species. They further prevent hydrogen peroxide degradation, chelate metal ions, and block the activity of enzymes.

2.6.5 Drug Delivery

Vaccines, proteins, enzymes, and molecules, along with other pharmaceutical agents, depend on advanced drug delivery systems to maximize their effectiveness. Traditional methods like suspensions, emulsions, and solutions lack selectivity, stability, and drug concentration stability (Chauhan *et al.,* 2011). These systems are designed for continuous drug release through oral and transdermal routes, reducing dosing frequency and improving patient protection from drug degradation. The second generation of drug delivery systems introduced eco-friendly nanomaterials, which enhance drug delivery efficiency, reduce side effects, and provide superior control. Nanoparticles also improve the solubility and bioavailability of medications. The third generation aims to characterize the physical and biological limitations of models, focusing on targeted delivery, minimizing toxicity, and optimizing drug release in nanoparticle-based systems.

2.7 CHALLENGES AND FUTURE PROSPECTS

Green synthesis is an eco-friendly approach for the production of nanoparticles, such as AuNPs, utilized in the delivery of drugs and cancer therapy. This environmentally friendly process employs plant extracts and microorganisms as reducing agents, providing an alternative to traditional chemical synthesis. Nevertheless, constraints limit stability, functionalization, and processability in biological and industrial settings. The improvement of stability and establishment of standardized industrial manufacturing processes for green-synthesized nanoparticles should occupy a high priority to open up the complete range of applications in medicine (Fazal *et al.*, 2014). Nanoparticles focus on synthesis optimization, control over nanoparticle properties, and scale-up to maintain quality in therapeutic and diagnostic applications.

2.8 CONCLUSION

The green production of gold nanoparticles represents a significant advancement in nanotechnology, merging aesthetics with the principles of materials science. The utilization of natural plant biomass as an agent of protection for reducing gold ions for the synthesis of biodegradable, and multipurpose nanoparticles, devoid of the toxicity associated with chemical reagents, is a principal distinction among green and chemical production methods. The green method directly addresses several environmental problems embodied in chemical synthesis by incorporating natural reducing agents, non-toxic solvents, and environmentally safe stabilizing agents, so upholding the cause of sustainability. The adaptability of these gold nanoparticles is seen in numerous applications, including medication delivery, remediation of the environment, diagnostics, and catalysis. The multidisciplinary aspects of biological, substances, and chemical sciences in green material synthesis foster the advancement of sustainable nanotechnologies and industrialized processes that are scalable and economically viable. Therefore, such an innovative synthesis has great prospects to meet the impending technological requirements, with the least risk to ecological well-being.

REFERENCES

Abhijith, K. S., and M. S. Thakur. "Application of Green Synthesis of Gold Nanoparticles for Sensitive Detection of Aflatoxin B1 Based on Metal Enhanced Fluorescence." *Analytical Methods* 4 (2012): 4250–4256.

Alaallah, N. J., E. Abd Alkareem, A. Ghaidan, and N. A. Imran. "Eco-Friendly Approach for Silver Nanoparticles Synthesis from Lemon Extract and Their Anti-Oxidant, Anti-Bacterial, and Anti-Cancer Activities." *Journal of the Turkish Chemical Society, Section A: Chemistry* 10, no. 1 (2023): 205–216.

Aljabali, A. A. A., Y. Akkam, M. Salim, A. Zoubi, K. M. Al-Batayneh, B. A. Id, O. A. Alrob, A. M. Alkilany, M. Benamara, and D. J. E. Id. "Synthesis of Gold Nanoparticles Using Leaf Extract of *Ziziphus zizyphus* and Their Antimicrobial Activity." *Nanomaterials* 8 (2018): 1–15.

Basavegowda, N., A. Sobczak-Kupiec, R. I. Fenn, and S. Dinakar. "Bioreduction of Chloroaurate Ions Using Fruit Extract *Punica granatum* (Pomegranate) for Synthesis of Highly Stable Gold Nanoparticles and Assessment of Its Antibacterial Activity." *Micro & Nano Letters* 8, no. 8 (2013): 400–404.

Begum, S. J., S. Pratibha, J. M. Rawat, D. Venugopal, P. Sahu, A. Gowda, and M. Jaremko. "Recent Advances in Green Synthesis, Characterization, and Applications of Bioactive Metallic Nanoparticles." *Pharmaceuticals* 15, no. 4 (2022): 455.

Bennur, T., Z. Khan, R. Kshirsagar, V. Javdekar, and S. Zinjarde. "Biogenic Gold Nanoparticles from the Actinomycete *Gordonia marae*: Application in Rapid Sensing of Copper Ions." *Sensors and Actuators B: Chemical* 233 (2016): 684–690.

Chauhan, A., S. Zubair, S. Tufail, A. Sherwani, M. Sajid, S. C. Raman, A. Azam, and M. Owais. "Fungus-Mediated Biological Synthesis of Gold Nanoparticles: Potential in Detection of Liver Cancer." *International Journal of Nanomedicine* 6 (2011): 2305–2319.

Das, S. K., A. R. Das, and A. K. Guha. "Gold Nanoparticles: Microbial Synthesis and Application in Water Hygiene Management." *Langmuir* 25 (2009): 8192–8199.

Driskell, J. D., C. A. Jones, S. M. Tompkins, and R. A. Tripp. "One-Step Assay for Detecting Influenza Virus Using Dynamic Light Scattering and Gold Nanoparticles." *Analyst* 136 (2011): 3083–3090.

Drummer, S., T. Madzimbamuto, and M. Chowdhury. "Green Synthesis of Transition-Metal Nanoparticles and Their Oxides: A Review." *Materials* 14, no. 10 (2021): 2700.

Fazal, S., A. Jayasree, S. Sasidharan, M. Koyakutty, S. V. Nair, and D. Menon. "Green Synthesis of Anisotropic Gold Nanoparticles for Photothermal Therapy of Cancer." *ACS Applied Materials & Interfaces* 6 (2014): 8080–8089.

Flieger, J., W. Flieger, J. Baj, and R. Maciejewski. "Antioxidants: Classification, Natural Sources, Activity/Capacity Measurements, and Usefulness for the Synthesis of Nanoparticles." *Materials* 14, no. 15 (2021): 4135.

Geethalakshmi, R., and D. Sarada. "Characterization and Antimicrobial Activity of Gold and Silver Nanoparticles Synthesized Using Saponin Isolated from *Trianthema decandra* L." *Industrial Crops and Products* 51 (2013): 107–115.

Ghosh, P., G. Han, M. De, C. K. Kim, and V. M. Rotello. "Gold Nanoparticles in Delivery Applications." *Advanced Drug Delivery Reviews* 60, no. 11 (2008): 1307–1315.

Giljohann, D. A., D. S. Seferos, W. L. Daniel, M. D. Massich, P. C. Patel, and C. A. Mirkin. "Gold Nanoparticles for Biology and Medicine." *Angewandte Chemie International Edition* 49, no. 19 (2010): 3280–3294.

Islam, N. U., K. Jalil, M. Shahid, A. Rauf, N. Muhammad, A. Khan, M. R. Shah, and M. A. Khan. "Green Synthesis and Biological Activities of Gold Nanoparticles Functionalized with *Salix alba*." *Arabian Journal of Chemistry* 12 (2019): 2914–2925.

Jeyaraj, M., S. Gurunathan, M. Qasim, M. H. Kang, and J. H. Kim. "A Comprehensive Review on the Synthesis, Characterization, and Biomedical Application of Platinum Nanoparticles." *Nanomaterials* 9 (2017).

Kanwar, R., J. Rathee, D. B. Salunke, and S. K. Mehta. "Green Nanotechnology-Driven Drug Delivery Assemblies." *ACS Omega* 4, no. 5 (2019): 8804–8815.

Khalil, M. M. H., E. H. Ismail, and F. El-Magdoub. "Biosynthesis of Au Nanoparticles Using Olive Leaf Extract: First Nano Updates." *Arabian Journal of Chemistry* 5 (2012): 431–437.

Kumar, S. A., Y. A. Peter, and J. L. Nadeau. "Facile Biosynthesis, Separation, and Conjugation of Gold Nanoparticles to Doxorubicin." *Nanotechnology* 19 (2008): 495101.

Kuppusamy, P., M. M. Yusoff, G. P. Maniam, and N. Govindan. "Biosynthesis of Metallic Nanoparticles Using Plant Derivatives and Their New Avenues in Pharmacological Applications—An Updated Report." *Saudi Pharmaceutical Journal* 24 (2016): 473–484.

Li, S., F. A. Al-Misned, H. A. El-Serehy, and L. Yang. "Green Synthesis of Gold Nanoparticles Using Aqueous Extract of *Mentha longifolia* Leaf and Investigation of Its Anti-Human Breast Carcinoma Properties in the *In Vitro* Condition." *Arabian Journal of Chemistry* 14, no. 2 (2021): 102931.

Lowry, G. V., R. J. Hill, S. Harper, A. F. Rawle, C. O. Hendren, and F. Klaessig. "Guidance to Improve the Scientific Value of Zeta Potential Measurements in Nano Environmental Health and Safety." *Environmental Science: Nano* 3 (2016): 953–965.

Luo, P., Y. Liu, Y. Xia, H. Xu, and G. Xie. "Aptamer Biosensor for Sensitive Detection of Toxin A of *Clostridium difficile* Using Gold Nanoparticles Synthesized by *Bacillus stearothermophilus*." *Biosensors and Bioelectronics* 54 (2014): 217–221.

Mahoney, J., C. Monroe, A. M. Swartley, M. G. Ucak-Astarlioglu, and C. A. Zoto. "Surface Analysis Using X-ray Photoelectron Spectroscopy." *Spectroscopy Letters* 53 (2020): 726–736.

Mishra, A., D. Pradhan, J. Halder, P. Biswasroy, V. K. Rai, D. Dubey, B. Kar, G. Ghosh, and G. Rath. "Metal Nanoparticles Against Multi-Drug-Resistant Bacteria." *Journal of Inorganic Biochemistry* 237 (2022): 111938.

Mosquera-Romero, S., J. Anaya-Garzon, C. Garcia-Timermans, J. Van Dorpe, A. Hoorens, N. Commenges-Bernole, and J. Varia. "Combined Gold Recovery and Nanoparticle Synthesis in Microbial Systems Using Fractional Factorial Design." *Nanomaterials* 13 (2022): 83.

Mukherjee, P., A. Ahmad, D. Mandal, S. Senapati, S. R. Sainkar, M. I. Khan, R. Ramani, R. Parischa, P. V. Jayakumar, M. Alam. "Bioreduction of AuCl4-Ions by the Fungus *Verticillium* sp. and Surface Trapping of the Gold Nanoparticles Formed." *Angewandte Chemie International Edition* 40 (2001): 3585–3588.

Nangia, Y., N. Wangoo, S. Sharma, J. S. Wu, V. Dravid, G. S. Shekhawat, and C. Raman Suri. "Facile Biosynthesis of Phosphate-Capped Gold Nanoparticles by a Bacterial Isolate *Stenotrophomonas maltophilia*." *Applied Physics Letters* 94 (2009): 233901.

Naveena, B., and S. Prakash. "Biological Synthesis of Gold Nanoparticles Using Marine Algae *Gracilaria corticata* and Its Application as a Potent Antimicrobial and Antioxidant Agent." *Asian Journal of Pharmaceutical and Clinical Research* 6 (2013): 179.

Pal, S. L., U. Jana, P. K. Manna, G. P. Mohanta, and R. Manavalan. "Nanoparticle: An Overview of Preparation and Characterization." *Journal of Applied Pharmaceutical Science* 1 (2011): 228–234.

Philip, D. "Biosynthesis of Au, Ag, and Au–Ag Nanoparticles Using Edible Mushroom Extract." *Spectrochimica Acta Part A: Molecular and Biomolecular Spectroscopy* 73 (2009): 374–381.

Poojary, M. M., P. Passamonti, and A. V. Adhikari. "Green Synthesis of Silver and Gold Nanoparticles Using Root Bark Extract of *Mammea Auriga*: Characterization, Process Optimization, and Their Antibacterial Activity." *BioNanoScience* 6 (2016): 110–120.

Rajan, A., V. Vilas, and D. Philip. "Studies on Catalytic, Antioxidant, Antibacterial, and Anticancer Activities of Biogenic Gold Nanoparticles." *Journal of Molecular Liquids* 212 (2015): 331–339.

Raveendran, P., J. Fu, and S. L. Wallen. "A Simple and 'Green' Method for the Synthesis of Au, Ag, and Au-Ag Alloy Nanoparticles." *Green Chemistry* 8, no. 1 (2006): 34–38.

Salam, H. A., P. Rajiv, M. Kamaraj, P. Jagadeeswaran, S. Gunalan, and R. Sivaraj. "Plants: Green Route for Nanoparticle Synthesis." *International Research Journal of Biological Sciences* 1, no. 5 (2012): 85–90.

Salomi, V., Abdulbari A. Alfaris, P. Saranraj, M. S. Swetha, and K. Gayathri. "Extraction of Fungal Pigment Melanin from *Aspergillus niger* and Analysis of Its Antimicrobial Activity." *European Chemical Bulletin* 12, no. 11 (2023): 431–438.

Saranraj, P., S. Dinesh Kumar, K. Manickan Kirandas, K. Gayathri, and B. Ana Lokeshwari. "Larvicidal and Ovicidal Activities of *Datura metel* L. (Family: Solanaceae) Solvent Extracts Against *Aedes aegypti*." *International Journal of Entomology Research* 9, no. 5 (2024): 53–56.

Sarkar, J., S. Ray, D. Chattopadhyay, A. Laskar, and K. Acharya. "Mycogenesis of Gold Nanoparticles Using a Phytopathogen *Alternaria alternata*." *Bioprocess and Biosystems Engineering* 35 (2012): 637–643.

Shankar, S. S., A. Ahmad, R. Pasricha, M. I. Khan, R. Kumar, and M. Sastry. "Immobilization of Biogenic Gold Nanoparticles in Thermally Evaporated Fatty Acid and Amine Thin Films." *Journal of Colloid and Interface Science* 274 (2004): 69–75.

Sublemontier, O., C. Nicolas, D. Aureau, M. Patanen, H. Kintz, X. Liu, M. A. Gaveau, J. L. Le Garrec, E. Robert, F. A. Barreda. "X-ray Photoelectron Spectroscopy of Isolated Nanoparticles." *Journal of Physical Chemistry Letters* 5 (2014): 3399–3403.

Sujitha, M. V., and S. Kannan. "Green Synthesis of Gold Nanoparticles Using Citrus Fruits (*Citrus limon*, *Citrus reticulata*, and *Citrus sinensis*) Aqueous Extract and Its Characterization." *Spectrochimica Acta Part A: Molecular and Biomolecular Spectroscopy* 102 (2013): 15–23.

Suman, T. Y., S. R. Radhika Rajasree, R. Ramkumar, C. Rajthilak, and P. Perumal. "The Green Synthesis of Gold Nanoparticles Using an Aqueous Root Extract of *Morinda citrifolia* L." *Spectrochimica Acta Part A: Molecular and Biomolecular Spectroscopy* 118 (2014): 11–16.

Tatarchuk, T., N. Danyliuk, A. Shyichuk, V. Kotsyubynsky, I. Lapchuk, and V. Mandzyuk. "Green Synthesis of Cobalt Ferrite Using Grape Extract: The Impact of Cation Distribution and Inversion Degree on the Catalytic Activity in the Decomposition of Hydrogen Peroxide." *Emergent Materials* 5 (2021): 89–103.

Timoszyk, A. "A Review of the Biological Synthesis of Gold Nanoparticles Using Fruit Extracts: Scientific Potential and Application." *Bulletin of Materials Science* 41 (2018): 154.

Vedha, V., Abdulbari A. Alfaris, Narayan D. Totewad, P. Saranraj, and K. Gayathri. "Mosquito Larvicidal Potential of Marine *Streptomyces griseus* Against *Aedes aegypti*." *International Journal of Entomology Research* 8, no. 11 (2023): 39–41.

Velmurugan, P., K. Anbalagan, M. Manosathyadevan, K. J. Lee, M. Cho, S. M. Lee, J. H. Park, S. G. Oh, K. S. Bang, and B. T. Oh. "Green Synthesis of Silver and Gold Nanoparticles Using *Zingiber officinale* Root Extract and Antibacterial Activity of Silver Nanoparticles Against Food Pathogens." *Bioprocess and Biosystems Engineering* 37 (2014): 1935–1944.

Vijayakumar, S., B. Vaseeharan, B. Malaikozhundan, N. Gopi, P. Ekambaram, R. Pachaiappan, P. Velusamy, K. Murugan, G. Benelli, and R. Suresh Kumar. "Therapeutic Effects of Gold Nanoparticles Synthesized Using *Musa paradisiaca* Peel Extract Against Multiple Antibiotic-Resistant *Enterococcus faecalis* Biofilms and Human Lung Cancer Cells (A549)." *Microbial Pathogenesis* 102 (2017): 173–183.

Vinay Gopal, J., M. Thenmozhi, K. Kannabiran, G. Rajakumar, K. Velayutham, and A. A. Rahuman. "Actinobacteria Mediated Synthesis of Gold Nanoparticles Using *Streptomyces* sp. VITDDK3 and Its Antifungal Activity." *Materials Letters* 93 (2013): 360–362.

Wang, A. Z., R. Langer, and O. C. Farokhzad. "Nanoparticle Delivery of Cancer Drugs." *Annual Review of Medicine* 63 (2012): 185–198.

Wang, Z., Q. Zhang, D. Kuehner, A. Ivaska, and L. Niu. "Green Synthesis of 1–2 nm Gold Nanoparticles Stabilized by Amine-Terminated Ionic Liquid and Their Electrocatalytic Activity in Oxygen Reduction." *Green Chemistry* 10, no. 9 (2008): 907–909.

Wu, X., F. Fang, B. Zhang, J. J. Wu, and K. Zhang. "Biogenic Silver Nanoparticles-Modified Forward Osmosis Membranes with Mitigated Internal Concentration Polarization and Enhanced Antibacterial Properties." *npj Clean Water* 5, no. 1 (2022).

Yamini, S. Y., V. K. Vinay, B. Bodaiah, B. B. Pavani, and P. Sudhakar. "Green Synthesis of Silver and Gold Nanoparticles Using *Shorea tumbuggaia* Bark Extract and Screening for Their Catalytic Activity." *International Journal of Advanced Science Research and Management* 4 (2019): 180–185.

Yang, N., L. WeiHong, and L. Hao. "Biosynthesis of Au Nanoparticles Using Agricultural Waste Mango Peel Extract and Its *In Vitro* Cytotoxic Effect on Two Normal Cells." *Materials Letters* 134 (2014): 67–70.

Yu, J., D. Xu, H. N. Guan, C. Wang, L. K. Huang, and D. F. Chi. "Facile One-Step Green Synthesis of Gold Nanoparticles Using *Citrus maxima* Aqueous Extracts and Its Catalytic Activity." *Materials Letters* 166 (2016): 110–112.

Yusefi, M., K. Shameli, O. S. Yee, S. Y. Teow, Z. Hedayatnasab, H. Jahangirian, T. J. Webster, and K. KuČa. "Green Synthesis of Fe3O4 Nanoparticles Stabilized by a *Garcinia mangostana* Fruit Peel Extract for Hyperthermia and Anticancer Activities." *International Journal of Nanomedicine* 16 (2021): 2515.

Zhang, X., X. He, K. Wang, and X. Yang. "Different Active Biomolecules Involved in Biosynthesis of Gold Nanoparticles by Three Fungus Species." *Journal of Biomedical Nanotechnology* 7 (2011): 245–254.

3 Plants in the Biogenic Synthesis of Medicinally Significant Silver Nanoparticles

Divya Sharma, Prangya Rath, Vartika Singh, Shweta Khanna, Harshita Jain, and Laxmi Kant Bhardwaj

3.1 INTRODUCTION

Nanotechnology (NT) advancements have ushered in a new era of pharmaceutical applications, with silver nanoparticles (AgNPs) emerging as essential agents due to their superior antibacterial, anti-inflammatory, and anticancer properties (Bhardwaj et al. 2023). Traditionally, the manufacture of AgNPs has been based mostly on physical and chemical methods, which might include dangerous compounds and complex procedures. In comparison, biogenic AgNP synthesis from plant extracts is a naïve and more sustainable approach (Rath et al. 2024). Flavonoids, polyphenols, and terpenoids are the secondary metabolites that are observed in plants and can behave as stabilizing agents in the formation of AgNPs (Sharma et al. 2009). This approach reduces the environmental impact of nanoparticles (NPs) and makes them more suitable for therapeutic applications (Siddiqi and Husen 2022).

Due to their unique characteristics, plant-mediated AgNPs have attracted great benefits in medicine. The use of AgNPs is increasing in different sectors such as food, pharma, healthcare, orthopedics, etc. They act as antibacterial and anticancer agents and lead to enhancing their effects of tumor killing of anticancer drugs (Chernousova and Epple 2013). These NPs have improved biological activity, which is critical for the expansion of innovative therapeutic medicines. For example, Vigneshwaran et al. (2007) have reported AgNPs from the extracts of *Azadirachta indica* (Neem) and *Aloe vera* and stated that they act against cancer cells and various pathogens. Inherent antioxidant properties of AgNPs promote the decrease of oxidative stress-related diseases (Sangeetha and Ramaswamy 2019).

This study focuses on the mechanism of the production of AgNPs, their medical applications, and future predictions. Several researchers have emphasized the potential of AgNPs in various therapeutic supports. The antibacterial and anticancer characteristics of plant extracts with AgNPs open new opportunities for medical intermediations. AgNPs, which are plant-derived, have shown advanced benefits and

DOI: 10.1201/9781003559900-3

biocompatibility in medicine delivery systems, injury dressings, and diagnostic tools (Dhandapani et al. 2022). Moreover, the ability of these NPs to target specific cellular routes and modify biological reactions suggests a substantial benefit in personalized cures. Chandraker et al. (2021) stated that as researchers continue their research on the interaction between AgNPs and plant-derived compounds, the advanced treatment approaches and applications are increasing, which can help in the future medical sciences. The current study describes these innovations and presents a complete summary of the innovative methods that play a significant role in plant-mediated AgNPs in progressive medicine.

3.2 BIOSYNTHESIS OF SILVER NANOPARTICLES (AGNPS) USING MEDICINAL PLANTS

Synthesis of AgNPs from medicinal plants has emerged as a green as well as sustainable alternative to conventional chemical and physical methods. The process employs the unique properties of medicinal plant extracts, which are a mixture of phytochemicals that can reduce silver ions (Ag^+) and stabilize the formed NPs (Mohammed et al. 2022). The use of medicinal plants not only minimizes the footprint on the environment but also improves the biological properties of AgNPs such that they become suitable for numerous applications in agriculture, medicine, and industry (Amaliyah et al. 2022).

Biosynthesis of AgNPs involves a cutback process where Ag^+ is reduced to metallic silver (Ag^0) in the presence of plant extracts. Medicinal plants have a group of secondary metabolites such as flavonoids, alkaloids, terpenoids, and phenolic acids. These compounds are both reducing and capping agents. The hydroxyl, carbonyl, and amine groups present in these phytochemicals play a significant role in reducing Ag+ and stabilizing the produced NPs (Rakib-Uz-Zaman et al. 2022). The choice of medicinal plants for the biosynthesis of AgNPs is significant, as varied plants yield different concentrations and types of phytochemicals, which influence the size, shape, and stability of the NPs (Table 3.1).

Plants such as *Azadirachta indica* (Neem), *Ocimum sanctum* (Tulsi), and *Cymbopogon citratus* (Lemongrass) have been extensively studied due to their rich phytochemical contents and well-documented healing properties (Rather et al. 2022). Neem leaves are rich in flavonoids and terpenoids, which are effective in reducing Ag+ and aiding in the formation of stable NPs (Kumar et al. 2016). The conventional protocol for the synthesis of AgNPs using medicinal plants involves several key steps. Plant materials, either fresh or dried, are collected, purified, and then subjected to extraction processes using solvents like water, ethanol, or methanol. The extracts are then filtered to remove any solid residue (Chinni et al. 2021).

A solution of silver nitrate ($AgNO_3$) in water is prepared and mixed with the plant extract under well-controlled conditions. The solution is generally stored at room temperature or slightly elevated temperatures to facilitate the process of cutback (Ebrahimzadeh et al. 2020). The verification of AgNP synthesis is done using various characterization methods, such as UV-Visible spectroscopy. It shows a clear Surface Plasmon Resonance (SPR) peak, which is generally found between 400 to 450 nm wavelengths. Further characterization using methods like Scanning Electron

TABLE 3.1
Synthesis of Silver Nanoparticles (AgNPs) Using Medical Plants

S. No.	Name of the Plants	Plant Part Used	Nanoparticle (NP) Size (nm)	References
1	*Scabiosa atropurpurea subsp. maritima*	Fruits	40 to 50	Essghaier et al. 2022
2	*Moringa oleifera*	Leaves	15.22 to 29.45	Mohammed et al. 2022
3	*Piper retrofractum*	Fruits	1 to 5	Amaliyah et al. 2022
4	*Cymbopogon citratus*	Leaves	47	Rakib-Uz-Zaman et al. 2022
5	*Cuphea carthagenensis*		10.65 ± 0.1	Rather et al. 2022
6	*Zingiber officinale*	Stems	10 to 20	Mehata 2021a, 2021b
7	*Coccinia indica*	Leaves	8 to 48	Chinni et al. 2021
8	*Kalanchoe pinnata*		40.8	Mehata 2021a, 2021b
9	*Prangos ferulacea*	Roots	79 to 200	Mavaei et al. 2020
10	*Alpinia ofcinarum*	Rhizome	20 to 80	Li et al. 2020
11	*Crataegus pentagyna*	Fruit	15 to 60	Ebrahimzadeh et al. 2020
12	*Shorea robusta*	Leaves	12 to 37	Shaikh et al. 2020
13	*Prosopis julifora*		30	Malini et al. 2020
14	*Gymnema sylvestre*		20 to 30	Rajkumar et al. 2020
15	*Lemon*	Peels	9.3 to 20.3	Nasr et al. 2019
16	*Azadirachta indica*	Leaves	11 to 35	Shaikh et al. 2018
17	*Camellia japonica*		12 to 25	Karthik et al. 2017
18	*Lantana camara*	Flowers	53.5	Kumar et al. 2016

Microscopy (SEM), Transmission Electron Microscopy (TEM), and X-ray Diffraction (XRD) provides useful information with regard to the size, morphology, and crystalline nature of the NPs (Rajkumar et al. 2020).

The biosynthesis of AgNPs depends on several parameters like the concentration of plant extract and $AgNO_3$, reaction mixture pH, temperature, and reaction time. Optimization of these parameters plays a vital role in controlling the size and morphology of the NPs. Higher dilutions of the plant extract may lead to the formation of smaller NPs because more reducing agents are present. Additionally, the pH level of the solution affects the ionization state of the phytochemicals and the reduction potential, thus changing the synthesis process (Hashemi et al. 2020).

AgNPs that are produced from the extracts of medicinal plants have shown numerous benefits as compared to established methods. It eliminates the application of dangerous chemicals and reduces ecological pollution. Sharifi-Rad et al. (2020) stated that the use of easily accessible plant sources decreases the expenditure linked with manufacturing. The biocompatibility of AgNPs is improved by the use of phytochemicals that are obtained from medicinal plants, and this improved biocompatibility is suitable for biomedical treatments. During production, the healing abilities of the plant extracts are transferred to the AgNPs and improve their properties to fight against cancer and microorganisms. Plant-based produced AgNPs show effective antimicrobial action against pathogens (Dangi et al. 2020).

These things make AgNPs extremely helpful in injury healing, in the development of an outer layer for medical apparatus, and in water cleaning. Erdogan et al. (2019) studied the properties of AgNPs and reported that these properties can activate apoptosis and delay cell proliferation, making AgNPs a strong agent for the treatment of cancer. These particles exhibit great efficacy as catalysts for many chemical reactions owing to their substantial surface area-to-volume ratio and active surface sites. Nasr et al. (2019) described the synthesis of AgNPs from medicinal plants and reported their catalytic properties that are useful in industrialized applications. Moreover, these particles have unique optical properties that can be applied in the detection of heavy metals and other pollutants.

The presence of the functional groups in plant extracts can improve the capability of target analytes to bind selectively. Although there are advantages, the procedure of biosynthesizing AgNPs using plants faces many barriers (Shaikh et al. 2018). The divergence in the chemical constituents of plant isolates can result in inaccuracies in the formation of NPs. Making certain that the extraction and synthesis procedures are regulated is necessary to ensure reproducibility. Additionally, doing research is required to fully understand the involved processes involved in the decline and stabilization accelerated by phytochemicals.

Karthik et al. (2017) stated that studies should pay attention to obtaining standardized procedures for the extraction and synthesis procedures to confirm reproducibility and reliability. Scientists are still examining the altered molecular procedures underlying the biosynthesis of AgNPs so as to have appropriate observation over the properties of the NPs. A wide-ranging toxicological assessment should be directed to confirm the protection of plant-mediated AgNPs for biomedical claims (Essghaier et al. 2022). In spite of several challenges, nonstop research connects the capacity to release new functions and recover the scalability and consistency of this eco-friendly creation technique. The incorporation of AgNPs that are plant-based into several industrial, medical, and ecological purposes heralds a modern era of natural NT (Tokas et al. 2024).

3.3 PROPERTIES OF BIOSYNTHESIS SILVER NANOPARTICLES (AGNPS)

3.3.1 Size and Shape

Species of algae, bacteria, fungi, and plants have the ability to biosynthesize AgNPs, which are greener as well as cheaper alternatives to chemical and physical processes. AgNPs' cytotoxicity has been observed to depend upon its chemical and physical properties, like chemistry of surfaces, dimensions, shapes, distributions, morphologies of particles, arrangements, coatings, accumulation, rate of disintegration, release efficiency of ions, particle reactivity, cell types, and reducing agent type (Zhang et al. 2011). Previous research studies have suggested that smaller-sized particles might be more hazardous because of their larger surface areas (Sriram et al. 2012). The presence of chemical or biological coverings on NPs outermost layer has significantly impacted AgNP toxicity (Suresh et al. 2012). AgNPs' surface charges could also impact the toxicity of cells. Positive charges on the exterior of these NPs have allowed them to be present in the bloodstream for a longer duration than negative charges, thereby making them suitable for administering anticancerous drugs (Schlinkert et al. 2015).

3.3.2 Stability

Biosynthesized AgNPs tend to exhibit high durability under various environmental conditions, chiefly due to encapsulation of the NPs using biomolecules in the synthesis step. The proteins form a capping layer surrounding the NPs, protecting them from oxidation or aggregation, which are the two most prevalent issues reported with chemically synthesized AgNPs. AgNPs biosynthesized by biological systems are influenced by various parameters such as pH, ionic strength, and temperature with respect to their stability. For instance, AgNPs derived from plant extracts generally remain stable for extended periods in neutral or weakly alkaline conditions. However, the occurrence of strongly acidic or alkaline conditions or highly ionized environments can lead to the aggregation or degradation of the NPs, which subsequently influences their ability to remain stable and, consequently, their efficiency in different applications (Mittal et al. 2013).

3.3.3 Optical Properties

Localized Surface Plasmon Resonance (LSPR) refers to the cooperative travel of the spread electrons within a metallic NP. This event can be accelerated by light. LSPR state is determined by various parameters, such as the NP's electrical characteristics, size and shape, temperature, and the dielectric environment. Biosynthesized AgNPs commonly display a distinct LSPR peak in the visible range of the electromagnetic spectrum, typically between 400 and 450 nm. By manipulating the size and form of the NPs during synthesis, one may precisely adjust the position of the LSPR band. This adjustable optical characteristic is very beneficial in biosensing, imaging, and photothermal therapy fields.

3.3.4 Anti-Bacterial Properties

The antibacterial action of AgNPs is based not only on their size but also on their form. AgNPs were manufactured using four diverse saccharides, resulting in a standard size of 25 nm. These NPs presented potent antibacterial and bactericidal effects counter to both gram-negative and gram-positive bacteria, including highly resistant strains like methicillin-resistant *Staphylococcus aureus*. As earlier stated, the efficacy of AgNPs is influenced not only by their size but also by their shape, as they interact with the gram-negative bacterium *E. coli* in a manner that depends on their shape (Pal et al. 2007).

3.3.5 Angiogenic Properties

Pathological angiogenesis is a characteristic feature of cancer and several ischemic and provocative disorders. Recently, numerous studies have stipulated supporting proof using both *in vivo and in vitro* models demonstrating that AgNPs have both anticancer and anti-angiogenic properties. Kalishwaralal et al. (2009) showed that biologically synthesized AgNPs have anti-angiogenic properties. They used Bovine Retinal Endothelial Cells (BRECs) as a model system and observed that after 24 hours of therapy with AgNPs at an intensity of 500 nm, there was an unwillingness to migrate and proliferate in BRECs. The activation of caspase-3 and DNA fragmentation are mechanisms that impede the angiogenic process produced by Vascular Endothelial Growth Factor (VEGF). Additionally, AgNPs inhibit the VEGF-driven PI3K/Akt pathway in BRECs (Kemp et al. 2009).

3.4 MECHANISM OF ANTIBACTERIAL, CYTOTOXIC, AND ANTIOXIDANT ACTIVITIES FOR MEDICINAL PLANT-DERIVED NANOPARTICLES (NPS)

Plant-mediated NPs synthesized and emerged in NT are cost-effective, eco-friendly, and beneficial approaches. Medicinal plants contain a large number of natural components with their unique properties. Additionally, greenly synthesized NPs possess biopotential and pharmacological properties such as antibacterial, antioxidant, antifungal, anticancerous, anti-inflammatory, and antiviral effects, and also have promising mixtures for further amendment during drug development (Alshameri and Owais 2022). In the biological method, different sections of the plant, such as leaves, bark, flowers, fruits, roots, seeds, callus, and peels, are used for the synthesis of medicinal plant-mediated NPs in varied shapes and sizes. They mainly help in the synthesis of NPs with improved biological properties.

Amino acids, proteins, enzymes, vitamins, organic acids, and other bioactive constituents are found in plant extracts that are mainly responsible for reduction, capping, and stabilization during green synthesis of NPs. Fourier Transform Infrared Spectroscopy (FTIR), Nuclear Magnetic Resonance (NMR), Gas Chromatography-Mass Spectrometry (GC-MS), High-Performance Liquid Chromatography (HPLC), and other testing methods are used to identify and purify the bioactive constituents that are responsible for pharmacological effects. Alkaloids, flavonoids, polyphenols, terpenoids, phenolic acids, and other metabolites are sensitive to the decrease of metal ions to zerovalent metals. The secondary metabolites flavonoids can be able to chelate metal ions due to the existence of carbonyl and hydroxyl functional groups, which act as capping and reducing agents during the production of metal NPs. Several researchers have reported the mechanism of AgNPs for antibacterial, antioxidant, and anticancer effects.

3.4.1 Antibacterial Mechanism of Silver Nanoparticles (AgNPs)

Various researchers reported the mode of action of AgNPs against bacterial growth. Despite the facts, the precise antibacterial mechanism of AgNPs is unknown. Though contact killing and ion-mediated killing methods are two types of methods that have been recognized as antibacterial actions. Antibacterial mechanisms of AgNPs are reported by direct contact and adhesion of AgNPs on the cell wall of bacteria and penetration into it. These accumulated NPs on bacterial cell membranes liberate the Ag^+. The continuous liberation of Ag^+ through NPs has the possibility to disrupt the cell membrane, and they can cause a leak of cellular content and bacterial death. These liberated Ag^+ are attached to the cell wall and disrupt the cytoplasmic membrane due to the electrostatic hold and affinity of sulfur proteins.

The permeability of the cytoplasmic membrane is increased by the attachment of Ag^+ and damages the cell envelope. After penetration and disruption of the bacterial cytoplasmic membrane, Ag^+ enters the cytoplasm of bacterial cells. The antibacterial activities of AgNPs are dependent on size. Nano-sized particles show effective antibacterial properties as compared to larger-sized NPs. More in number, nanosized NPs easily enter the cytoplasm and interact with intercellular structures and biomolecules such as proteins, enzymes, genetic material, etc. These interactions of AgNPs can damage and rupture intercellular structures like cell organelles, resulting in the death of the microbes.

Another method of antibacterial action of AgNPs is the ion-mediated killing method by Reactive Oxygen Species (ROS) production, which causes the death of cells. ROS production in the cell by Ag^+ can inhibit respiratory enzymes. In normal cells, ROS formation might be inhibited by antioxidant compounds such as glutathione. Increased ROS formation leads to reduced adenosine triphosphate (ATP), apoptosis and lipid peroxidation, DNA damage, and microbial death. The biosynthesized AgNPs report an antibacterial mode of action initiated by the delivery of Ag^+ from AgNPs. The AgNPs have a high surface area and liberate Ag^+ in numbers with high concentrations in and out of the cells. These Ag^+ ions interact with the sulfhydryl group of enzymes and proteins, leading to inhibition of ATP formation.

Likewise, Ag^+ ions can inhibit respiratory chain enzymes, interact with DNA, increase cellular oxidative stress, and prevent cell division. The intracellular ROS eventually damages the cytoplasmic membrane and leads to dysfunctional mitochondria. Intracellular ROS can also cause the alteration of DNA due to phosphorus and sulfur are key components of DNA, during interaction with Ag^+, disrupting DNA replication and cell multiplication. Moreover, the denaturation of cytoplasmic ribosomes by interfering with Ag^+ can inhibit the process of protein synthesis (Bapat et al. 2018; Shanmuganathan et al. 2019).

3.4.2 In Vitro Cytotoxicity and Antioxidant Activity Mechanism of Silver Nanoparticles (AgNPs)

Over the last few years, numerous researchers have conveyed that plant-mediated AgNPs might be treated as anticancer therapeutic agents due to their potentially significant properties of cytotoxic and antioxidant activity. The World Health Organization (WHO) reported that cancer is the second highest common cause of human death, following cardiovascular disease. Hence, the production of anticancer mediators is essential. The AgNPs contain significant properties of anticancer. The greenly synthesized AgNPs contain potent anticancer properties and are actively involved in cancer treatment, which can be more effective and a substitute for current cancer treatment methodologies (Ahmad et al. 2019). Cancerous cells and oncogenes may be treated and protected from malignant cells with the help of these AgNPs. The greenly synthesized AgNPs have bioactive components and possess antioxidant and anticancer properties of plants. Varied types of AgNPs have shown different anticancer mechanisms against cancerous cells based on the precursor of plants. They also comprise apoptotic cell death, membrane damage, misfolded proteins, suppression of angiogenesis, DNA damage, and other factors (Walimbe et al. 2022).

The most acceptable action of AgNPs against the cancerous cell is the activation of the caspase-3 enzyme, then intracellular glutathione oxidizes and damages the genetic content (Figure 3.1). The Ag^+ is released from AgNPs, which are highly toxic. These Ag^+ ions interact with the membrane and disrupt the RNA polymerase activity. Due to this action, ROS is also generated, with the result of intracellular oxidative stress and ultimately cell damage. The ROS generation leads to disequilibrium in the activity of the protein by this caspase, and caspase enzymes are activated and cause necrosis. Behind this, the main reason is that cells are damaged and ultimately killed by the toxicity of AgNPs in a dose-dependent manner. The mode of action is also dependent on the shape and size of AgNPs showed a significant toxicity effect. Nanosized NPs showed effective results and easily reached the cytoplasm. Additionally, high surface area AgNPs are released in high concentrations of Ag^+ (Li et al. 2020). Generally, it is not possible to explain the general mechanism of AgNPs due to their phenomenon being extremely dependent on physicochemical properties like size, surface chemistry, etc. A large number of organic molecules of plant-mediated AgNPs interact with the surface and easily penetrate the cell membrane. AgNPs strongly reduce the ATP content of the cell, which ultimately causes mitochondrial damage and increases the generation of ROS in a dose-dependent manner (Mohanta et al. 2017). The AgNPs worked against cytotoxicity; their mode of action is schematically represented in Figure 3.1.

The most widely accepted mechanism by researchers, ROS, is formed by the reaction of AgNPs with cancerous cells and damaging their cell membranes. The formation of intracellular ROS eventually disrupts the equilibrium of the cell by interfering with biomolecules and enzymes, leading to dysfunctional mitochondria. Cell membrane disruption prevents the multiplication of cancerous cells. The mechanisms of AgNPs against cancerous cells are represented in Figure 3.2.

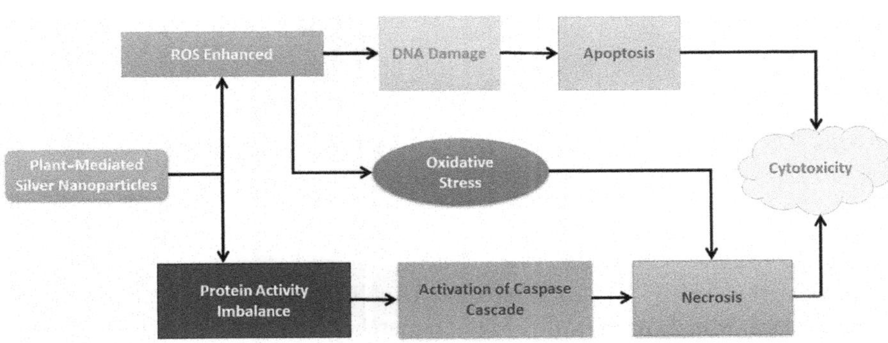

FIGURE 3.1 Cytotoxicity mechanism of silver nanoparticles.

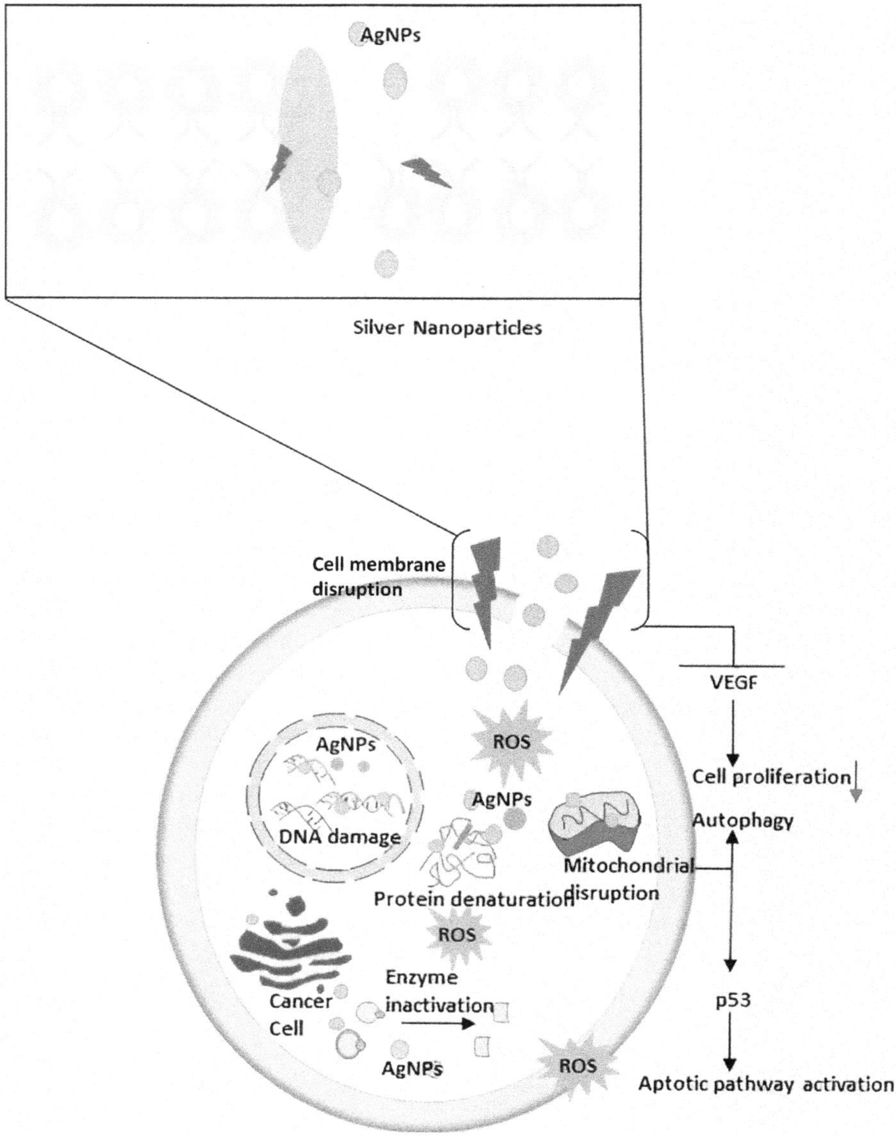

FIGURE 3.2 The schematic diagram shows the anticancer mechanism of plant-mediated silver nanoparticles (AgNPs).

Plant-mediated AgNPs are also effective against breast cancer cell lines MDA-MB231 and MCF-7. The biosynthesized AgNPs showed a significant role against anticancerous cells therapeutically. Many researchers reported the reduced production of breast cancer cells by some *in vitro* activities. The greenly synthesized AgNPs from the leaves of the plant *Syzygium jambolanum* showed their efficiency against breast cancer cell MCF-7 with effective inhibitory concentration. The plant-mediated AgNPs are cost-effective, eco-friendly, non-toxic, easy, rapid, and viable substitutes to traditional methods (Figure 3.2). *Dactylifera* seed extract-mediated AgNPs showed high cytotoxic activity and anticancerous action against the cancer cell line A-549 of lung adenocarcinoma cells in humans (Madakka et al. 2021).

AgNPs lead to ROS generation, apoptosis, and necrosis, which could be able to inhibit cancer cells. Majorly, AgNPs increase the rate of programmed cell death due to the up- and downregulation of genetic material like p53 genes. In cancer cells, the cell cycle can be arrested by treatment with AgNPs. Different cancer cells exhibited sub-G1 phase block and apoptosis after being treated with AgNPs. Moreover, with reduced tumor cell angiogenesis and migration, AgNPs can avoid isolated metastasis (Mishra et al. 2021). Some studies revealed the significance of anti-proliferative characteristics of biosynthesized AgNPs. However, such action may be due to the synergetic influence of both nano-sized silver and the bioactive phyto-complexes attached to the external surface of the NPs (Mohanta et al. 2017). The anticancer properties of AgNPs have been shown to be safe and efficient anticancer drugs. The greenly synthesized AgNPs possess cytotoxicity action against cancerous cells.

3.4.3 Antioxidant Mechanism and Therapeutic Applications of Silver Nanoparticles (AgNPs)

Antioxidants possess the ability to slow down the rate of the autoxidation process. This is achieved by either interacting with chain-carrying free radicals or by reducing the initiation process. Various mechanisms have been reported to contribute to the antioxidative activities of NPs. One of those is through the catalase-mimic (CAT-mimic) mechanism observed in metal NPs. According to research, AgNPs could convert H_2O_2 into H_2O and O_2 on their surfaces, thereby exhibiting antioxidant properties. Consequently, green-synthesized AgNPs possess these antioxidant characteristics. Plant-mediated AgNPs have been reported to be advantageous because of their cost-effectiveness, eco-friendliness, biocompatibility, and significant therapeutic properties. These NPs impart safe, easy, reliable, and precise treatment options. Furthermore, the plant metabolites present in the extract have been researched to serve as reducing agents for Ag^+ ions as well as capping agents for AgNPs, thus eliminating the need for toxic chemicals. Previous research studies have suggested that the presence of phytochemicals reduces cytotoxicity, genotoxicity, as well as immunotoxicity because of their antioxidant properties. They have also been reported to enhance the therapeutic efficacy of AgNPs without causing side effects (Magdolenova et al. 2014).

3.5 CONCLUSION AND FUTURE PERSPECTIVES

Biosynthesized AgNPs are promising nanomaterials (NMs) with broad and powerful characteristics for biological applications. Using biological sources like plant extracts and microorganisms makes the production process eco-friendly and gives AgNPs medicinal properties due to these bioactive compounds. Many research studies have reported the successful usage of biosynthesized AgNPs for antibacterial and antifungal uses, as well as for medicinal purposes due to their effectiveness as anti-cancerous, anti-inflammatory agents. Biosynthesized AgNPs have also been used for industrial purposes, agricultural purposes, etc. Their properties depend upon the size, shape, source, concentration, reaction process, etc. To maximize their therapeutic potential in clinical practice, more biocompatibility, long-term safety, and a mechanism of action are needed. Among numerous synthetic approaches for AgNPs, biological approaches seem to be rapid, simple, non-toxic, and green approaches that can generate well-defined sizes and morphology under boosted situations for translational research. They are gaining attention due to their cost-effectiveness, sustainability, and environmentally friendly applications. Therefore, a green chemistry methodology for the synthesis of AgNPs exhibits much promise. This study has highlighted the biosynthesis of AgNPs, their properties, their use as antibacterial and antifungal agents, their application as anticancerous agents, as well as their mode of action. However, more studies are required to overcome challenges, as well as to study their other properties, synergistic applications, and wide applications in industries and pharmaceuticals.

ACKNOWLEDGMENT

The authors are deeply grateful to their respective institutes for providing the platform to do this study.

REFERENCES

Ahmad, S., Munir, S., Zeb, N., Ullah, A., Khan, B., Ali, J., . . . and S. Ali. 2019. Green nanotechnology: A review on green synthesis of silver nanoparticles—An ecofriendly approach. *International Journal of Nanomedicine* 5087–5107.

Alshameri, A. W., and M. Owais. 2022. Antibacterial and cytotoxic potency of the plant-mediated synthesis of metallic nanoparticles Ag NPs and ZnO NPs: A review. *OpenNano 8*: 100077.

Amaliyah, S., Sabarudin, A., Masruri, M., and S. B. Sumitro. 2022. Characterization and antibacterial application of biosynthesized silver nanoparticles using Piper retrofractum Vahl fruit extract as bioreductor. *Journal of Applied Pharmaceutical Science 12*(3): 103–114.

Bapat, R. A., Chaubal, T. V., Joshi, C. P., Bapat, P. R., Choudhury, H., Pandey, M., . . . and P. Kesharwani. 2018. An overview of application of silver nanoparticles for biomaterials in dentistry. *Materials Science and Engineering: C 91*: 881–898.

Bhardwaj, L. K., Rath, P., and M. Choudhury. 2023. A comprehensive review on the classification, uses, sources of nanoparticles (NPs) and their toxicity on health. *Aerosol Science and Engineering 7*(1): 69–86.

Chandraker, S. K., Ghosh, M. K., Lal, M., and R. Shukla. 2021. A review on plant-mediated synthesis of silver nanoparticles, their characterization and applications. *Nano Express 2*(2): 022008.

Chernousova, S., and M. Epple. 2013. Silver as antibacterial agent: Ion, nanoparticle, and metal. *Angewandte Chemie International Edition 52*(6): 1636–1653.

Chinni, S. V., Gopinath, S. C., Anbu, P., Fuloria, N. K., Fuloria, S., Mariappan, P., . . . and S. Samuggam. 2021. Characterization and antibacterial response of silver nanoparticles biosynthesized using an ethanolic extract of Coccinia indica leaves. *Crystals 11*(2): 97.

Dangi, S., Gupta, A., Gupta, D. K., Singh, S., and N. Parajuli. 2020. Green synthesis of silver nanoparticles using aqueous root extract of Berberis asiatica and evaluation of their antibacterial activity. *Chemical Data Collections 28*: 100411.

Dhandapani, P., Santhoshkumar, M., Narenkumar, J., AlSalhi, M. S., Kumar, P. A., Devanesan, S., . . . and A. Rajasekar. 2022. Bio-approach: Preparation of RGO-AgNPs on cotton fabric and interface with sweat environment for antibacterial activity. *Bioprocess and Biosystems Engineering 45*(11): 1825–1837.

Ebrahimzadeh, M. A., Naghizadeh, A., Amiri, O., Shirzadi-Ahodashti, M., and S. Mortazavi-Derazkola. 2020. Green and facile synthesis of Ag nanoparticles using Crataegus pentagyna fruit extract (CP-AgNPs) for organic pollution dyes degradation and antibacterial application. *Bioorganic Chemistry 94*: 103425.

Erdogan, O., Abbak, M., Demirbolat, G. M., Birtekocak, F., Aksel, M., Pasa, S., and O. Cevik. 2019. Green synthesis of silver nanoparticles via Cynara scolymus leaf extracts: The characterization, anticancer potential with photodynamic therapy in MCF7 cells. *PloS One 14*(6): e0216496.

Essghaier, B., Toukabri, N., Dridi, R., Hannachi, H., Limam, I., Mottola, F., . . . and M. Abdelkarim. 2022. First report of the biosynthesis and characterization of silver nanoparticles using scabiosa atropurpurea subsp. Maritima fruit extracts and their antioxidant, antimicrobial and cytotoxic properties. *Nanomaterials 12*(9): 1585.

Hashemi, Z., Ebrahimzadeh, M. A., Biparva, P., Mortazavi-Derazkola, S., Goli, H. R., Sadeghian, F., . . . and A. Rafiei. 2020. Biogenic silver and zero-Valent iron nanoparticles by Feijoa: Biosynthesis, characterization, cytotoxic, antibacterial and antioxidant activities. *Anti-Cancer Agents in Medicinal Chemistry (Formerly Current Medicinal Chemistry-Anti-Cancer Agents) 20*(14): 1673–1687.

Kalishwaralal, K., Banumathi, E., Pandian, S. R. K., Deepak, V., Muniyandi, J., Eom, S. H., and S. Gurunathan. 2009. Silver nanoparticles inhibit VEGF induced cell proliferation and migration in bovine retinal endothelial cells. *Colloids and Surfaces B: Biointerfaces 73*(1): 51–57.

Karthik, R., Govindasamy, M., Chen, S. M., Cheng, Y. H., Muthukrishnan, P., Padmavathy, S., and A. Elangovan. 2017. Biosynthesis of silver nanoparticles by using Camellia japonica leaf extract for the electrocatalytic reduction of nitrobenzene and photocatalytic degradation of Eosin-Y. *Journal of Photochemistry and Photobiology B: Biology 170*: 164–172.

Kemp, M. M., Kumar, A., Mousa, S., Dyskin, E., Yalcin, M., Ajayan, P., . . . and S. A. Mousa. 2009. Gold and silver nanoparticles conjugated with heparin derivative possess anti-angiogenesis properties. *Nanotechnology 20*(45): 455104.

Kumar, B., Smita, K., and L. Cumbal. 2016. Biosynthesis of silver nanoparticles using Lantana camara flower extract and its application. *Journal of Sol-Gel Science and Technology 78*: 285–292.

Li, J. F., Liu, Y. C., Chokkalingam, M., Rupa, E. J., Mathiyalagan, R., Hurh, J., . . . and D. C. Yang. 2020. Phytosynthesis of silver nanoparticles using rhizome extract of Alpinia officinarum and their photocatalytic removal of dye under UV and visible light irradiation. *Optik 208*: 164521.

Madakka, M., Jayaraju, N., and N. Rajesh. 2021. Evaluating the antimicrobial activity and antitumor screening of green synthesized silver nanoparticles compounds, using Syzygium jambolanum, towards MCF7 cell line (Breast cancer cell line). *Journal of Photochemistry and Photobiology 6*: 100028.

Magdolenova, Z., Collins, A., Kumar, A., Dhawan, A., Stone, V., and M. Dusinska. 2014. Mechanisms of genotoxicity. A review of in vitro and in vivo studies with engineered nanoparticles. *Nanotoxicology 8*(3): 233–278.

Malini, S., Kumar, S. V., Hariharan, R., Bharathi, A. P., Devi, P. R., and E. Hemananthan. 2020. Antibacterial, photocatalytic and biosorption activity of chitosan nanocapsules embedded with Prosopis juliflora leaf extract synthesized silver nanoparticles. *Materials Today: Proceedings 21*: 828–832.

Mavaei, M., Chahardoli, A., Shokoohinia, Y., Khoshroo, A., and A. Fattahi. 2020. One-step synthesized silver nanoparticles using isoimperatorin: evaluation of photocatalytic, and electrochemical activities. *Scientific Reports 10*(1): 1762.

Mehata, M. S. 2021a. Green route synthesis of silver nanoparticles using plants/ginger extracts with enhanced surface plasmon resonance and degradation of textile dye. *Materials Science and Engineering: B 273*: 115418.

Mehata, M. S. 2021b. Green synthesis of silver nanoparticles using Kalanchoe pinnata leaves (life plant) and their antibacterial and photocatalytic activities. *Chemical Physics Letters 778*: 138760.

Mishra, V., Nayak, P., Singh, M., Tambuwala, M. M., Aljabali, A. A., Chellappan, D. K., and K. Dua. 2021. Pharmaceutical aspects of green synthesized silver nanoparticles: A boon to cancer treatment. *Anti-Cancer Agents in Medicinal Chemistry (Formerly Current Medicinal Chemistry-Anti-Cancer Agents) 21*(12): 1490–1509.

Mittal, A. K., Chisti, Y., and U. C. Banerjee. 2013. Synthesis of metallic nanoparticles using plant extracts. *Biotechnology Advances 31*(2): 346–356.

Mohammed, A. A., Mohamed, A., El-Naggar, N. E. A., Mahrous, H., Nasr, G. M., Abdella, A., . . . and A. S. Ali. 2022. Antioxidant and antibacterial activities of silver nanoparticles biosynthesized by Moringa Oleifera through response surface methodology. *Journal of Nanomaterials 2022*(1): 9984308.

Mohanta, Y. K., Panda, S. K., Jayabalan, R., Sharma, N., Bastia, A. K., and T. K. Mohanta. 2017. Antimicrobial, antioxidant and cytotoxic activity of silver nanoparticles synthesized by leaf extract of Erythrina suberosa (Roxb.). *Frontiers in Molecular Biosciences 4*: 14.

Nasr, H. A., Nassar, O. M., El-Sayed, M. H., and A. A. Kobisi. 2019. Characterization and antimicrobial activity of lemon peel mediated green synthesis of silver nanoparticles. *International Journal of Biology and Chemistry 12*(2): 56–63.

Pal, S., Tak, Y. K., and J. M. Song. 2007. Does the antibacterial activity of silver nanoparticles depend on the shape of the nanoparticle? A study of the gram-negative bacterium Escherichia coli. *Applied and Environmental Microbiology 73*(6): 1712–1720.

Rajkumar, P. V., Prakasam, A., Rajeshkumar, S., Gomathi, M., Anbarasan, P. M., and R. Chandrasekaran. 2020. Green synthesis of silver nanoparticles using Gymnema sylvestre leaf extract and evaluation of its antibacterial activity. *South African Journal of Chemical Engineering 32*(1): 1–4.

Rakib-Uz-Zaman, S. M., Hoque Apu, E., Muntasir, M. N., Mowna, S. A., Khanom, M. G., Jahan, S. S., . . . and K. Khan. 2022. Biosynthesis of silver nanoparticles from Cymbopogon citratus leaf extract and evaluation of their antimicrobial properties. *Challenges 13*(1): 18.

Rath, P., Bhardwaj, L. K., Yadav, P., and A. K. Bhardwaj. 2024. A synthesis of biogenic nanoparticles (NPs) for the treatment of wastewater and its application: A review. *Biogenic Wastes-Enabled Nanomaterial Synthesis: Applications in Environmental Sustainability* 127–148.

Rather, M. A., Deori, P. J., Gupta, K., Daimary, N., Deka, D., Qureshi, A., . . . and M. Mandal. 2022. Ecofriendly phytofabrication of silver nanoparticles using aqueous extract of Cuphea carthagenensis and their antioxidant potential and antibacterial activity against clinically important human pathogens. *Chemosphere 300*: 134497.

Sangeetha, K., and U. Ramaswamy. 2019. Green synthesis and characterization of silver nanoparticles using aqueous extract of microalga Desmococcus Olivaceus and its biological activities. In *Biotechnology and Biological Sciences* (pp. 337–342). CRC Press.

Schlinkert, P., Casals, E., Boyles, M., Tischler, U., Hornig, E., Tran, N., . . . and A. Duschl. 2015. The oxidative potential of differently charged silver and gold nanoparticles on three human lung epithelial cell types. *Journal of nanobiotechnology 13*: 1–18.

Shaikh, W. A., Chakraborty, S., and R. U. Islam. 2018. UV-assisted photo-catalytic degradation of anionic dye (Congo red) using biosynthesized silver nanoparticles: A green catalysis. *Desalination and Water Treatment 130*: 232–242.

Shaikh, W. A., Chakraborty, S., and R. U. Islam. 2020. Photocatalytic degradation of rhodamine B under UV irradiation using Shorea robusta leaf extract-mediated bio-synthesized silver nanoparticles. *International Journal of Environmental Science and Technology 17*(4): 2059–2072.

Shanmuganathan, R., Karuppusamy, I., Saravanan, M., Muthukumar, H., Ponnuchamy, K., Ramkumar, V. S., and A. Pugazhendhi. 2019. Synthesis of silver nanoparticles and their biomedical applications—A comprehensive review. *Current Pharmaceutical Design 25*(24): 2650–2660.

Sharifi-Rad, J., Rayess, Y. E., Rizk, A. A., Sadaka, C., Zgheib, R., Zam, W., . . . and N. Martins. 2020. Turmeric and its major compound curcumin on health: Bioactive effects and safety profiles for food, pharmaceutical, biotechnological and medicinal applications. *Frontiers in Pharmacology 11*: 550909.

Sharma, V. K., Yngard, R. A., and Y. Lin. 2009. Silver nanoparticles: Green synthesis and their antimicrobial activities. *Advances in Colloid and Interface Science 145*(1–2): 83–96.

Siddiqi, K. S., and A. Husen. 2022. Plant response to silver nanoparticles: A critical review. *Critical Reviews in Biotechnology 42*(7): 973–990.

Sriram, M. I., Kalishwaralal, K., Barathmanikanth, S., and S. Gurunathani. 2012. Size-based cytotoxicity of silver nanoparticles in bovine retinal endothelial cells. *Nanoscience Methods 1*(1): 56–77.

Suresh, A. K., Pelletier, D. A., Wang, W., Morrell-Falvey, J. L., Gu, B., and M. J. Doktycz. 2012. Cytotoxicity induced by engineered silver nanocrystallites is dependent on surface coatings and cell types. *Langmuir 28*(5): 2727–2735.

Tokas, R., Bhardwaj, L. K., Kumar, N., and T. Jindal. 2024. Nanotechnology for sustainable development and future: A review. In *Green and Sustainable Approaches Using Wastes for the Production of Multifunctional Nanomaterials* (pp. 221–233). Elsevier.

Vigneshwaran, N., Ashtaputre, N. M., Varadarajan, P. V., Nachane, R. P., Paralikar, K. M., and R. H. Balasubramanya. 2007. Biological synthesis of silver nanoparticles using the fungus Aspergillus flavus. *Materials letters 61*(6): 1413–1418.

Walimbe, K. G., Dhawal, P. P., and S. A. Kakodkar. 2022. Anticancer potential of biosynthesized silver nanoparticles: A review. *European Journal of Biology and Biotechnology 3*(2): 10–20.

Zhang, Q., Li, N., Goebl, J., Lu, Z., and Y. Yin. 2011. A systematic study of the synthesis of silver nanoplates: Is citrate a "magic" reagent? *Journal of the American Chemical Society 133*(46): 18931–18939.

4 Phytosynthesis of Medicinally Significant Non-Noble Metal Nanoparticles

Arun Ghuge, Megha Sapkota, Indarchand Gupta, and Sandeep Sangle

4.1 INTRODUCTION

The present era belongs to the field of nanotechnology. This technology majorly involves the role of nanoparticles (NPs). Nanoparticles are particles with a size range less than 1nm. These particles are formed from various biodegradable particles including natural and synthetic polymers (Goh *et al.*, 2016). They exhibit distinctive shape features due to their ultra tiny size (Altammar, 2023). With the advent of nanotechnology, the study of NPs has drawn great interest in different fields of research and technology due to their size-dependent physical and chemical properties. In the last decade, much attention has been given to NPs because they possess unique chemical and physical properties that differ significantly from their bulk counterparts. NPs have very high surface-to-volume ratios. Nanoparticles, with their larger surface area relative to bulk materials, enhance catalytic properties and chemical reactivity, which helps make them already widely used in the biomedical field (Abbasi *et al.*, 2023). Research on NPs has been stimulated by considering their possible applications in different fields of sciences, such as chemical, energy, electronics, biomedical, and space domains. On the global scale, NPs are already being used in cosmetics, medications, sunscreens, etc. (Berta *et al.*, 2021). Nanoparticles have significant applications in various fields such as environmental sciences, agriculture, food, biomedical, etc. (Rassaei *et al.*, 2011; Zahra *et al.*, 2020; Islam *et al.*, 2022; Chen *et al.*, 2023). The need for sustainable and environmentally friendly means for synthesis of nanoparticles has significantly grown focus on the field of nanotechnology (Osman *et al.*, 2024).

Synthesis of nanoparticles is an important step. They can be synthesized majorly by physical, chemical, and biological methods. Among them biosynthesis of nanoparticles is a greener approach. Nanoparticle biosynthesis using natural extracts has many advantages over the chemical route (Selvakesavan and Franklin, 2021). It has become a priority for researchers to turn away from toxic and hazardous chemicals and focus on using safer alternatives. Simultaneously, one- or zero-cost biosynthesis of NPs,

DOI: 10.1201/9781003559900-4

accessible and scalable methods, renewable natural resources, and elimination of toxic wastes are goals striving for the interest of researchers. Plant extracts meet these goals because they are abundant, renewable, and environmentally friendly. Among the many green synthesis approaches, plant-mediated NPs synthesis is preferred by researchers due to the capability of plants to reduce and stabilize Mn^{+} ions, short processing time, and versatility of plant extracts. The initial cost is low, and biosynthesis occurs at ambient temperature, saving energy. In the synthesis process of metallic NPs, phytochemicals in the plant/extracts convert metal ions into form of NPs while preventing agglomeration, which serves as both a reducing and stabilizing agent for NPs. These plants possess secondary metabolites such as flavonoids, terpenoids, alkaloids, vitamins, polysaccharides, etc., responsible for synthesizing NPs (Thatyana *et al.*, 2023). Phytochemicals play a crucial role in synthesizing noble metal NPs that are highly effective catalysts and have significant electrical properties. Plant-mediated NPs have higher stability than the chemically synthesized NPs and vary in shape and size (Javed *et al.*, 2021).

4.2 METHODS OF NANOPARTICLE SYNTHESIS

The primary methods of nanoparticles synthesis are categorized into chemical, physical, and biological.

4.2.1 Chemical Methods

4.2.1.1 Precipitation Based

It is the simplest method in which precipitate is formed from super saturated solution. A precursor solution of desired metal ions is reacted by the precipitating agents like bases, reducing agents or salts that induce the precipitation reaction (Jain, Kumar and Meena, 2022). On addition of precipitating agent, a nucleation occurs. These nuclei further grow by additional attachment of ions and molecules leading to the formation of nanoparticles (Thanh, Maclean and Mahiddine, 2014). The formed particles are washed and dried to eliminate any impurities or byproducts. Drying typically includes oven drying or freeze drying. The precipitate is often aged to enhance the crystallinity and even distribution of the particle size (Nyabadza *et al.*, 2023).

4.2.1.2 Sol-Gel

This is a wet-chemical technique widely used for synthesis of a variety of nanoparticles, specifically metal oxides. The precursors are hydrolysed and condensed to form the metal hydroxide. The process involves formation of colloidal suspension followed by gelation. The gel is then dried and calcinated to obtain nanoparticles. The gel is often aged to increase the homogeneity and crystallinity (Bokov *et al.*, 2021; Choudhary, 2024).

4.2.1.3 Hydrothermal and Solvothermal

It involves heating the reaction mixture in a sealed vessel at very high temperature and pressure. The hydrothermal method uses water as the solvent and involves chemical reactions above the boiling point of water (Dinadayalane *et al.*, 2022). It uses organic solvents, such as ethanol, and involves chemical reactions at relatively high temperatures (Li, Wu and Wu, 2015).

4.2.1.4 Microwave-Assisted Synthesis

A promising technique for producing nanoparticles, particularly silver nanoparticles, is microwave-assisted synthesis. The process raises the temperature of a reaction mixture above its boiling point by means of microwave irradiation. Effective internal heating is produced by the direct coupling of the microwave radiation with the mixture's ions and dipoles (Desai *et al.*, 2024).

4.2.2 Physical Methods

4.2.2.1 Gas Phase Condensation

This technique involves evaporation, sputtering, and laser ablation. In evaporation, the material is evaporated in a vacuum. Vapors then condensed back to form nanoparticles (Nande *et al.*, 2023). The sputtering technique involves bombarding the target surface with ions which in turn releases atoms. These atoms can be condensed back to form nanoparticles (Wender *et al.*, 2013). In the laser ablation technique, a laser beam is imparted over the target. This causes vaporization of material which is then condensed to obtain nanoparticles (Wang *et al.*, 2023).

4.2.2.2 Ball Milling

The target material is placed in the rotating drum with the balls made up of steel or ceramic. As there is a rotation of the drum, the balls collide with each other, thereby causing a breakdown of target material into small particles (Tiwari *et al.*, 2022).

4.2.2.3 Ultrasonic Cavitation

The material to be treated is put into a liquid and high-frequency sound waves are added. The sound pressure causes the bubbles that are created by the sound waves in the liquid to expand and contract. The material may be broken down into nanoparticles by the strong localized heating, pressure, and shear forces produced when these bubbles break down (Carlton, 2019).

4.2.2.4 Biological Methods of Nanoparticle Synthesis

Biological means of nanoparticle synthesis have emerged as a promising alternative to the traditional and inefficient methods of nanoparticle synthesis.

4.2.2.4.1 Microbial Synthesis

Various microbes including bacteria, fungi, yeast, virus, actinomycetes and microalgae have shown satisfactory results in nanoparticles synthesis. Microbes have a natural tendency to grow in diverse habitats. Bacterial species like *Bacillus licheniformis*, *Pseudomonas deceptionensis*, *Pyrococcus furiosus*, *Pyrobaculum islandicum*, *Pseudomonas aeruginosa,* etc., have been recently explored for the synthesis of silver (Ag), gold (Au), iron (Fe), nickel (Ni), zinc (Zn), and other important NPs (Singh *et al.*, 2016; Adebayo *et al.*, 2021; Sandhu and Goel, 2023). The synthesis of nanoparticles using microbes can be done in both intracellular and extracellular ways (Das *et al.*, 2014; Shah *et al.*, 2015).

4.2.2.4.2 Phytosynthesis

Plant-mediated synthesis of nanoparticles can be done by reducing different parts of plant like leaves, roots, stems, etc. *Brassica juncea* (mustard greens) and *Medicago sativa* (alfalfa) have been found to accumulate 50nm sized silver nanoparticles at a significant level, amounting to 13.6% of their total weight, when cultivated using silver nitrate as a growth substrate (Harris and Bali, 2008).

4.3 ADVANTAGES OF PHYTOSYNTHESIS OF NANOPARTICLES

Phytosynthesis eliminates the use of harmful chemicals that reduces the environmental impact (Kharissova *et al.*, 2019). Since it is biodegradable, it reduces the environmental concerns (Soni *et al.*, 2021). Phytosynthesis is economically viable, as biomaterials are widely available at relatively cheaper costs. Phytosynthesis is highly sustainable as it utilizes the biomaterial likely to be wasted (Janssen *et al.*, 2014). It involves fewer and less complex steps in synthesis of nanoparticles over conventional chemical synthesis methods (Nyabadza *et al.*, 2023). The shape, size, and properties of the nanoparticles synthesized, can be controlled by adjusting the concentration of the extract, reaction, and temperature (Almatroudi, 2020).

Conventional methods of NP synthesis are largely dependent on the use of toxic materials which makes them unsuitable for the use. These hazardous effects can be seen specifically in biomedical and environmental realms (Kulkarni *et al.*, 2023; Kumari *et al.*, 2023; Osman *et al.*, 2024). Taking into consideration these issues, phytosynthesis or biosynthesis of nanoparticles using plant-based materials has proved to be a viable green strategy. The use of green chemistry principles in nanotechnology has led to different plant-mediated pathways for the synthesis of nanoparticles including those made of non-noble metals like CuNPs, Fe_2O_3, ZnNPs, and NiNPs. Plant-based synthesis of nanoparticles results in large-scale production in a short time (Prabhu, 2018; Rawat, Singh and Kaur, 2018). The use of plant-based extracts as a capping or reducing agent offers a wide range of advantages over conventional methods. It is a highly reliable, cost-effective, eco-friendly, and simple technique that can be easily scaled up for large-scale production (Huston *et al.*, 2021; Singh *et al.*, 2023). Also, the inherent antimicrobial and anticancerous properties of the plants can be used to enhance the biological properties of synthesized nanoparticles (Rawat, Singh and Kaur, 2018). Nanomaterials have emerged as a revolutionary driving force in medicine due to their tiny structures, larger surface area, and ability to be modified at the molecular level. Nanomaterials are utilized as the most effective drug delivery system (Afzal *et al.*, 2022).

4.4 PHYTOSYNTHESIS OF NON-NOBLE METAL NANOPARTICLES

As mentioned earlier, phytosynthesis is an ecofriendly method for synthesis of nanoparticles. This method makes use of plant extracts for the synthesis. The bioactive components from plants are extracted from the plants and reacted with the salt.

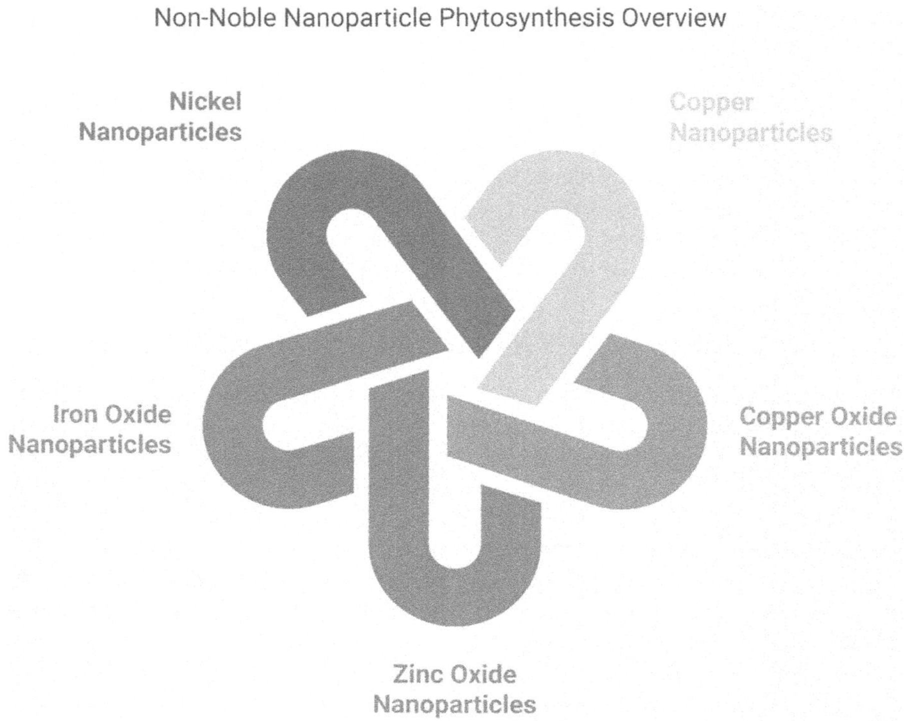

FIGURE 4.1 Overview of non-noble nanoparticle synthesis.

The reaction can be observed in the form of colour change. The resultant product can be filtered or centrifuged to obtained nanoparticles (Nzilu *et al.*, 2023). Various plant parts such as root, shoot, and leaves can be used for the synthesis of non-noble metal nanoparticles viz. CuNPs, Fe_2O_3, ZnNPs, and NiNPs (Bali *et al.*, 2006; Tiwari *et al.*, 2021). An overview of phytosynthesis of non-noble nanoparticles is illustrated in Figure 4.1.

4.4.1 Phytosynthesis of Copper and Copper Oxide Nanoparticles

CuNPs and copper oxide nanoparticles (CuONPs) have gained significant attention due to their unique properties and large range of applications. Their phytosynthesis is an eco-friendly method. During this process, Cu^{2+} get reduced to Cu^0 (Tiwari *et al.*, 2021). For instance *Aloe vera* and *Ferula macrocolea* has been demonstrated to synthesize CuNPs (Karimi and Mohsenzadeh, 2015; Shiravand and Azarbani, 2017; Fahiminia *et al.*, 2019). The synthesis was confirmed by characterizing them by UV-Vis spectrophotometry, FTIR, XRD, and EDX. *Euphorbia cheiradenia Boiss* is a traditionally important medicinal plant. It has been employed for synthesis of CuNPs. *Oldenlandia corymbosa L.* leaf extract has been found to be useful for the synthesis of CuNPs (Yulizar *et al.*, 2018). A simple and low-cost synthesis of CuNPs was demonstrated to be synthesized by using tea extract (Ismail, 2020; Gomaa *et al.*, 2021). Recently, fruit extract of *Piper longum* was exploited for the synthesis of CuNPs and it was shown to have antimicrobial applications (Kiranmayee *et al.*, 2023).

Plants were also employed for the synthesis of CuONPs. Various plants such as *Acanthospermum hispidum L.* (Pansambal *et al.*, 2017), *Phyllanthus reticulatus*, *Conyza bonariensis* (Potbhare *et al.*, 2019), and *Aloe vera* (Karimi and Mohsenzadeh, 2015; Sharma *et al.*, 2021) have been successfully used for this purpose. *Citrus aurantifolia* enzyme extract has been reported to synthesize CuONPs, which showed potential anticancer activity, implying their biomedical applications (Vincent *et al.*, 2022). Moreover, an aqueous extract of brown algae *Turbinaria* species was also reported to synthesize CuNPs and CuONPs, having promising antibacterial activity (Raja Rajamanikkam *et al.*, 2024). All these studies suggest the use of phytosynthesis for CuNPs and CuONPs, which could be used for many applications.

4.4.2 Phytosynthesis of Zinc Oxide Nanoparticles (ZnO NPs)

Zinc oxide is a largely inert compound. It possesses properties like large surface area, antimicrobial activity, and biocompatibility. This makes it suitable for a variety of uses. The green synthesis of ZnO NPs generally involves the use of plant extracts as reducing and stabilizing agents. Plant parts such as leaves, stems, and flowers have been utilized for this purpose. Plant extracts from some plants like Neem and *Aloe vera* have shown the reduction of the zinc salts into zinc oxide. It involves use of reducing and capping agents for the synthesis of zinc oxide nanoparticles (Kharissova *et al.*, 2019). Leaf extracts of *Cassia fistula* and *Melia azedarach* have the property of reducing the zinc acetate to ZnO NPs and thereby are considered to be the potential candidate for nanoparticle synthesis (Naseer *et al.*, 2020). The synthesized ZnO NPs were confirmed by SEM, XRD, and FTIR. Alyammani and group has utilised leaf extract of *Phlomis* for the synthesis of ZnONPs (Alyamani *et al.*, 2021). They were found to have the potential to antimicrobial properties, suggesting the viability of various plant extracts for nanoparticles synthesis.

4.4.3 Phytosynthesis of Iron and Iron Oxide Nanoparticles (Fe_2O_3NPs)

Recently, iron oxide nanoparticles have gained significant attention due to their properties, like superparamagnetism, easy synthesis process, surface to volume ratio, etc. Phytosynthesis employs plants or their derivatives for the synthesis of the iron oxide nanoparticles. Bioactive compounds from plants or other living organisms are extracted and they are used for the process of reduction for iron oxide nanoparticles synthesis. This plant-based bioactive compound acts as a reducing stabilizing agent. This acts upon purification and leads to the formation of the iron nanoparticles (Ali *et al.*, 2016). Eucalyptus and oak leaf extracts have been exploited for the synthesis of iron oxide nanoparticles. These plants were indicated to exhibit a high amount of arsenic adsorption capacity (Kamath, Chandra and Jeppu, 2020). Similarly, the role of *Ficus carica* was highlighted for the synthesis of nanoparticles by its extract (Patil, 2020). Other plants such as *Thymus vulgaris*, *Rosa damascene*, and *Urtica dioica* have been used to produce FeNP (Vijayaram *et al.*, 2024). Various plants such as *Eichhornia crassipes*, *Lantana camera*, and *Mimosa pudica* were found to produce FeNPs having the diameter of 20–60nm. Thus generated FeNPs were found to be of irregular and aggregated quasi-spherical shape (Vijayaram *et al.*, 2024). Various phytochemicals such as poluphenols, flavonoids, and aminoacids were demonstrated to synthesize FeNPs, having prominent antioxidant activity (Kirti, Kamsonlian and Agarwal, 2022). Such study demonstrates the effectiveness of locally available plants and their varieties for nanoparticle synthesis.

4.4.4 Phytosynthesis of Nickel (NiNPs) and Nickel Oxide Nanoparticles (NiONPs)

As that of other non-noble metal nanoparticles, NiNPs can also be synthesized by using various types of plants. However, phytosynthesis of NiNPs has gained significant attention in recent years due to its eco-friendly and sustainable nature. Bioactive compounds from living material like plants and certain microorganisms acts as a reducing agent. This reduces nickel containing compounds to NiNPs. This method offers a sustainable and eco-friendly methods of nanoparticle synthesis (Hassan *et al.*, 2023; Hussain *et al.*, 2023). Sudhasree and group have performed a comparative study and showed that the green synthesized nickel nanoparticles have better monodispersity, reduced size, and lower toxicity as compared to the chemically synthesized nickel nanoparticles (Tailor *et al.*, 2023). NiONPs has been reported to be synthesized by extract of *Calendula officinalis* and *Solanum*. Similarly, anthraquinones in leaf extract of *Moringa oleifera* was found to play a role as reducing agent to synthesize nickel NiNPs (Mamuru and Jaji, 2015; Tailor *et al.*, 2023). Accordingly, there are studies showing the role of phytochemicals to act as reducing, stabilizing, and capping agents (Mamuru and Jaji, 2015). *Aegle marmelos* Correa leaf extract also has been exploited for phytofabrication of NiNPs (Angajala, Ramya and Subashini, 2014).

4.5 BIOMEDICAL APPLICATIONS OF NON-NOBLE NANOPARTICLES (NNPS)

Nanotechnology has ushered in a new era for biomedical applications. The engineered nanoparticles for biomedical purposes have proved to play a pivotal role. Non-noble metals boast a range of biomedical applications due to their unique physiochemical properties, and hence they share common properties as anticancer, antibacterial, and antifungal agents and are helpful in targeted drug delivery.

4.5.1 Targeted Drug Delivery

Targeted drug delivery aims at enhancing the therapeutic potency, while minimizing the side effects. Traditional methods of drug administration often lack specificity. This often led to the systemic distribution of administered drugs, inducing damage to healthy tissues. Nanoparticles are engineered to selectively deliver therapeutic drugs at the targeted side. This improves the effectiveness of the localized treatment. The surface of the nanoparticles can be modified to increase the biocompatibility and drug loading capacity. The mechanism of the action is done by functionalizing the nanoparticle with the ligands like antibodies or proteins that bind selectively to the biomarkers present on the site of infection. Once bounded, nanoparticles facilitate the cellular assimilation by various endocytic pathways. This approach to drug delivery not only increases the concentration of the drug at the target site but also reduces the risk of offset distribution of therapeutic drugs (Singh and Lillard, 2009; Osman *et al.*, 2024).

4.5.2 Cancer Therapy

Despite several advancements of cancer therapies like chemotherapy and radiation, concern still exists. Challenges remain concerning drug toxicity, drug resistance, and enhanced drug solubility. Many therapeutics used in cancer treatment suffer from poor solubility and bioavailability. Nanoparticles like lipid-based nanoparticles enhance the solubility of hydrophobic compounds. This boots the therapeutic indices of the drugs. Nanoparticles can further be modified to deliver multiple therapeutic drugs simultaneously. This allows for a combination of therapies that target different cancer-causing cells by different mechanisms. This can be particularly effective in overcoming the drug resistance as multiple pathways are attacked concomitantly. Nanoparticles are not only used for delivering the therapeutic agents at the site but can also be used for diagnostic imaging mediators. They can create a platform for theragnostic i.e. therapeutics and diagnostics, such as injecting magnetic beads at the site of a tumour by using external magnetic fields for real-time tracking of treatment response (Berta *et al.*, 2021; Sun *et al.*, 2023).

4.5.3 Antibacterial Activity

The rise of antibiotic-resistant microbes has necessitated the development of novel strategies. Nanoparticles possess unique antimicrobial activity. They exert antimicrobial activity following different mechanisms, including generation of ROS (reactive oxygen species), causing disruption of the microbial cell membrane and causing interference with the metabolic pathways (Urnukhsaikhan *et al.*, 2021). Antibacterial activity of CuONPs was found to be shape-dependent causing DNA and cell membrane damage (Laha *et al.*, 2014) in pathogenic bacteria.

4.5.4 Antifungal Activity

Similar to the antibacterial activity, nanoparticles also offer solutions to mitigate fungal infections. It is important in patients with compromised immune systems. NNPs like copper oxide and zinc oxide have been shown to have strong antifungal properties against a range of harmful fungus (AlHarethi *et al.*, 2024). Combining NNPs with conventional antifungal medicines may increase their effectiveness and lower dosages, perhaps reducing the negative effects of higher drug concentrations.

4.6 CONCLUSION

The developing realm of nanotechnology offers a sustainable and green approach to nanoparticle synthesis. Phytosynthesis has emerged as an intriguing method that makes use of the reducing and capping qualities of plant extracts to yield metal nanoparticles with special qualities. By harnessing the intrinsic properties of the plant biomolecules, photosynthesis further eliminates the use of toxic chemicals and harsh solvent for the synthesis of nanoparticles. This also reduces the detrimental effects on the environment. Additionally, the compatibility of the plant-derived nanoparticles increases the potential of these molecules to be used for various purposes including drug delivery, imaging, and other therapies.

Even while phytosynthesis has advanced significantly, there are still several constraints to overcome. Standardization of the synthesis process, its optimization for the reaction conditions and scaling up of the synthesis are the key study areas for future research. Furthermore, exploration for a comprehensive understanding of the underlying mechanisms of the nanoparticle formation and their interaction with biological systems is essential. Despite several challenges, the outlook for phytosynthesis shows great potential. By further exploring the different plant varieties and their bioactive compounds, scientists can create a unique and cutting-edge nanocomposite with tailored attributes. As the need for regenerative and ecofriendly approaches increases, photosynthesis presents an appealing solution to address the demand of the dynamic world.

REFERENCES

Abbasi, R. *et al.* (2023) 'Structural parameters of nanoparticles affecting their toxicity for biomedical applications: a review', *Journal of Nanoparticle Research*, 25(3), p. 43. Available at: https://doi.org/10.1007/s11051-023-05690-w.

Adebayo, E.A. *et al.* (2021) 'Fungi as veritable tool in current advances in nanobiotechnology', *Heliyon*, 7(11), p. e08480. Available at: https://doi.org/10.1016/j.heliyon.2021.e08480.

Afzal, O. *et al.* (2022) 'Nanoparticles in drug delivery: from history to therapeutic applications', *Nanomaterials*, 12(24), p. 4494. Available at: https://doi.org/10.3390/nano12244494.

AlHarethi, A.A. *et al.* (2024) 'Zinc oxide and copper oxide nanoparticles as a potential solution for controlling Phytophthora infestans, the late blight disease of potatoes', *Discover Nano*, 19(1), p. 105. Available at: https://doi.org/10.1186/s11671-024-04040-6.

Ali, A. *et al.* (2016) 'Synthesis, characterization, applications, and challenges of iron oxide nanoparticles', *Nanotechnology, Science and Applications*, 9, pp. 49–67. Available at: https://doi.org/10.2147/NSA.S99986.

Almatroudi, A. (2020) 'Silver nanoparticles: synthesis, characterisation and biomedical applications', *Open Life Sciences*, 15(1), pp. 819–839. Available at: https://doi.org/10.1515/biol-2020-0094.

Altammar, K.A. (2023) 'A review on nanoparticles: characteristics, synthesis, applications, and challenges', *Frontiers in Microbiology*, 14. Available at: https://doi.org/10.3389/fmicb.2023.1155622.

Alyamani, A.A. *et al.* (2021) 'Green fabrication of zinc oxide nanoparticles using phlomis leaf extract: characterization and *in vitro* evaluation of cytotoxicity and antibacterial properties', *Molecules*, 26(20), p. 6140. Available at: https://doi.org/10.3390/molecules26206140.

Angajala, G., Ramya, R. and Subashini, R. (2014) '*In vitro* anti-inflammatory and mosquito larvicidal efficacy of nickel nanoparticles phytofabricated from aqueous leaf extracts of Aegle marmelos Correa', *Acta Tropica*, 135, pp. 19–26. Available at: https://doi.org/10.1016/j.actatropica.2014.03.012.

Bali, R. *et al.* (2006) 'The synthesis of metallic nanoparticles inside live plants', in *2006 International Conference on Nanoscience and Nanotechnology*. IEEE. Available at: https://doi.org/10.1109/ICONN.2006.340592.

Berta, L. *et al.* (2021) 'A review on plant-mediated synthesis of bimetallic nanopartoicles, characterisation and their biological applications', *Materials*, 14(24), p. 7677. Available at: https://doi.org/10.3390/ma14247677.

Bokov, D. *et al.* (2021) 'Nanomaterial by sol-gel method: synthesis and application', in *Advances in Materials Science and Engineering*. Edited by Z. Wang. Available at: https://doi.org/10.1155/2021/5102014.

Carlton, J.S. (2019) 'Cavitation', in *Marine Propellers and Propulsion*. Elsevier, pp. 217–260. Available at: https://doi.org/10.1016/B978-0-08-100366-4.00009-2.

Chen, J. *et al.* (2023) 'Emerging nanoparticles in food: sources, application, and safety', *Journal of Agricultural and Food Chemistry*, 71(8), pp. 3564–3582. Available at: https://doi.org/10.1021/acs.jafc.2c06740.

Choudhary, S. (2024) 'A review of the sol-gel process and its application', *International Education and Research Journal*, 10(7). Available at: https://doi.org/10.21276/IERJ24449856325648.

Das, V.L. *et al.* (2014) 'Extracellular synthesis of silver nanoparticles by the Bacillus strain CS 11 isolated from industrialized area', *3 Biotech*, 4(2), pp. 121–126. Available at: https://doi.org/10.1007/s13205-013-0130-8.

Desai, P.P. *et al.* (2024) 'Versatile strategies for multifaceted nanoparticle synthesis—an overview', in *Nanotechnology and In Silico Tools*. Elsevier, pp. 155–168. Available at: https://doi.org/10.1016/B978-0-443-15457-7.00023-X.

Dinadayalane, T. *et al.* (2022) 'Structures, properties, and applications of nitrogen-doped graphene', pp. 211–248. Available at: https://doi.org/10.1016/B978-0-12-819514-7.00010-5.

Fahiminia, M. *et al.* (2019) 'Phytosynthesis of Cu/rGO using *Euphorbia cheiradenia Boiss* extract and study of its ability in the reduction of organic dyes and 4-nitrophenol in aqueous medium', *IET Nanobiotechnology*, 13(2), pp. 202–213. Available at: https://doi.org/10.1049/iet-nbt.2018.5175.

Goh, E.T. *et al.* (2016) 'Accelerated wound healing using nanoparticles', in *Nanoscience in Dermatology*. Elsevier, pp. 287–306. Available at: https://doi.org/10.1016/B978-0-12-802926-8.00023-9.

Gomaa, Alshimaa *et al.* (2021) 'Green synthesis of copper nanoparticles: synthesis, characterization and their application: about future', *Egyptian Journal of Chemistry*. Available at: https://doi.org/10.21608/ejchem.2021.87059.4209.

Harris, A.T. and Bali, R. (2008) 'On the formation and extent of uptake of silver nanoparticles by live plants', *Journal of Nanoparticle Research*, 10(4), pp. 691–695. Available at: https://doi.org/10.1007/s11051-007-9288-5.

Hassan, Md.R. *et al.* (2023) 'Synthesis and applications of nickel nanoparticles (NiNPs)-comprehensive review', *Journal of Ultra Chemistry*, 19(1), pp. 9–37. Available at: https://doi.org/10.22147/juc/190102.

Hussain, S. *et al.* (2023) 'Green synthesis of nickel oxide nanoparticles using *Acacia nilotica* leaf extracts and investigation of their electrochemical and biological properties', *Journal of Taibah University for Science*, 17(1). Available at: https://doi.org/10.1080/16583655.2023.2170162.

Huston, M. *et al.* (2021) 'Green synthesis of nanomaterials', *Nanomaterials*, 11(8), p. 2130. Available at: https://doi.org/10.3390/nano11082130.

Islam, F. *et al.* (2022) 'Exploring the journey of zinc oxide nanoparticles (ZnO-NPs) toward biomedical applications', *Materials*, 15(6), p. 2160. Available at: https://doi.org/10.3390/ma15062160.

Ismail, M.I.M. (2020) 'Green synthesis and characterizations of copper nanoparticles', *Materials Chemistry and Physics*, 240, p. 122283. Available at: https://doi.org/10.1016/j.matchemphys.2019.122283.

Jain, R., Kumar, S. and Meena, S.K. (2022) 'Precipitating agent (NaOH and NH_4OH) dependent magnetic properties of cobalt ferrite nanoparticles', *AIP Advances*, 12(9). Available at: https://doi.org/10.1063/5.0098157.

Janssen, P.J.D. *et al.* (2014) 'Photosynthesis at the forefront of a sustainable life', *Frontiers in Chemistry*, 2. Available at: https://doi.org/10.3389/fchem.2014.00036.

Javed, R. *et al.* (2021) 'Comparison of chemically and biologically synthesized nanoparticles for the production of secondary metabolites, and growth and development of plants', pp. 303–329. Available at: https://doi.org/10.1016/bs.coac.2021.02.002.

Kamath, V., Chandra, P. and Jeppu, G.P. (2020) 'Comparative study of using five different leaf extracts in the green synthesis of iron oxide nanoparticles for removal of arsenic from water', *International Journal of Phytoremediation*, 22(12), pp. 1278–1294. Available at: https://doi.org/10.1080/15226514.2020.1765139.

Karimi, J. and Mohsenzadeh, S. (2015) 'Rapid, green, and eco-friendly biosynthesis of copper nanoparticles using flower extract of *Aloe vera*', *Synthesis and Reactivity in Inorganic, Metal-Organic, and Nano-Metal Chemistry*, 45(6), pp. 895–898. Available at: https://doi.org/10.1080/15533174.2013.862644.

Kharissova, O.V. *et al.* (2019) 'Greener synthesis of chemical compounds and materials', *Royal Society Open Science*, 6(11), p. 191378. Available at: https://doi.org/10.1098/rsos.191378.

Kiranmayee, M. *et al.* (2023) 'Green synthesis of *Piper nigrum* copper-based nanoparticles: in silico study and ADMET analysis to assess their antioxidant, antibacterial, and cytotoxic effects', *Frontiers in Chemistry*, 11. Available at: https://doi.org/10.3389/fchem.2023.1218588.

Kirti, Kamsonlian, S. and Agarwal, V.P. (2022) 'Review on synthesis of plant-mediated green iron nanoparticles and their application for decolorization of dyes', *Materials Today: Proceedings* [Preprint]. Available at: https://api.semanticscholar.org/CorpusID:254331021.

Kulkarni, D. *et al.* (2023) 'Biofabrication of nanoparticles: sources, synthesis, and biomedical applications', *Frontiers in Bioengineering and Biotechnology*, 11. Available at: https://doi.org/10.3389/fbioe.2023.1159193.

Kumari, S. *et al.* (2023) 'A comprehensive review on various techniques used for synthesizing nanoparticles', *Journal of Materials Research and Technology*, 27, pp. 1739–1763. Available at: https://doi.org/10.1016/j.jmrt.2023.09.291.

Laha, D. *et al.* (2014) 'Shape-dependent bactericidal activity of copper oxide nanoparticle mediated by DNA and membrane damage', *Materials Research Bulletin*, 59, pp. 185–191. Available at: https://doi.org/10.1016/j.materresbull.2014.06.024.

Li, J., Wu, Q. and Wu, J. (2015) 'Synthesis of nanoparticles via solvothermal and hydrothermal methods', in *Handbook of Nanoparticles*. Cham: Springer International Publishing, pp. 1–28. Available at: https://doi.org/10.1007/978-3-319-13188-7_17-1.

Mamuru, S.A. and Jaji, N. (2015) 'Voltammetric and impedimetric behaviour of phytosynthesized nickel nanoparticles', *Journal of Nanostructure in Chemistry*, 5(4), pp. 347–356. Available at: https://doi.org/10.1007/s40097-015-0166-x.

Nande, A. *et al.* (2023) 'Exploring the world of functional materials', in *Functional Materials from Carbon, Inorganic, and Organic Sources*. Elsevier, pp. 1–19. Available at: https://doi.org/10.1016/B978-0-323-85788-8.00014-8.

Naseer, M. *et al.* (2020) 'Green route to synthesize Zinc Oxide Nanoparticles using leaf extracts of *Cassia fistula* and *Melia azadarach* and their antibacterial potential', *Scientific Reports*, 10(1), p. 9055. Available at: https://doi.org/10.1038/s41598-020-65949-3.

Nyabadza, A. *et al.* (2023) 'A review of physical, chemical and biological synthesis methods of bimetallic nanoparticles and applications in sensing, water treatment, biomedicine, catalysis and hydrogen storage', *Advances in Colloid and Interface Science*, 321, p. 103010. Available at: https://doi.org/10.1016/j.cis.2023.103010.

Nzilu, D.M. *et al.* (2023) 'Green synthesis of copper oxide nanoparticles and its efficiency in degradation of rifampicin antibiotic', *Scientific Reports*, 13(1), p. 14030. Available at: https://doi.org/10.1038/s41598-023-41119-z.

Osman, A.I. *et al.* (2024) 'Synthesis of green nanoparticles for energy, biomedical, environmental, agricultural, and food applications: a review', *Environmental Chemistry Letters*, 22(2), pp. 841–887. Available at: https://doi.org/10.1007/s10311-023-01682-3.

Pansambal, S. *et al.* (2017) 'Phytosynthesis and biological activities of fluorescent CuO nanoparticles using *Acanthospermum hispidum* L. extract', *Journal of Nanostructures*, 7(3), pp. 165–174. Available at: https://doi.org/10.22052/jns.2017.03.001.

Patil, S.P. (2020) '*Ficus carica* assisted green synthesis of metal nanoparticles: a mini review', *Biotechnology Reports*, 28, p. e00569. Available at: https://doi.org/10.1016/j.btre.2020.e00569.

Potbhare, A.K. *et al.* (2019) 'Phytosynthesis of nearly monodisperse CuO nanospheres using *Phyllanthus reticulatus/Conyza bonariensis* and its antioxidant/antibacterial assays', *Materials Science and Engineering: C*, 99, pp. 783–793. Available at: https://doi.org/10.1016/j.msec.2019.02.010.

Prabhu, N.N. (2018) 'Green synthesis of iron oxide nanoparticles (IONPs) and their nanotechnological applications', *Journal of Bacteriology & Mycology: Open Access*, 6(4). Available at: https://doi.org/10.15406/jbmoa.2018.06.00215.

Raja Rajamanikkam, S.C.R. *et al.* (2024) 'Green synthesis of copper and copper oxide nanoparticles from brown algae *Turbinaria* species' aqueous extract and its antibacterial properties', *Cureus* [Preprint]. Available at: https://doi.org/10.7759/cureus.57366.

Rassaei, L. *et al.* (2011) 'Nanoparticles in electrochemical sensors for environmental monitoring', *TrAC Trends in Analytical Chemistry*, 30(11), pp. 1704–1715. Available at: https://doi.org/10.1016/j.trac.2011.05.009.

Rawat, M., Singh, J. and Kaur, H. (2018) 'A uncanny potential of plants for metal nanoparticles synthesis', *Journal of Nanomedicine Research*, 7(2). Available at: https://doi.org/10.15406/jnmr.2018.07.00191.

Sandhu, A. and Goel, A. (2023) 'Biosynthesis of nanoparticles by micro-organisms and its applications', *Journal of Young Pharmacists*, 15(3), pp. 430–440. Available at: https://doi.org/10.5530/jyp.2023.15.58.

Selvakesavan, R.K. and Franklin, G. (2021) 'Prospective application of nanoparticles green synthesized using medicinal plant extracts as novel nanomedicines', *Nanotechnology, Science and Applications*, 14, pp. 179–195. Available at: https://doi.org/10.2147/NSA.S333467.

Shah, M. *et al.* (2015) 'Green synthesis of metallic nanoparticles via biological entities', *Materials*, 8(11), pp. 7278–7308. Available at: https://doi.org/10.3390/ma8115377.

Sharma, S. *et al.* (2021) 'Synthesis and characterization of CuO nanoparticles by *Aloe barbadensis* leaves', *Quantum Journal of Engineering, Science and Technology*, 2(5), pp. 1–9.

Shiravand, S. and Azarbani, F. (2017) 'Phytosynthesis, characterization, antibacterial and cytotoxic effects of copper nanoparticles', *Green Chemistry Letters and Reviews*, 10(4), pp. 241–249. Available at: https://doi.org/10.1080/17518253.2017.1360401.

Singh, H. *et al.* (2023) 'Revisiting the green synthesis of nanoparticles: uncovering influences of plant extracts as reducing agents for enhanced synthesis efficiency and its biomedical applications', *International Journal of Nanomedicine*, 18, pp. 4727–4750. Available at: https://doi.org/10.2147/IJN.S419369.

Singh, P. *et al.* (2016) 'Biological synthesis of nanoparticles from plants and microorganisms', *Trends in Biotechnology*, 34(7), pp. 588–599. Available at: https://doi.org/10.1016/j.tibtech.2016.02.006.

Singh, R. and Lillard, J.W. (2009) 'Nanoparticle-based targeted drug delivery', *Experimental and Molecular Pathology*, 86(3), pp. 215–223. Available at: https://doi.org/10.1016/j.yexmp.2008.12.004.

Soni, V. *et al.* (2021) 'Sustainable and green trends in using plant extracts for the synthesis of biogenic metal nanoparticles toward environmental and pharmaceutical advances: a review', *Environmental Research*, 202, p. 111622. Available at: https://doi.org/10.1016/j.envres.2021.111622.

Sun, L. *et al.* (2023) 'Smart nanoparticles for cancer therapy', *Signal Transduction and Targeted Therapy*, 8(1), p. 418. Available at: https://doi.org/10.1038/s41392-023-01642-x.

Tailor, G. *et al.* (2023) 'A review on green route synthesized nickel nanoparticles: biological and photo-catalytic applications', *Results in Chemistry*, 6, p. 101195. Available at: https://doi.org/10.1016/j.rechem.2023.101195.

Thanh, N.T.K., Maclean, N. and Mahiddine, S. (2014) 'Mechanisms of nucleation and growth of nanoparticles in solution', *Chemical Reviews*, 114(15), pp. 7610–7630. Available at: https://doi.org/10.1021/cr400544s.

Thatyana, M. *et al.* (2023) 'Advances in phytonanotechnology: a plant-mediated green synthesis of metal nanoparticles using Phyllanthus plant extracts and their antimicrobial and anticancer applications', *Nanomaterials*, 13(19), p. 2616. Available at: https://doi.org/10.3390/nano13192616.

Tiwari, S. *et al.* (2021) 'An overview of the phytosynthesis of various metal nanoparticles', *3 Biotech*, 11(11), p. 478. Available at: https://doi.org/10.1007/s13205-021-03014-0.

Tiwari, S.K. *et al.* (2022) 'Methods for the conversion of biomass waste into value-added carbon nanomaterials: recent progress and applications', *Progress in Energy and Combustion Science*, 92, p. 101023. Available at: https://doi.org/10.1016/j.pecs.2022.101023.

Urnukhsaikhan, E. *et al.* (2021) 'Antibacterial activity and characteristics of silver nanoparticles biosynthesized from *Carduus crispus*', *Scientific Reports*, 11(1), p. 21047. Available at: https://doi.org/10.1038/s41598-021-00520-2.

Vijayaram, S. *et al.* (2024) 'Applications of green synthesized metal nanoparticles—a review', in *Biological Trace Element Research*. Springer, pp. 360–386. Available at: https://doi.org/10.1007/s12011-023-03645-9.

Vincent, J. *et al.* (2022) 'Biogenic synthesis of copper-based nanomaterials using plant extracts and their applications: current and future directions', *Nanomaterials*, 12(19), p. 3312. Available at: https://doi.org/10.3390/nano12193312.

Wang, S. *et al.* (2023) 'Surface functionalization of metal and metal oxide nanoparticles for dispersion and tribological applications—a review', *Journal of Molecular Liquids*, 389, p. 122821. Available at: https://doi.org/10.1016/j.molliq.2023.122821.

Wender, H. *et al.* (2013) 'Sputtering deposition of nanoparticles onto liquid substrates: recent advances and future trends', *Coordination Chemistry Reviews*, 257(17–18), pp. 2468–2483. Available at: https://doi.org/10.1016/j.ccr.2013.01.013.

Yulizar, Y. *et al.* (2018) 'Plants extract mediated synthesis of copper (II) oxide nanoparticles using *Oldenlandia corymbosa* L. leaf', in, p. 020097. Available at: https://doi.org/10.1063/1.5064094.

Zahra, Z. *et al.* (2020) 'Exposure route of TiO_2 NPs from industrial applications to wastewater treatment and their impacts on the agro-environment', *Nanomaterials*, 10(8), p. 1469. Available at: https://doi.org/10.3390/nano10081469.

5 Plant-Associated Nanotechnology in Biosensoring and Medicinal Applications

Harpreet Kaur, Paramjeet Singh, Abhishek Kumar Mishra, and Ambar Bongirwar

5.1 INTRODUCTION

Nanotechnology involves various scientific and engineering fields, including organic chemistry, molecular biology, and molecular engineering, that concentrate on the manipulation of fundamental substances such as atoms or molecules at the nanoscale, particularly within a size range of 1 to 100 nanometres (Bhushan 2017). The field of nanotechnology has yielded extensively researched technological advancements, particularly in the domains of healthcare and medicine (Malik, Muhammad, and Waheed 2023), food and packaging, agriculture, and biosensoring (Malekzad et al. 2017). Recently, plant-associated nanotechnology has garnered significant attention as an alternative to methods utilizing chemically synthesized nanoparticles in the areas of agriculture, precision medicine (Pandey and Karmous 2024), and medical imaging and diagnostics. A growing number of researchers are diligently exploring innovative nanotechnology-driven approaches linked to plant materials to create diverse drug delivery systems (Kant et al. 2024) and biosensors that are both safe for use and sustainable. Extracellular vesicles like ginseng-derived nanoparticles, isolated from *Panax ginseng* plant (Cao et al. 2019), have demonstrated significant potential as safer nano-drugs in cancer immunotherapy. In addition to their applications in medical imaging, the remarkable electrical, optical, and electrochemical properties of various nanoparticles make them highly appropriate for signal transduction in biosensor technologies (Zhang, Guo, and Cui 2009). Numerous metallic nanoparticles, such as gold (Siddiqi and Husen 2017) or silver, synthesized from plant-associated nanotechnology, have exhibited the potential to significantly enhance electron transfer rates and become a candidate for biosensors. Gold nanoparticles exhibit exceptional effectiveness in optical detection techniques for certain antigens due to signals enhanced by surface plasmon resonance. Nanoparticles with varying degrees of refinement can be engineered with two prominent manufacturing approaches, categorized as Bottom-up and Top-down. The metal reduction method involves reduction of metallic ions (solution

DOI: 10.1201/9781003559900-5

form) by incorporating organic and inorganic reducing agents and stabilizing agents (Peralta-Videa et al. 2016). The Bottom-up approach incorporates Self-assembly, and Positional-assembly as manufacturing methodologies for generating molecules, crystals, and molecular devices, etc. Physical and chemical interactions among atoms and molecules tend to arrange them in an orderly fashion, which is referred to as Self-assembly. However, this slow process can be accelerated by applying an external electric or magnetic field to produce nanoparticles in an industrial setup. Conversely, Positional-assembly is not suitable for synthesizing nanostructures at an industrial level because it requires atomic-level refinement in measuring the starting materials (atoms/molecules). The Top-down synthetic approach, a physically driven method, incorporates various chemomechanical interventions (electrochemical method or ultrasonication) to minimize the size of bulk material, eventually transforming it into nano-sized particles for multi-functional uses. However, both physical and chemical approaches for synthesizing nanoparticles at an industrial scale pose several challenges (Gour and Jain 2019) due to indiscriminate use of chemicals during synthesis and production of harmful metabolites as side products. Thus, the concept of green synthesis of nanoparticles has attracted substantial attention among the scientific community as compared to traditional methods due to their sustainability, cost-effectiveness, and scalability. The replacement of perilous solvents with safer solvents like water and ionic liquids renders them ideal for usage in the field of biosensoring and pharmaceutical applications. However, a comprehensive framework is needed to ascertain the safety and effectiveness of the nanoparticles synthesized by green approaches (Albrecht, Evans, and Raston 2006) before their incorporation in medicine and diagnostics.

5.2 PLANT-MEDIATED NANOPARTICLES

In the last few decades, biological entities from nature, such as algae, plants, microorganisms, and fungi, have been harnessed to create environmentally benign metallic nanoparticles. Among these, plant-mediated metallic nanoparticles have emerged as a significant innovation in the field of medicine and diagnostics for treating various diseases. Plant-mediated nanoparticles can be synthesized through a process known as green synthesis, as depicted in Figure 5.1. This eco-friendly approach significantly reduces the reliance on hazardous chemicals, often generously utilized in conventional synthesis methods.

5.2.1 Green Synthesis of Plant-Mediated Nanoparticles

i. Preparation of plant extract: The bioactive compounds flavonoids, phenolic acids, alkaloids, and terpenoids are extracted from different parts of medicinal plants by various methods like maceration, infusion, digestion, decoction, and percolation.
ii. Synthesis and purification of nanoparticles: The plant extract is added to the salt solution of metal (zinc acetate, copper acetate, copper nitrate) and mixed thoroughly with a stirrer. This mixture is kept at ambient temperature, and

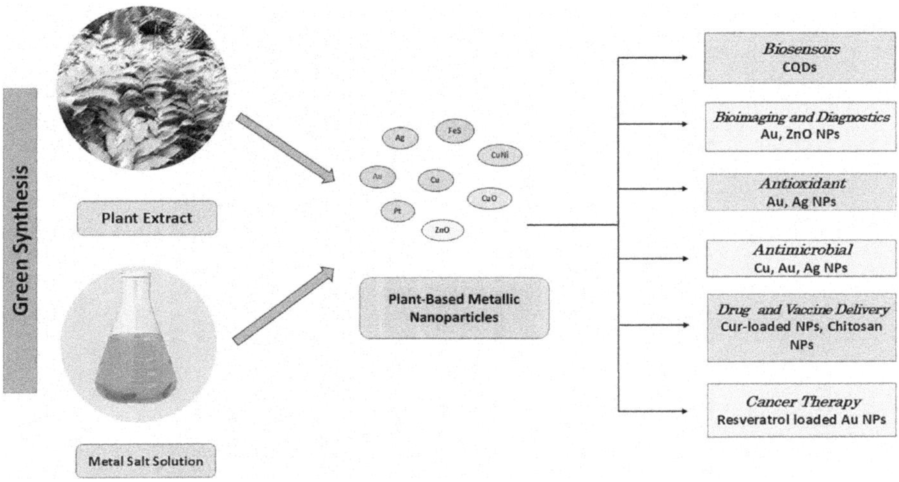

FIGURE 5.1 Green synthesis of nanoparticles and their applications in biomedical sciences.

over time, the salt solution of metal is reduced to metallic nanoparticles, evidenced by the colour transition of the solution. The nanoparticles can be separated and purified using ultracentrifugation at high speeds for several minutes.

iii. Characterization: In order to confirm the synthesis of nanoparticles, different analytical methods are used to analyse their crystal structure and to assess their size and shape. The various analytical techniques employed include UV-Visible spectroscopy, Fourier transform infrared spectroscopy (FTIR), Raman spectroscopy, Nuclear magnetic resonance (NMR), X-ray photoelectron spectroscopy (XPS), Dynamic light scattering (DLS), Transmission electron microscopy (TEM), Scanning electron microscopy (SEM), and X-ray diffraction (XRD), among others.

5.2.2 Types of Plant-Mediated Nanoparticles

Numerous metal nanoparticles have been synthesized through the reduction of metal ions to their respective nanoparticle forms. In the context of plant-mediated nanoparticle synthesis, the plants do not serve as the direct source of the nanoparticle material, rather, they play a significant role in the synthesis process. Plants are utilized as reducing, stabilizing, or capping agents during the fabrication of these nanoparticles. There exist several examples of synthesis of plant-mediated nanoparticles from literature, like metal nanoparticles (AgNPs, ZnO NPs), quantum dots (CdS QDs), carbon-based nanoparticles (graphene oxide NPs, Carbon Nanotubes), cerium oxide nanoparticles, and selenium nanoparticles. Table 5.1 illustrates comprehensive information about properties, mechanisms, and applications of various plant-mediated nanoparticles.

TABLE 5.1
Comprehensive Analysis of Nanoparticles, Their Morphology, Interactions, and Applications

Nanoparticle	Plant Source	Family	Size (nm)	Morpho-logy	Biological Assays	Mechanism	Application	Reference
CI-Au	*Clerodendrum inerme* (Leaves)	Acanthaceae	5.82	Spherical	DPPH free radical scavenging assay	Oxidative damage from their high redox potential, and ROS generation	Anti-oxidant, Anti-bacterial agents, Wound dressings	(Khan, Shahid, and Lee 2020)
CI-Ag	*Clerodendrum inerme* (Leaves)	Lamiaceae	5.54	Spherical	Standard well diffusion method	Oxidative damage from their high redox potential, and ROS generation	Anti-bacterial, Anti-inflammatory agents, Disinfecting medical devices	(Khan, Shahid, and Lee 2020)
Cu	*Moringa oleifera* (Leaves)	Moringaceae	35.8–49.2	Amorphous	DPPH free radical scavenging assay, Resazurin Microtiter assay	ROS generation and structural damage to microbial cells	Anti-bacterial, Anti-oxidant, Anti-fungal	(Das et al. 2020)
Se	*Catharanthus roseus* (flower)	Apocynaceae	23.2±6.06	Hollow spherical	–	Increase the level of ROS and decrease the mitochondrial potential and glutathione level, leading to inhibition of HepG2 cell proliferation	Anti-cancer	(Deepa and Ganesan 2015)
Pd	*Tabernaemon-tanadivaricata*	Apocynaceae	20–50	Spherical	Alpha-amylase activity	Inhibition of carbohydrate-hydrolyzing enzymes	Anti-microbial, Anti-diabetic	(Vinodhini, Vithiya, and Prasad 2022)
Pt	*Cordia myxa L.(leaves)*	Boraginaceae	40	Spherical	MTT assay	Inducing apoptosis	Anti-cancer, photothermal therapy	(Abed, Mishaal Mohammed, and Khalaf 2022)

ZnO	*Andrographis paniculata (Leaves)*	Acanthaceae	57	Hexagonal	DPPH free radical scavenging assay, Alpha-amylase inhibition activity	–	Anti-oxidant, Anti-diabetic, Anti-inflammatory	(Rajakumar et al. 2018)
Cp-Ag	*Cucumis prophetarum (Leaves)*	Cucurbitaceae	30–50	Ellipsoidal	DPPH assay, ABTS assay, MTT assay	–	Antioxidant, Anti-bacterial Anticancer	(Hemlata et al. 2020)

5.3 PLANT-MEDIATED NANOPARTICLES IN BIOSENSORS AND DIAGNOSTICS

Biosensors are sophisticated analytical devices (Pejcic, Marco, and Parkinson 2006) that integrate a biological sensing element with a transducer to detect and quantify specific substances known as analytes. They are extensively employed across diverse fields, including healthcare, food safety, personalized medicine, and biotechnology. A typical biosensor comprises a biological recognition element, which identifies the specific analyte through mechanisms such as enzymatic reactions, antibody binding, nucleic acid hybridization, or microbial cell activity. This is paired with a transducer, which converts the biological interaction into measurable signals. Additionally, an intricate electronic network amplifies and converts signals to digital form and displays them in many forms, like graphs, tables, or images. Gold nanoparticles have found applications in optical biosensors (Ciobotaru et al. 2023) for the detection of copper, aqueous ammonia, etc., in water, and biological samples. Furthermore, gold nanoparticle-based electrochemical biosensors have made possible the detection of amino acids, antibiotics, and glucose in biospecimens. This technology has evolved considerably from the inception of the first commercial biosensor for glucose detection to the advent of plant-mediated nanoparticles-based biosensors in bioimaging and diagnosis.

5.3.1 Carbon Quantum Dots (CQDs) in Biosensors

Carbon quantum dots (CQDs) integrated in biosensors are spherical nanoparticles surrounded by a coating that protects them from the extracellular signalling or enzymatic activities. Owing to their unique optical properties and up-conversion photoluminescence, carbon quantum dots (Pourmadadi et al. 2023) have recorded progress in translational research in areas of imaging and drug delivery. These nanoparticles can be synthesized through Bottom-up and Top-down synthesis approaches. A recent study by Fang et al. reported the bioimaging studies of gold nanoparticles synthesized by seed-derived carbon quantum dots. The fluorescence images of *E. Coli* bacterial strain (Fang et al. 2024) were captured and exhibited the compatibility of CQDs with bacterial cells. The CQDs displayed passive diffusion through the cell membrane and accumulation in the cytoplasm. These findings established them as a potential detector for imaging different types of bacteria. A comparative study between microscale disc electrode and the gold microdisc array electrode bound to CQDs/AuNPs nanohybrid materials reported a sensitivity in glucose detection (Buk and Pemble 2019) that is 13 times greater with the latter configuration. This compact glucose biosensor was way more accurate than the previously reported planar biosensor by the researchers. Carbon quantum dots carrying different functional groups on their surface have found applications in bioimaging and detecting intracellular metallic ions (Ding, Zhu, and Tian 2014).

5.3.2 ZnO Nanoparticles in Biosensors

Some noble metals and 2D nanostructures are well established in the manufacturing of biosensors with high detection sensitivity. Metal oxides, particularly zinc oxide

(ZnO) nanoparticles, exhibit a swift exchange of electrons between electrodes and signaling molecules in biological systems, making them a prospective candidate for use in biosensors. Plant-mediated synthesis of ZnO nanoparticles has been popularized in the past few years (Figure 5.1). The polyphenols, flavonoids, and terpenoids present in *Moringa Oleifera* extract facilitate the formation of ZnO nanoparticles (Manikandan et al. 2024) by acting as reducing/capping agents. The incorporation of these nanoparticles in electrochemical biosensors has made it possible to detect some neurochemicals like dopamine.

By integrating enzyme selectivity with the catalytic properties of nanoparticles, biosensors with enhanced sensitivity have also been reported. A highly sensitive biosensor has been developed, utilizing a glassy carbon electrode. This innovative design incorporates hybrid ZnO nanoparticles embedded nitrogen-doped carbon sheets derived from peach extract, and immobilized with glucose oxidase for the detection of glucose in blood serum (Muthuchamy et al. 2018).

5.4 MEDICINAL APPLICATIONS OF PLANT-MEDIATED NANOPARTICLES

In the field of medicine and diagnostics, plant-associated nanotechnology has offered healthcare professionals new methods for safely treating terminal diseases compared to ancient approaches. These applications include antimicrobial activities, drug delivery systems, and bioimaging as illustrated in Figure 5.1.

5.4.1 Antioxidative Activity

The oxidation of foods, particularly lipids, leads to the generation of reactive oxygen and nitrogen species, and other harmful chemicals that can negatively impact health. Antioxidant activity involves enzymes such as superoxide dismutase and other antioxidants like mycothiol, which work to scavenge these free radicals and protect the body from tissue injury, accelerated cellular death, and oxidative stress. Some recent research in the field of green synthesis of nanoparticles has delivered significant results about antioxidative activities in terms of DPPH free radical inhibition. A few findings worth mentioning include gold and silver nanoparticles derived from *Clerodendrum inerme leaf extract* (Khan, Shahid, and Lee 2020) and copper nanoparticles (Das et al. 2020) derived from *Moringa oleifera* leaves (Table 5.1). Vera et al. have demonstrated in their studies that % yield of ZnO nanoparticles is much more when synthesized using *A. chilensis* extract (Vera et al. 2023) as compared to other plant extracts. This generalizes the highest antioxidant activities in the case of ZnO nanoparticles derived from *A. chilensis* extract.

5.4.2 Antimicrobial Activity

Plant-mediated nanoparticles display significant antimicrobial properties against microbial organisms. The probable mechanisms of antimicrobial activity constitute cell wall disruption of the microbe and are determined by minimum inhibitory concentration (MIC) or minimum bactericidal concentration (MBC) against different strains of pathogens. *Moringa oleifera* has had exceptional medicinal value since ancient times and is still being researched extensively for synthesizing copper nanoparticles (Das et al. 2020) to establish their antifungal and antibacterial activities. Estimation of total phenolic content during the studies established the involvement of polyphenols in the reduction of metallic nanoparticles. The minimum inhibitory concentration values for copper nanoparticles were dropped exceptionally in the case of *Candida albicans and Candida glabrata*, establishing them as more potent antifungal agents. Another pivotal study (Khan, Shahid, and Lee 2020) about plant-mediated synthesis of gold and silver nanoparticles (CI-Au and CI-Au), incorporating *Clerodendrum inerme* leaf extract, exhibited elevated antioxidant properties and enhanced antibacterial activity against *B. subtilis*, *S. aureus*, *Klebsiella*, and *E. coli* microbial strains as compared to chemically synthesized nanoparticles. In order to validate these activities, comprehensive evaluations were performed using well diffusion assays, colorimetric assays, DPPH free radical scavenging assays, and MTT assays (Table 5.1). Concentration of *C. inerme-derived* nanoparticles falling between 1–50 μM exhibited a non-hemolytic effect as compared to their chemically synthesized counterparts. This can be attributed to the diminished toxicity of nanoparticles owing to their surface modification by *C. inerme.*

5.4.3 Antiviral Activity

Viruses are differently shaped pathogens encasing a viral genome (RNA/DNA) inside a protein coat. Innumerable complicated health conditions like AIDS, COVID-19, Influenza, Hepatitis, and many more are caused by different viruses. Several plant-mediated nanoparticles generously exerted antiviral activities in many instances, as reported in the literature. Multiple communications about silver and other nanoparticles synthesized from plant extracts have been reported in the last few decades as potential antiviral options. Green-synthesized silver nanoparticles derived from *M. oleifera* seeds (Sujitha et al. 2015) exhibited in vitro antiviral activity against DEN-2 infecting Vero cells. The antiviral activity was determined through the 50% tissue culture infectious dose ($TCID_{50}$) assay and exhibited cytopathic effects by showing a significant drop in viral titre after treatment with plant-mediated nanoparticles. Furthermore, plaque assay also substantiated the antiviral activity by exhibiting reduced viral yield during the studies. Similarly, silver nanoparticles derived from *A. paniculata* (Sharma et al. 2019) exerted antiviral activity against chikungunya virus (CHIKV). The antiviral effect was corroborated by undertaking MTT cell viability test and suggested maximum inhibition of viral infection by silver nanoparticles derived from *A. paniculata* compared to NPs derived from *P. niruri and T. cordifolia.*

5.4.4 Anti-Inflammatory Activity

Phytoconstituents like polyphenols, terpenoids, tannins, flavonoids, and saponins have established anti-inflammatory activities as summarized in the vast literature. Henceforth, nanoparticles synthesized from a plant-mediated approach exert double the efficacy of anti-inflammatory action as therapeutics (Gonfa et al. 2023). The underlying mechanism of this activity limits the release of cytokines that promote inflammation. Various reports have even established significantly higher inflammatory activities of plant-mediated nanoparticles as compared to the activity of plant extracts. *A. Paniculata* extract-mediated synthesis of ZnO nanoparticles reported by Rajakumar et al. (Rajakumar et al. 2018) has established their potent anti-inflammatory potential in therapeutic and nutraceuticals. Maximum inhibition activity (Table 5.1) was determined for diclofenac (reference), ZnO NPs, and *A. Paniculata* leaf extract. The inhibitory potential of diclofenac and ZnO NPs was almost similar, while that of *A. Paniculata* leaf extract showed maximum inhibition at higher concentrations.

5.4.5 Drug Delivery System

Phytochemicals, owing to their size and high molecular weight, face the challenge of passage through lipid bilayers. This limitation leads to lesser absorption and poor bioavailability of these constituents. This concern underscores the necessity for developing innovative drug delivery systems that enhance the effectiveness of bioactive molecules in treatment regimens. Although the utilization of nanoparticles as carriers in drug delivery systems gained prominence in the 1990s, the eco-friendly approach of employing plant-mediated nanoparticles for sustained and controlled drug release remains relatively nascent. A study demonstrated the encapsulation of curcumin within Poly (lactic-co-glycolic acid) nanoparticles (Chang et al. 2013) induced apoptotic cell death in the CAL27 cell line resistant to cisplatin. These nanoparticles significantly improved the cellular uptake of curcumin, thereby increasing cytotoxicity against oncogenic cells. This finding reinforces the potential of plant-mediated drug delivery systems as viable therapeutic options for the treatment of cisplatin-resistant human oral cancer.

Numerous reports (Chandrakala, Aruna, and Angajala 2022) highlight the efficacy of greener, plant-mediated nanoparticles, such as silver nanoparticles synthesized from *Brassica oleracea* and palladium nanoparticles derived from *Sapium sebiferum* leaf extract, for treatment in a safer manner. These nanoparticles are administered through transdermal routes, like AgNP-loaded antimicrobial wound dressing patches. Plant-mediated nanoparticles are natural entities having significant advantages over other drug delivery systems, like liposomes and exosomes. These are easily tunable depending upon the site of action and thus are more specific. Plant-mediated synthesis of ZnO nanoparticles carrying many anti-cancer drugs has aided in generating cytotoxicity in malignant cells and established them as effective drug delivery systems. The drug delivery systems worth mentioning include doxorubicin-loaded zinc oxide nanoparticles with *B. flabellifer* fruit extract (Vimala et al. 2014), and paclitaxel on chitosan-coated zinc oxide nanoparticles derived from *C. sinensis* leaf extract (Akbarian et al. 2020).

5.4.6 Cancer Therapy

The multifaceted advantages associated with plant-mediated nanoparticles make them significant components for next-generation oncological treatments. Metallic nanoparticles can be used as a carrier for the delivery of anti-cancer drugs (Chandrakala,

Aruna, and Angajala 2022). Silver nanoparticles exert anti-cancer effects by triggering the production of reactive oxygen species (ROS). Silver nanoparticles synthesized using the aqueous leaf extract of *Cucumis prophetarum* (Hemlata et al. 2020) have presented significant antibacterial activity against both *Staphylococcus aureus* (a Gram-positive bacterium) and *Salmonella typhi* (a Gram-negative bacterium). Furthermore, the MTT assay has established their antiproliferative efficacy in various cancer cell lines (Table 5.1). The IC_{50} value of silver nanoparticles was least for MCF-7 as compared to other cell lines. It generalizes that these nanoparticles show maximum potency towards MCF-7. These nanoparticles have exhibited significant efficacy as potent agents, exhibiting antioxidant, antibacterial, and anticancer properties. Biosynthesized silver nanoparticles (AgNPs) (Sukirtha et al. 2012) from *Melia azedarach* exhibited dose-dependent cytotoxicity against HeLa cells, with a lethal dose 50 (LD_{50}) value of 300 µg/mL. The induction of apoptosis was confirmed through fluorescence staining techniques. In another study, Zhang et al. reported the green synthesis of gold nanoparticles loaded with resveratrol (Zhang et al. 2019), presenting a remarkable anti-cancer effect compared to resveratrol alone, as evidenced by both pre-clinical studies. Various techniques, including MTT assay, flow cytometry, immunohistochemistry, and western blot analysis, were utilized to investigate the antitumor activity of resveratrol-loaded gold nanoparticles (Res-GNPs) in malignant hepatic cells.

5.4.7 Bioimaging

Bioimaging refers to diagnostic techniques incorporating sensors/probes to study physiological and chemical processes inside living organisms in a non-invasive manner. Nanoparticles are a highly significant component of bioimaging as they function as a contrasting agent in different techniques, namely magnetic resonance imaging (MRI), ultrasound, and computed tomography (CT), etc. However, the potential toxicity associated with chemically synthesized nanoparticles highlights an important concern about their use in bioimaging. The size and shape of the nanoparticles are crucial characteristics for their intended use as contrast agents in biological investigations. Recently, the paradigm of research has shifted towards plant-mediated synthesis of metallic nanoparticles. This approach renders surface functionalization to nanoparticles, making them safer for use in biomedical applications. There are references about the significant use of plant-mediated nanoparticles as both diagnostic and therapeutic agents for treating cancer. This revolutionary field of medicine is known as theranostics. In this approach, APIs can be coated with contrast agents, resulting in significant advantages. This methodology enhances our understanding of drug delivery mechanisms, kinetics, and the overall effectiveness of the API. Plant extract-synthesized gold nanoparticles have paved their way in cancer diagnostics and treatment regimens. Studies (Bharadwaj et al. 2021) have summarized that barley leaf-mediated gold nanoparticles presented better X-ray absorption capacity as compared to conventional contrast agents. In another reference, there is a mention of the synthesis of 4-allyl-2-methoxyphenolate-capped iron oxide nanoparticles from *P. dioica* leaf extract (Kharey et al. 2019). The synthesized nanoparticles were extensively characterized and assessed for cell viability and cytotoxicity by performing MTT assay. These studies inferred the potential of these nanoparticles as a novel contrast agent in biomedical imaging.

German physicist Wilhelm Röntgen discovered X-rays in 1895, which set the foundation of modern bioimaging. Since their inception, iodinated X-ray contrast agents have been employed to obtain high-resolution images in X-ray imaging. However, these agents can sometimes induce severe side effects, such as contrast-induced nephropathy. In response to this issue, extensive research has been undertaken. Recently, plant-mediated nanoparticles have emerged as alternatives to traditional iodinated X-ray contrast agents. A recent study reported that gold nanoparticles synthesized from barley leaf extract (Xue et al. 2021) exhibited superior image quality and higher contrast at lower doses in CT scans during in vitro studies, compared to conventional contrast agents.

5.5 CONCLUSION AND FUTURE DIRECTIONS

Plant-associated nanotechnology, especially the application of metallic nanoparticles in medicine and drug delivery systems, is still evolving, and its full potential remains to be fully realized. As diseases continue to change in both nature and severity, there is an increasing demand for extensive research that emphasizes plant-associated approaches over traditional methods. Recent advancements in bioimaging have provided us with tools that can help manage life-threatening conditions such as cancer and cardiovascular diseases. Combined with nanomedicines derived from green synthesis, these innovations have significantly decreased mortality rates in many previously untreatable conditions. Potential applications of plant-mediated nanoparticles include medicine, agriculture, food packaging, sensors, imaging, and biotechnology. However, several challenges must be addressed to shift the focus from traditional to plant-associated nanotechnology. A significant challenge in the production and commercialization of plant-mediated nanoparticles is scalability. The reproducibility and purity of plant extracts from even a single species can vary considerably under different sets of conditions, leading to unprecedented discrepancies in their use for therapeutics and diagnostics and thus raising regulatory concerns. To ensure the exponential development of green synthesis in the future, research must concentrate on the precise and reproducible synthesis of nanoparticles. Another ongoing debate regarding their long-term effects on human health and the environment warrants further investigation. Thus, future endeavors should focus on making plant-associated nanotechnology approach safer and more affordable, particularly in realms of vaccine development, chemotherapy, immunotherapy, and medical imaging.

5.6 DECLARATION OF GENERATIVE AI AND AI-ASSISTED TECHNOLOGIES IN THE WRITING PROCESS

The authors have used Grammarly to ensure that the English text is grammatically correct during the writing process. The manuscript was subsequently reviewed and edited. The authors take full responsibility for the content of the publication.

REFERENCES

Abed, Ali S., Ahmed Mishaal Mohammed, and Yousif H. Khalaf. "Novel photothermal therapy using platinum nanoparticles in synergy with near-infrared radiation (NIR) against human breast cancer MCF-7 cell line." *Results in Chemistry* 4 (2022): 100591. https://www.sciencedirect.com/science/article/pii/S2211715622003101

Akbarian, Maedeh, Soleiman Mahjoub, Seyed Mohammad Elahi, Ebrahim Zabihi, and Hamed Tashakkorian. "Green synthesis, formulation and biological evaluation of a novel ZnO nanocarrier loaded with paclitaxel as drug delivery system on MCF-7 cell line." *Colloids and Surfaces B: Biointerfaces* 186 (2020): 110686. https://www.sciencedirect.com/science/article/abs/pii/S0927776519308306

Albrecht, Matthew A., Cameron W. Evans, and Colin L. Raston. "Green chemistry and the health implications of nanoparticles." *Green chemistry* 8, no. 5 (2006): 417–432. https://pubs.rsc.org/en/content/articlelanding/2006/gc/b517131h/unauth

Bharadwaj, Kaushik Kumar, Bijuli Rabha, Siddhartha Pati, Tanmay Sarkar, Bhabesh Kumar Choudhury, Arpita Barman, Dorothy Bhattacharjya et al. "Green synthesis of gold nanoparticles using plant extracts as beneficial prospect for cancer theranostics." *Molecules* 26, no. 21 (2021): 6389. https://www.mdpi.com/1420-3049/26/21/6389

Bhushan, Bharat. "Introduction to nanotechnology." In *Springer handbook of nanotechnology* (2017): 1–19. https://link.springer.com/chapter/10.1007/978-3-662-54357-3_1

Buk, Vuslat, and Martyn E. Pemble. "A highly sensitive glucose biosensor based on a micro disk array electrode design modified with carbon quantum dots and gold nanoparticles." *Electrochimica Acta* 298 (2019): 97–105. https://www.sciencedirect.com/science/article/abs/pii/S001346861832766X

Cao, Meng, Huaijiang Yan, Xuan Han, Ling Weng, Qin Wei, Xiaoyan Sun, Wuguang Lu et al. "Ginseng-derived nanoparticles alter macrophage polarization to inhibit melanoma growth." *Journal for Immunotherapy of Cancer* 7 (2019): 1–18. https://link.springer.com/article/10.1186/s40425-019-0817-4

Chandrakala, V., Valmiki Aruna, and Gangadhara Angajala. "Review on metal nanoparticles as nanocarriers: Current challenges and perspectives in drug delivery systems." *Emergent Materials* 5, no. 6 (2022): 1593–1615. https://link.springer.com/article/10.1007/s42247-021-00335-x

Chang, Pei-Ying, Shu-Fen Peng, Chao-Ying Lee, Chi-Cheng Lu, Shih-Chang Tsai, Tzong-Ming Shieh, Tian-Shung Wu, Ming-Gene Tu, Michael Yuanchien Chen, and Jai-Sing Yang. "Curcumin-loaded nanoparticles induce apoptotic cell death through regulation of the function of MDR1 and reactive oxygen species in cisplatin-resistant CAR human oral cancer cells." *International Journal of Oncology* 43, no. 4 (2013): 1141–1150. https://www.spandidos-publications.com/ijo/43/4/1141

Ciobotaru, Iulia Corina, Daniela Oprea, Constantin Claudiu Ciobotaru, and Teodor Adrian Enache. "Low-cost plant-based metal and metal oxide nanoparticle synthesis and their use in optical and electrochemical (bio) sensors." *Biosensors* 13, no. 12 (2023): 1031. https://www.mdpi.com/2079-6374/13/12/1031

Das, Prince Edwin, Imad A. Abu-Yousef, Amin F. Majdalawieh, Srinivasan Narasimhan, and Palmiro Poltronieri. "Green synthesis of encapsulated copper nanoparticles using a hydroalcoholic extract of Moringa oleifera leaves and assessment of their antioxidant and antimicrobial activities." *Molecules* 25, no. 3 (2020): 555. https://www.mdpi.com/1420-3049/25/3/555

Deepa, B., and V. Ganesan. "Bioinspiredsynthesis of selenium nanoparticles using flowers of Catharanthus roseus (L.) G. Don. and Peltophorum pterocarpum (DC.) Backer ex Heyne–a comparison." *International Journal of ChemTechnol Research* 7 (2015): 725–733. https://www.cabidigitallibrary.org/doi/full/10.5555/20153091493

Ding, Changqin, Anwei Zhu, and Yang Tian. "Functional surface engineering of C-dots for fluorescent biosensing and in vivo bioimaging." *Accounts of Chemical Research* 47, no. 1 (2014): 20–30. https://pubs.acs.org/doi/abs/10.1021/ar400023s

Fang, Xuan-Wei, Hao Chang, Tsunghsueh Wu, Chen-Hao Yeh, Fu-Li Hsiao, Tsung-Shine Ko, Chiu-Lan Hsieh, Mei-Yao Wu, and Yang-Wei Lin. "Green synthesis of carbon quantum dots and carbon quantum dot-gold nanoparticles for applications in bacterial imaging and catalytic reduction of aromatic nitro compounds." *ACS Omega* 9, no. 22 (2024): 23573–23583. https://pubs.acs.org/doi/full/10.1021/acsomega.4c00833

Gonfa, Yilma Hunde, Fekade Beshah Tessema, Archana Bachheti, Nishant Rai, Mesfin Getachew Tadesse, A. Nasser Singab, Kundan Kumar Chaubey, and Rakesh Kumar Bachheti. "Anti-inflammatory activity of phytochemicals from medicinal plants and their nanoparticles: A review." *Current Research in Biotechnology* 6 (2023): 100152. https://www.sciencedirect.com/science/article/pii/S2590262823000345

Gour, Aman, and Narendra Kumar Jain. "Advances in green synthesis of nanoparticles." *Artificial Cells, Nanomedicine, and Biotechnology* 47, no. 1 (2019): 844–851. https://www.tandfonline.com/doi/full/10.1080/21691401.2019.1577878

Hemlata, Prem Raj Meena, Arvind Pratap Singh, and Kiran Kumar Tejavath. "Biosynthesis of silver nanoparticles using *Cucumis prophetarum* aqueous leaf extract and their antibacterial and antiproliferative activity against cancer cell lines." *ACS Omega* 5, no. 10 (2020): 5520–5528. https://pubs.acs.org/doi/full/10.1021/acsomega.0c00155

Kant, Shashi, Harpreet Kaur, Abhishek Kumar Mishra, Ranadheer Reddy Challa, Bhaskar Vallamkonda, Pankaj Popli, and Rajan Swami. "Harmonizing Nature and Technology: Lipidic Nano-phytotherapeutics in the Management of Atopic Dermatitis." *Revista Brasileira de Farmacognosia* 34, no. 5 (2024): 948–968. https://link.springer.com/article/10.1007/s43450-024-00570-3

Khan, Shakeel Ahmad, Sammia Shahid, and Chun-Sing Lee. "Green synthesis of gold and silver nanoparticles using leaf extract of Clerodendrum inerme; characterization, antimicrobial, and antioxidant activities." *Biomolecules* 10, no. 6 (2020): 835. https://www.mdpi.com/2218-273X/10/6/835

Kharey, Prashant, Surjendu Bikash Dutta, M. Manikandan, I. A. Palani, S. K. Majumder, and Sharad Gupta. "Green synthesis of near-infrared absorbing eugenate capped iron oxide nanoparticles for photothermal application." *Nanotechnology* 31, no. 9 (2019): 095705. https://iopscience.iop.org/article/10.1088/1361-6528/ab56b6/meta

Malekzad, Hedieh, Parham Sahandi Zangabad, Hamed Mirshekari, Mahdi Karimi, and Michael R. Hamblin. "Noble metal nanoparticles in biosensors: Recent studies and applications." *Nanotechnology Reviews* 6, no. 3 (2017): 301–329. https://www.degruyter.com/document/doi/10.1515/ntrev-2016-0014/html

Malik, Shiza, Khalid Muhammad, and Yasir Waheed. "Emerging applications of nanotechnology in healthcare and medicine." *Molecules* 28, no. 18 (2023): 6624. https://www.mdpi.com/1420-3049/28/18/6624

Manikandan, M., G. Lakshmi Priya, E. Manikandan, and Vithyasaahar Sethumadhavan. "ZnO nanoparticle-enhanced electrochemical sensor utilizing moringa oleifera leaf extract for real-time dopamine detection." *Journal of Inorganic and Organometallic Polymers and Materials* (2024): 1–10. https://link.springer.com/article/10.1007/s10904-023-02955-8

Muthuchamy, Nallal, Raji Atchudan, Thomas Nesakumar Jebakumar Immanuel Edison, Suguna Perumal, and Yong Rok Lee. "High-performance glucose biosensor based on green synthesized zinc oxide nanoparticle embedded nitrogen-doped carbon sheet." *Journal of Electroanalytical Chemistry* 816 (2018): 195–204. https://www.sciencedirect.com/science/article/abs/pii/S1572665718302340

Pandey, Ashish, and Ines Karmous. "Exploring the potential of plant-based nanotechnology in cancer immunotherapy: benefits, limitations, and future perspectives." *Biological Trace Element Research* (2024): 1–18. https://link.springer.com/article/10.1007/s12011-024-04266-6

Pejcic, Bobby, Roland De Marco, and Gordon Parkinson. "The role of biosensors in the detection of emerging infectious diseases." *Analyst* 131, no. 10 (2006): 1079–1090. https://pubs.rsc.org/en/content/articlehtml/2006/an/b603402k

Peralta-Videa, Jose R., Yuxiong Huang, Jason G. Parsons, Lijuan Zhao, Laura Lopez-Moreno, Jose A. Hernandez-Viezcas, and Jorge L. Gardea-Torresdey. "Plant-based green synthesis of metallic nanoparticles: scientific curiosity or a realistic alternative to chemical synthesis?" *Nanotechnology for Environmental Engineering* 1 (2016): 1–29. https://link.springer.com/article/10.1007/s41204-016-0004-5

Pourmadadi, Mehrab, Erfan Rahmani, Maryam Rajabzadeh-Khosroshahi, Amirmasoud Samadi, Razieh Behzadmehr, Abbas Rahdar, and Luiz Fernando Romanholo Ferreira. "Properties and application of carbon quantum dots (CQDs) in biosensors for disease detection: A comprehensive review." *Journal of Drug Delivery Science and Technology* 80 (2023): 104156. https://www.sciencedirect.com/science/article/abs/pii/S1773224723000084

Rajakumar, Govindasamy, Muthu Thiruvengadam, Govindarasu Mydhili, Thandapani Gomathi, and Ill-Min Chung. "Green approach for synthesis of zinc oxide nanoparticles from Andrographis paniculata leaf extract and evaluation of their antioxidant, antidiabetic, and anti-inflammatory activities." *Bioprocess and Biosystems Engineering* 41 (2018): 21–30. https://link.springer.com/article/10.1007/s00449-017-1840-9

Sharma, Vikrant, Sulochana Kaushik, Pooja Pandit, Divya Dhull, Jaya Parkash Yadav, and Samander Kaushik. "Green synthesis of silver nanoparticles from medicinal plants and evaluation of their antiviral potential against chikungunya virus." *Applied Microbiology and Biotechnology* 103 (2019): 881–891. https://link.springer.com/article/10.1007/s00253-018-9488-1#article-info

Siddiqi, Khwaja Salahuddin, and Azamal Husen. "Recent advances in plant-mediated engineered gold nanoparticles and their application in biological system." *Journal of Trace Elements in Medicine and Biology* 40 (2017): 10–23. https://www.sciencedirect.com/science/article/abs/pii/S0946672X16303819

Sujitha, Vasu, Kadarkarai Murugan, Manickam Paulpandi, Chellasamy Panneerselvam, Udaiyan Suresh, Mathath Roni, Marcello Nicoletti et al. "Green-synthesized silver nanoparticles as a novel control tool against dengue virus (DEN-2) and its primary vector Aedes aegypti." *Parasitology Research* 114 (2015): 3315–3325. https://link.springer.com/article/10.1007/s00436-015-4556-2

Sukirtha, Raman, Kandula Manasa Priyanka, Jacob Joe Antony, Soundararajan Kamalakkannan, Ramar Thangam, Palani Gunasekaran, Muthukalingan Krishnan, and Shanmugam Achiraman. "Cytotoxic effect of Green synthesized silver nanoparticles using Melia azedarach against in vitro HeLa cell lines and lymphoma mice model." *Process Biochemistry* 47, no. 2 (2012): 273–279. https://www.sciencedirect.com/science/article/abs/pii/S1359511311004004

Vera, Joelis, Wence Herrera, Edward Hermosilla, Marcela Díaz, Javiera Parada, Amedea B. Seabra, Gonzalo Tortella, Héctor Pesenti, Gustavo Ciudad, and Olga Rubilar. "Antioxidant activity as an indicator of the efficiency of plant extract-mediated synthesis of zinc oxide nanoparticles." *Antioxidants* 12, no. 4 (2023): 784. https://www.mdpi.com/2076-3921/12/4/784

Vimala, Karuppaiya, Shenbagamoorthy Sundarraj, Manickam Paulpandi, Srinivasan Vengatesan, and Soundarapandian Kannan. "Green synthesized doxorubicin loaded zinc oxide nanoparticles regulates the Bax and Bcl-2 expression in breast and colon carcinoma." *Process Biochemistry* 49, no. 1 (2014): 160–172. https://www.sciencedirect.com/science/article/abs/pii/S1359511313005680

Vinodhini, S., B. Scholastica Mary Vithiya, and T. Augustine Arul Prasad. "Green synthesis of palladium nanoparticles using aqueous plant extracts and its biomedical applications." *Journal of King Saud University-Science* 34, no. 4 (2022): 102017. https://www.sciencedirect.com/science/article/pii/S1018364722001987

Xue, NiChang, ChaoHui Zhou, ZhongYun Chu, LiNa Chen, and NengQin Jia. "Barley leaves mediated biosynthesis of Au nanomaterials as a potential contrast agent for computed tomography imaging." *Science China Technological Sciences* 64, no. 2 (2021): 433–440. https://link.springer.com/article/10.1007/s11431-019-1546-3#article-info

Zhang, Duoduo, Jie Zhang, Jie Zeng, Zhubo Li, Hua Zuo, Chengzhi Huang, and Xiaoyan Zhao. "Nano-gold loaded with resveratrol enhance the anti-hepatoma effect of resveratrol in vitro and in vivo." *Journal of Biomedical Nanotechnology* 15, no. 2 (2019): 288–300. https://www.ingentaconnect.com/contentone/asp/jbn/2019/00000015/00000002/art00006

Zhang X., Q. Guo, and D. Cui. "Recent advances in nanotechnology applied to biosensors." *Sensors* 9, no. 02 (2009 Feb 17): 1033–1053. https://www.mdpi.com/1424-8220/9/2/1033

6 Nanotechnology in Enhancing Herb-Based Medicines' Bioavailability and Activity

Alka Singh, Shalu Verma, Gauree Kukreti, Chetan Singh Chauhan, Sumit Durgapal, and Pankaj Kumar Chaurasia

6.1 INTRODUCTION

Herbal medicines are becoming increasingly popular around the world for treating a range of illnesses. They frequently provide the primary source of healthcare for a significant portion of the population in impoverished nations as they are cost-efficient. India has a long history of traditional medicine, primarily herb-based therapeutic modalities like Ayurveda, Unani, Homoeopathy, and Sidha. Furthermore, there is a tendency towards using natural herbal items as ingredients in food and drink, as well as in the production of contemporary medications and nutritional supplements.

Traditionally, humans have utilised plants and herbs to treat, mitigate, and cure several diseases. The drugs or active constituents obtained from plants are often called herbal medicines, phytomedicines, or secondary metabolites. They consist of mixtures of potent plant metabolites with medicinal, curative, and healing properties. The presence of active chemicals is necessary for the efficiency of different kinds of medicinal plants (Garg et al., 2024). The pharmacological assets of several herbs are utilised in treating long-term degenerative illnesses, including cancer and diabetes to various viral, fungal, and bacterial infections. The majority of the components of biologically active phytochemicals like terpenoids, tannins, and flavonoids have drawbacks, such as poor solubility, large molecular size, poor distribution, poor permeability, and absorption and relatively low target specificity, which leads to low bioavailability and reduced biological activity (Ahmad et al., 2021).

Furthermore, several of these phytochemical substances have low stability and are sensitive to acidic environments, and substantial quantities are needed to elicit their activity. These limitations hampered their clinical applicability. It can be best explained with an example of a turmeric plant (*Curcuma longa*) containing a yellow polyphenolic chemical called curcumin, which has antioxidant, anti-inflammatory, anti-cancer, and neuroprotective properties. Its hydrophobic structure, low water solubility, and limited absorption, however, limit its therapeutic use and cause rapid

DOI: 10.1201/9781003559900-6

metabolism and elimination, which leaves the body at pharmacologically negligible levels (Saka et al., 2021). Another example is of Puerarin, a primary bioactive component of Pueraria lobata root with pharmacological properties, including vasodilation, cardioprotection, neuroprotection, antioxidant, anticancer, anti-inflammation, pain relief, bone formation, alcohol inhibition, and attenuation of insulin resistance (Raman et al., 2022). However, the applicability of Puerarin is restricted by its low solubility. In the past few decades, researchers have been focusing on the improvement of such limitations by the use of nanotechnological approaches which overcome the problem of solubility and permeability in by carrier drug delivery with reduced targeted drug delivery (Alexander et al., 2016).

By acting as drug carriers, delivery systems based on nanotechnology can get around some of the drawbacks of herbal medicines, such as boosting the bioavailability and bioactivity of phytochemicals. Applying nanotechnology to phytochemical ingredients can be a promising and creative strategy that can improve the effectiveness of herbal medicines' phytotherapy. The objective of numerous researchers is to create a safe and effective drug delivery system. Interest in herbal medicine formulations has increased recently due to advancements in nanotechnology. Numerous delivery system strategies have been put forth, including phytosomes, solid lipid nanoparticles (SLN), nanostructured lipid carriers, polymeric nanoparticles, nanoemulsions, etc. Since various medications' pharmacokinetic characteristics have been altered and improved by nanoparticles, the application of nanotechnology is anticipated to boost bioavailability of herbal medicines (Baishya et al., 2020).

6.2 NANOTECHNOLOGY

The creation of various functional systems at the molecular level with enhanced features and distinct physical, electrical, and optical properties—typically in the 1–100 nm range—is known as nanotechnology. These nanoparticles come in a variety of forms, such as dendrimers, carbon nanotubes, micellar systems, polymeric nanoparticles, hard-phospholipid nanoparticles, nano-emulsions, etc. The particle size of the nanoscale system is 0.1 micrometres, also known as submicrometre. This feature increases the level of development and research interest in this type of nanotechnology by providing a number of advantages pertaining to many factors, including the mode of administration and improved therapeutic effects. Since these nanoscale systems have the ability to improve efficacy, reduce dosages, and minimise negative effects, many research have combined herbal medicine with nanotechnology (Bonifacio et al., 2014).

The pharmacological effect of any drug depends on both its therapeutic effectiveness and its bioavailability at the site of administration. A number of phytopharmaceuticals are characterised by low water solubility, leading to poor absorption through membranes and, consequently, low bioavailability when taken orally. Therefore, the design and development of effective pharmaceutical formulations for the delivery of specific phytoconstituents is of utmost importance (Chen et al., 2024).

Technological advancements are developing novel delivery systems for herbal medicine to increase the therapeutic efficacy and bioavailability of medications with limited water solubility. Phytosomes/herbosomes, liposomes, nanostructured

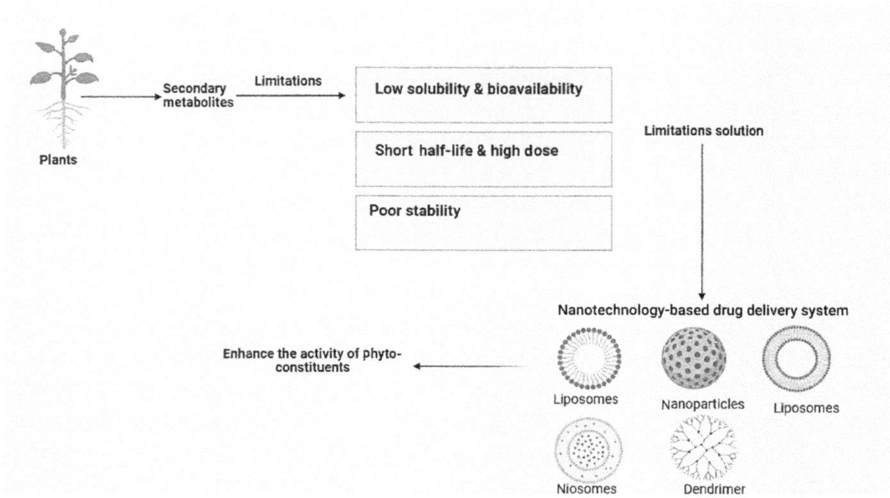

FIGURE 6.1 Nanotechnological approaches used for bioavailability enhancement of herbs.

lipid carriers (NLCs), nanoemulsions (NE), polymeric nanoparticles (NPs), dendrimers, and micelles are among the potential techniques in development as shown in Figure 6.1. By changing the pharmacokinetics of currently prescribed drugs, these technologies can improve the delivery of herbal products to their intended location and greatly boost their bioavailability (Chauhan et al., 2024). Herbal medicine delivery systems based on nanotechnology have a lot of promise and unique qualities, such as the capacity to successfully turn less soluble, poorly absorbed, or unstable ingredients into pharmaceuticals. As a result, these delivery methods are thought to be a promising way to reduce the difficulties associated with using herbal remedies and increase their efficacy (Desai, 2024).

6.3 DRUG DELIVERY SYSTEM FOR PHYTOCHEMICAL COMPOUNDS BASED ON NANOTECHNOLOGICAL APPROACHES

Determining an appropriate pharmaceutical formulation for delivery is crucial when developing new treatments. Therefore, for diseases to be effectively prevented and treated, phytomolecule administration is essential. Growing interest in creating NDDS to improve the pharmacological and therapeutic qualities of herbal medications is a result of new technology. The pursuit of knowledge and technology to address the limitations of herbal medicine for both therapeutic benefit and bioavailability enhancement is ongoing (Ali et al., 2022).

Nanoparticles derived from active pharmaceutical ingredients in medicinal plants include formulations such as nanosuspensions and nano-eutectics. In addition, drug nanocarriers come in various dosage forms, including liposomes, nanoparticles, nanoemulsions, and colloidal polymers (Dubey et al., 2022). Recent advancements in nanocarrier research and development have significantly improved through modifications to carrier surfaces and the addition of surface-active agents like transferrin, folate, low-density lipoprotein, peptides, lectins, and epidermal growth factor. These enhancements have greatly contributed to better drug targeting, increased bioavailability, and improved solubility. Moreover, innovative nano-based formulations of Chinese medicinal products, including eutectics and inorganic carriers, as well as phospholipid complexes and suspension gels, have been reported, emphasising the ongoing trend in developing nano-dosage forms aimed at boosting targeting efficiency, bioavailability, and solubility (Chavda et al., 2022).

6.4 ETHOSOME

Lipid-based vesicular structures called ethosomes are composed of phospholipids and ethanol. It is a non-invasive carrier that uses alcohol (ethanol) to improve penetration into the skin's deeper layers and maintain systemic circulation. The alcohol in the vesicular system interacts with the ethosomal lipids and the lipid bilayers of the stratum corneum (SC), to permit the soft and malleable ethosomes to penetrate. In this way, ethosomes proved a better, promising vesicular system for drug delivery as compared to the conventional drug delivery systems. It offers high deformability as well as good encapsulation efficiency, improved bioavailability, and better biocompatibility. All these characteristic features are utilised in the formulation of ethosomal transdermal drug delivery resulting in improved stability and efficacy. Many phytoconstituents like quercetin, curcumin, ammonium glycyrrhizinate, and oxymatrine are available in the form of ethosomal topical formulation with improved therapeutic effects used for the treatment of dermatitis, eczema, psoriasis, rheumatoid arthritis, etc. (Teli et al., 2024).

6.5 TRANSFEROSOMES

The specially designed vesicular formulations known as transferosomes are ultra foldable, elastic carrier systems resembling liposomes. They are fabricated by using phospholipids, surfactants, water, and ethanol. They can topically deliver hydrophobic and hydrophilic moieties. Larger molecules can be transported by these vesicular carriers through the SC which leads to improved bioavailability and therapeutic efficacy (Vimala et al., 2021).

Currently, a wide range of illnesses are treated with phytoconstituents, but most of them have limited therapeutic value due to poor permeability and large molecular size. The improvement of bio-efficacy in the transportation of phytoconstituents is therefore thought to be significantly hindered by bioavailability. The soft nature of transferosomes increases the membrane flexibility for transdermal phytochemical administration for both local and systemic drug activity.

Transfersomes are speculated to work in two ways: (1) as drug carriers that load and move drug molecules into the skin; and (2) as penetration enhancers that enter the SC and change the intercellular lipid layers, which makes it easier for unbound drug molecules to move into and through the SC. For example, a transfersome formulation of capsaicin has been produced, and it exhibits better topical absorption than pure capsaicin.

6.6 NIOSOME

Niosomes are hydrated vesicular systems that are intended to deliver medications to particular target locations. They are composed of nonionic surfactants mixed with cholesterol or phospholipids. These vesicles have lamellar structures made up of amphiphilic molecules that enclose an aqueous compartment. Both hydrophilic and hydrophobic medications can be delivered with them; the pharmaceuticals are enclosed in the outer lipid layer or the interior hydrophilic area. Because niosomes are stable over long periods of time in a variety of circumstances, biodegradable, biocompatible, and non-toxic, they may encapsulate large amounts of material in a comparatively tiny volume. They are also adaptable carrier systems that can be applied transdermally, intramuscularly, intravenously, orally, ocularly, or pulmonarily (Yeligar et al., 2023). Nonionic surfactants, which are used to improve the solubility of poorly water-soluble medications and hence increase their bioavailability, produce the bilayer lipid structures that make up niosomes. Niosomes have a longer shelf life, are more stable, and can administer drugs in a controlled or sustained manner for longer than liposomes. Through transdermal application, they can improve the bioavailability of medications like methotrexate, etoposide, and podophyllotoxin by increasing the permeability and fluidity of biological membranes.

6.7 LIPOSOME

Liposomes are lipidic vesicles with two layers that carry cholesterol and phospholipids. They range in size from 2.5 to 25 nm. Their capacity to encapsulate different materials and their structural adaptability are the clear advantages. Liposomes can encapsulate medications with widely different lipophilicity or solubility, either at the bilayer interface or trapped in the phospholipid bilayer's aqueous core. Natural lipid-based liposomes are biodegradable, nonimmunogenic, and physiologically inactive, and have a low inherent toxicity (Gorain et al., 2023).

Because liposome-encapsulated medications are made to be carried without breaking down quickly, they have less adverse effects. Increasingly, liposomes are being used in herbal products to deliver particular drugs for the treatment or prevention of different illnesses. In terms of their toxicity, biodistribution, pharmacokinetic characteristics, and relative stability, several formulations have been created and studied. Moreover, liposomes can efficiently transport medications to their intended locations by means of endocytosis or fusion. Liposomes' instability in plasma is their main drawback, despite the fact that they have several advantages, such as safety and biocompatibility. Liposome Herbasec is a commercial liposomal formulation having standardised extracts effective in enhancing the activities of its herbal ingredients. This novel approach improves the controlled drug delivery by extending the drug release of the product (Zeng et al., 2021).

6.8 DENDRIMER

Three-dimensional, unimolecular, hyperbranched polymeric structures are called dendrimers. With dimensions ranging from 2 to 20 nm, dendrimers are incredibly tiny. They are promising in gene delivery, medication delivery, and drug targeting. Their unimolecular micellar characteristics are a result of their structure's hydrophilic exteriors and interiors which make it suitable to encapsulate a wide variety of drug moieties (Puglia et al., 2017). Dendrimers are non-immunogenic, non-toxic, and biodegradable in nature. Abderrezak et al. investigated the antibacterial effect of dendrimers containing bacterial cellulose on a variety of bacterial strains and found increased antibacterial effect. A wide variety of dendrimer formulation containing quercetin, camptothecin, paclitaxel, silybin, etc., are available for the treatment of cancer and osteoporosis. Dendrimers have thus been employed as a platform for various ligands and natural compounds (Jeetah et al., 2014).

6.9 NANOPARTICLE

Nanoparticles (NP) are the nanosized drug delivery system with particle size ranging from 1 to 1000 nanometres. They are fabricated by using a wide variety of polymers, enabling them to encapsulate hydrophilic and hydrophobic moieties for target drug delivery. Their potential as drug carriers includes the capacity to improve bioavailability and therapeutic index, mask the taste of compounds and facilitate controlled release. A drug may be attached to the NP matrix, dissolved, trapped, or encapsulated. The therapeutic efficacy of the encapsulated drugs is enhanced by NPs through pharmacokinetic alteration, toxicity reduction, regulated release over prolonged periods, and protection from enzymatic degradation (Kumar et al., 2022b). NPs also improve the oral bioavailability of poorly soluble drug substances by adhering to the capillary walls and increasing tissue uptake following parenteral delivery (More et al., 2021). Furthermore, substances including artemisinin, berberine, quercetin, and glycyrrhizic acid have been added to NPs to increase their efficacy and bioavailability (Javed et al., 2020).

6.10 PHYTOSOME

"Phyto" means plants or herbs, and "some" means a structure that resembles a cell. Particle sizes in the nanoscale range (less than 100 nm) comprise the phospholipid complex system of herbal extracts or phytoconstituents that make up phytosomes, which are used as controlled-and-sustained-release delivery systems. In a nonpolar solvent, a standardised extract or phytoconstituent reacts with a stoichiometric quantity of phospholipid (phosphatidylcholine) to produce this technology. Standardised extracts or phytoconstituents are encapsulated within phospholipids using patented phytosome technology, which produces molecular complexes that improve the compounds' permeability and bioavailability, especially for those with low water solubility and a strong propensity to aggregate (Alharbi et al., 2021).

Only phospholipid makes up the core of the phytosomal system, and different phospholipids, such as hydrogenated soy phosphatidylcholine (HSPC), dipalmitoylphosphatidylcholine, and distearoylphosphatidylcholine, are employed for this function (Mahato et al., 2021). The goal of phospholipid-based delivery systems is to increase the bioavailability of herbal extracts and phytoconstituents. Numerous plant extracts (including milk thistle, ginkgo, and green tea) and phytochemicals (such silybin and curcumin) have been effectively treated with phytosomes, with noteworthy outcomes in pharmacokinetic studies conducted on both humans and animals. Compounds such as quercetin, curcumin, EA, naringenin, AP, resveratrol, mangiferin, and GA have been successfully used to increase therapeutic efficacy and bioavailability in a number of studies utilising phytoconstituent-phospholipid complexes (Kang et al., 2023).

6.11 MICELLES

Micelles are lipid molecules that, in aqueous solutions, arrange themselves into a spherical shape. The size of polymeric micelles ranges from 10 to 100 nanometres, and they are usually extremely narrow. Medications' solubility, stability, and biomembrane permeability are all improved, which increases the medications' bioavailability in the micellar environment. Drug diffusion rate, partition coefficient, micelle stability, and the copolymer's rate of biodegradation are some of the variables that affect drug release from micelles. Since polymeric micelles made from amphiphilic block copolymers have special properties including low toxicity and high solubility, they show great promise as drug delivery vehicles for a variety of anticancer drugs (Krishnaswamy, 2024).

They can also lessen the P-glycoprotein efflux effect, which gives them a different mode of action from the medications they contain. The effective therapeutic use of powerful chemotherapy drugs, such as doxorubicin and paclitaxel, is limited by their high toxicity; however, this therapeutic window can be expanded by using such delivery systems. According to preclinical research, when compared to paclitaxel given alone, the maximum plasma concentration (Cmax) in tumours and the area under the curve (AUC) were three times higher in colon-26-bearing CDF1 mice. In conclusion, there is growing hope for the use of polymeric micelle systems in cancer treatment, and these systems are becoming more and more significant in oncology (Kebebe et al., 2018).

6.12 SOLID LIPID NANOPARTICLES AND NLCS

Stearic acid, cetyl alcohol, and tripalmitin are among the lipids that are utilised to create solid lipid nanoparticles (SLNs). Both at body temperature and at room temperature, these lipids are solid. SLNs have several advantages, such as improved bioavailability, biocompatibility, resistance to deterioration, and ease of use on an industrial scale (Kumar et al., 2022a). Nanostructured lipid carriers (NLCs) blend liquid and solid lipids to form a less-than-ideal lipid particle matrix, which increases drug-loading capacity and decreases drug ejection during storage. Reduced drug leakage, improved drug loading and release profiles, more solid content dispersions, and the feasibility of producing final dose forms like tablets and capsules are some advantages of NLCs (Gugleva et al., 2021).

Numerous herbal substances are subjected to these methods. In contrast to pure α-asarone, α-asarone-loaded SLNs improved oral bioavailability and tissue uptake, according to pharmacokinetic research conducted in male rats. Comparing curcumin-loaded SLNs to curcumin solutions, research revealed notable increases in oral bioavailability, with enhancements ranging from 39 to 155 times depending on the dose. In a different investigation, quercetin SLNs showed delays in Tmax and mean residence time in plasma, but their relative bioavailability was 571.4% when compared to a quercetin solution. All things considered, SLNs and NLCs show potential as carrier systems for enhancing the absorption of medications that are poorly soluble (Gupta et al., 2022).

6.13 MICROEMULSIONS AND NANOEMULSIONS

Isotropic blends of oils, surfactants, and cosolvents, microemulsions and nanoemulsions (NEs) are distinguished by their thermodynamic stability and the existence of oil droplets covered with an emulsifier in an aqueous medium. These droplets fall into one of two categories: water-in-oil or oil-in-water. The size of these droplets depends on the homogenisation technique used and the makeup of the system. Following oral treatment, NEs quickly spread throughout gastrointestinal fluids, generating 20–200 nm nanodroplets with the help of digestive motility (Mohammadabadi and Mozafari, 2018). When these methods are used, the stability of herbal medications is improved, permeability through the skin and mucosa is increased, and sustained-release effects are made possible. Several herbal substances have been studied using these methods, including camptothecin and oil from Brucea javanica. Research showed that the use of an oral NE formulation of paclitaxel resulted in a significant improvement in its terminal elimination half-life, reaching 5.79 hours as opposed to 1.22 hours for the intravenous formulation, indicating a sustained release profile. In a mouse model, this formulation also increased the absolute oral bioavailability by up to 55.9% (Ma et al., 2019).

6.14 FUTURE PROSPECTS AND CHALLENGES OF NANOTECHNOLOGY IN HERBAL DRUG DELIVERY

Due to a variety of physicochemical challenges, poor absorption frequently limits the use of phytomedicines. Innovative delivery methods seek to increase the solubility of lipophilic plant secondary metabolites in order to increase their penetration and bioavailability. A significant amount of active chemicals must be delivered for herbal medicines to be effective, and two important parameters that impact these molecules' capacity to pass across biological membranes are lipid solubility and molecular size. Low-water-soluble and lipophilic medications' bioavailability can be increased by using oral lipid-based formulations, which are becoming more and more popular due to their capacity to improve solubility, ease gastrointestinal absorption, and ease food-related absorption problems.

Many advantages are offered by nanotechnology-based delivery systems for natural components, including increased stability, pharmacological activity, bioavailability, and solubility as well as less dosage needs. The use of herbs is promoted as a beneficial alternative therapy in these systems. But they also face constraints like high production costs, scalability issues, and a lack of safety and toxicity data, which draw attention to possible hazards related to these technologies. Further complicating their utilisation are stability problems with nanoparticles, including their propensity to leak when they come into touch with blood components. Since drug funding in European countries is dependent on cost-effectiveness, the high production costs of nanosystems can limit their market accessibility, making it difficult for costly nanomedicines to be accepted.

6.15 CONCLUSION

Since ancient times, herbal medicines have been used extensively around the world. Both physicians and patients agree that herbal remedies have greater therapeutic value than modern drugs since they have fewer side effects. By reducing toxicity and the requirement for repeated administration to overcome non-compliance, as well as by improving bioavailability, novel drug delivery systems help to increase therapeutic value. A lot of research is being done on herbal remedies in order to incorporate them into novel drug delivery systems. Therefore, using nanotechnology in the creation of herbal medicines is one potential strategy to improve their therapeutic effects. Further investigation on the toxicity and safety is necessary to ensure their reliability and efficacy in treating a variety of ailments. Enhancing public health worldwide requires the adoption of delivery technologies for phytochemical compounds based on nanotechnology. Although herbal medications are becoming more and more popular, problems like low solubility, poor absorption, restricted pharmacological action, and vulnerability to physical and chemical instability frequently limit their clinical efficacy. In addition to having poor pharmacokinetic characteristics, numerous compounds produced from plants also have low bioavailability conditions. For these phytomolecules to be effective as medicines, their half-life is essential. Herbal remedies are therefore becoming more and more popular. Novel delivery system developments allow for the creation of new formulations that alter the physical properties of phytochemicals to improve bioavailability. Solid lipid nanoparticles (SLNs), phytosomes, liposomes, niosomes, transfersomes, and nanoemulsions (NEs) are phospholipid-based delivery systems that have demonstrated superior efficacy over traditional techniques. By lowering dosages and enhancing the pharmacological activity and bioavailability of herbal medications, these novel techniques may improve therapeutic outcomes.

REFERENCES

Ahmad, Razi, Sukriti Srivastava, Shubhrima Ghosh, and Sunil Kumar Khare. "Phytochemical delivery through nanocarriers: A review." *Colloids and Surfaces B: Biointerfaces* 197 (2021): 111389. https://doi.org/10.1016/j.colsurfb.2020.111389

Alexander, Amit, Ravish J. Patel, Swarnlata Saraf, and Shailendra Saraf. "Recent expansion of pharmaceutical nanotechnologies and targeting strategies in the field of phytopharmaceuticals for the delivery of herbal extracts and bioactives." *Journal of Controlled Release* 241 (2016): 110–124. https://doi.org/10.1016/j.jconrel.2016.09.017

Alharbi, Waleed S., Fahad A. Almughem, Alshaimaa M. Almehmady, Somayah J. Jarallah, Wijdan K. Alsharif, Nouf M. Alzahrani, and Abdullah A. Alshehri. "Phytosomes as an emerging nanotechnology platform for the topical delivery of bioactive phytochemicals." *Pharmaceutics* 13, no. 9 (2021): 1475. https://doi.org/10.3390/pharmaceutics13091475

Ali, Tayyab, Fatma Hussain, Muhammad Naeem, Ajmal Khan, and Ahmed Al-Harrasi. "Nanotechnology approach for exploring the enhanced bioactivities and biochemical characterization of freshly prepared nigella sativa L. nanosuspensions and their phytochemical profile." *Frontiers in Bioengineering and Biotechnology* 10 (2022): 888177. https://doi.org/10.3389/fbioe.2022.888177

Baishya, Rinku, Jyoti L. Hati Boruah, Manob J. Bordoloi, Deepak Kumar, and Pratap Kalita. "Novel drug delivery system in phytochemicals: Modern era of ancient science." *Herbal Medicine in India: Indigenous Knowledge, Practice, Innovation and its Value* (2020): 175–189. https://doi.org/10.1007/978-981-13-7248-3_12

Bonifacio, Bruna Vidal, Patricia Bento da Silva, Matheus Aparecido dos Santos Ramos, Kamila Maria Silveira Negri, Taís Maria Bauab, and Marlus Chorilli. "Nanotechnology-based drug delivery systems and herbal medicines: A review." *International Journal of Nanomedicine* (2014): 1–15. https://doi.org/10.2147/IJN.S52634

Chen, Yuhao, Yuying Tang, Yuanbo Li, Yukui Rui, and Peng Zhang. "Enhancing the efficacy of active pharmaceutical ingredients in medicinal plants through nanoformulations: A promising field." *Nanomaterials* 14, no. 19 (2024): 1598. https://doi.org/10.3390/nano14191598

Chauhan, Divya, Pavan K. Yadav, Nazneen Sultana, Arun Agarwal, Saurabh Verma, Manish K. Chourasia, and Jiaur R. Gayen. "Advancements in nanotechnology for the delivery of phytochemicals." *Journal of Integrative Medicine* (2024). https://doi.org/10.1016/j.joim.2024.04.005

Chavda, Vivek P., Aayushi B. Patel, Kavya J. Mistry, Suresh F. Suthar, Zhuo-Xun Wu, Zhe-Sheng Chen, and Kaijian Hou. "Nano-drug delivery systems entrapping natural bioactive compounds for cancer: Recent progress and future challenges." *Frontiers in Oncology* 12 (2022): 867655. https://doi.org/10.3389/fonc.2022.867655

Desai, Nivas. "Herbal medicines and nanotechnology-based drug delivery systems." *Biomaterials Connect* 2, no. 1 (2024): 1–10. https://doi.org/10.69709/BIOMATC.2024.101090

Dubey, Sunil Kumar, Shraddha Parab, Vaishnav Pavan Kumarr Achalla, Avinash Narwaria, Swapnil Sharma, B. H. Jaswanth Gowda, and Prashant Kesharwani. "Microparticulate and nanotechnology mediated drug delivery system for the delivery of herbal extracts." *Journal of Biomaterials Science, Polymer Edition* 33, no. 12 (2022): 1531–1554. https://doi.org/10.1080/09205063.2022.2065408

Garg, Ruchika, Kiran Dobhal, and Alka Singh. "Utilization of medicinal herbal plants in the management of respiratory conditions." (2024). https://doi.org/10.5772/intechopen.1008176

Gorain, Bapi, Varnita Karmakar, Biswatrish Sarkar, Monika Dwivedi, Janelle Tsui Lyn Leong, Jing Hen Toh, Even Seah et al. "Biomacromolecule-based nanocarrier strategies to deliver plant-derived bioactive components for cancer treatment: A recent review." *International Journal of Biological Macromolecules* (2023): 126623. https://doi.org/10.1016/j.ijbiomac.2023.126623

Gugleva, Viliana, Nadezhda Ivanova, Yoana Sotirova, and Velichka Andonova. "Dermal drug delivery of phytochemicals with phenolic structure via lipid-based nanotechnologies." *Pharmaceuticals* 14, no. 9 (2021): 837. https://doi.org/10.3390/ph14090837

Gupta, Manish Kumar, Vipul Sansare, Birendra Shrivastava, Santosh Jadhav, and Prashant Gurav. "Comprehensive review on use of phospholipid based vesicles for phytoactive delivery." *Journal of Liposome Research* 32, no. 3 (2022): 211–223. https://doi.org/10.1080/08982104.2021.1968430

Javed, Md Noushad, Ekta Singh Dahiya, Abdallah Mohammad Ibrahim, Md Sabir Alam, Firdos Alam Khan, and Faheem Hyder Pottoo. "Recent advancement in clinical application of nanotechnological approached targeted delivery of herbal drugs." *Nanophytomedicine: Concept to Clinic* (2020): 151–172. https://doi.org/10.1007/978-981-15-4909-0_9

Jeetah, Roubeena, Archana Bhaw-Luximon, and Dhanjay Jhurry. "Nanopharmaceutics: phytochemical-based controlled or sustained drug-delivery systems for cancer treatment." *Journal of Biomedical Nanotechnology* 10, no. 9 (2014): 1810–1840. https://doi.org/10.1166/jbn.2014.1884

Kang, Chenglong, Jianwen Wang, Ruotong Li, Jianing Gong, Kuanrong Wang, Yuxin Wang, Zhenghua Wang, Ruzhe He, and Fengyun Li. "Smart targeted delivery systems for enhancing antitumor therapy of active ingredients in traditional Chinese medicine." *Molecules* 28, no. 16 (2023): 5955. https://doi.org/10.3390/molecules28165955

Krishnaswamy, Suma. "Phytopharmaceutical biotechnology: Integration of botany, pharmacology and plant biotechnology to deliver the best therapeutic potential of herbs." In *Concepts in Pharmaceutical Biotechnology and Drug Development*, pp. 437–464. Singapore: Springer Nature Singapore, 2024. https://doi.org/10.1007/978-981-97-1148-2_20.

Kebebe, Dereje, Yuanyuan Liu, Yumei Wu, Maikhone Vilakhamxay, Zhidong Liu, and Jiawei Li. "Tumor-targeting delivery of herb-based drugs with cell-penetrating/tumor-targeting peptide-modified nanocarriers." *International Journal of Nanomedicine* (2018): 1425–1442. https://doi.org/10.2147/IJN.S156616

Kumar, Parveen, Nishant Yadav, Benu Chaudhary, Vivek Jain, Vishal M. Balaramnavar, Khalid Saad Alharbi, Sattam Khulaif Alenezi et al. "Promises of phytochemical based nano drug delivery systems in the management of cancer." *Chemico-Biological Interactions* 351 (2022a): 109745. https://doi.org/10.1016/j.cbi.2021.109745

Kumar, Rajeev, Mohd A. Mirza, Punnoth Poonkuzhi Naseef, Mohamed Saheer Kuruniyan, Foziyah Zakir, and Geeta Aggarwal. "Exploring the potential of natural product-based nanomedicine for maintaining oral health." *Molecules* 27, no. 5 (2022b): 1725. https://doi.org/10.3390/molecules27051725

Ma, Zhe, Yuqi Fan, Yumei Wu, Dereje Kebebe, Bing Zhang, Peng Lu, Jiaxin Pi, and Zhidong Liu. "Traditional Chinese medicine-combination therapies utilizing nanotechnology-based targeted delivery systems: A new strategy for antitumor treatment." *International Journal of Nanomedicine* (2019): 2029–2053. https://doi.org/10.2147/IJN.S197889

Mahato, Mrityunjoy, Sanjukta Patra, and Manashjit Gogoi. "Herbal nanocarriers for cancer therapy." *Nanopharmaceuticals: Principles and Applications Vol. 2* (2021): 41–75. https://doi.org/10.1007/978-3-030-44921-6_2

Mohammadabadi, M. R., and M. R. Mozafari. "Enhanced efficacy and bioavailability of thymoquinone using nanoliposomal dosage form." *Journal of Drug Delivery Science and Technology* 47 (2018): 445–453. https://doi.org/10.1016/j.jddst.2018.08.019

More, Mahesh P., Sagar R. Pardeshi, Chandrakantsing V. Pardeshi, Gaurav A. Sonawane, Mahesh N. Shinde, Prashant K. Deshmukh, Jitendra B. Naik, and Abhijeet D. Kulkarni. "Recent advances in phytochemical-based nano-formulation for drug-resistant cancer." *Medicine in Drug Discovery* 10 (2021): 100082. https://doi.org/10.1016/j.medidd.2021.100082

Puglia, Carmelo, Maria Rosaria Lauro, Giorgia Giusy Tirendi, Giuseppina Emanuela Fassari, Claudia Carbone, Francesco Bonina, and Giovanni Puglisi. "Modern drug delivery strategies applied to natural active compounds." *Expert Opinion on Drug Delivery* 14, no. 6 (2017): 755–768. https://doi.org/10.1080/17425247.2017.1234452

Raman, Subashini, Vikneswaran Murugaiyah, and Thaigarajan Parumasivam. "Andrographis paniculata dosage forms and advances in nanoparticulate delivery systems: An overview." *Molecules* 27, no. 19 (2022): 6164. https://doi.org/10.3390/molecules27196164

Saka, Raju, and Naveen Chella. "Nanotechnology for delivery of natural therapeutic substances: A review." *Environmental Chemistry Letters* 19, no. 2 (2021): 1097–1106. https://doi.org/10.1007/s10311-020-01103-9

Teli, Divya, Riyansi Satasia, Vishwa Patel, Ruchi Nair, Rudri Khatri, Dhvani Gala, Pankti C. Balar et al. "Nature meets technology: Harnessing nanotechnology to unleash the power of phytochemicals." *Clinical Traditional Medicine and Pharmacology* (2024): 200139. https://doi.org/10.1016/j.ctmp.2024.200139

Vimala, Karuppaiya, and Soundarapandian Kannan. "Phyto-drug conjugated nanomaterials enhance apoptotic activity in cancer." *Advances in Protein Chemistry and Structural Biology* 125 (2021): 275–305. https://doi.org/10.1016/bs.apcsb.2020.12.003

Yeligar, Rachana R., Khomendra Kumar Sarwa, Manisha Chandrakar, and Manmohan Singh Jangde. "Nanotechnology-based delivery of genistein to overcome physicochemical hindrance and enhance therapeutic response in skin cancer." *Bionanoscience* 13, no. 3 (2023): 1339–1358. https://doi.org/10.1007/s12668-023-01118-w

Zeng, Yujun, Yufan Xiang, Ruilong Sheng, Helena Tomás, João Rodrigues, Zhongwei Gu, Hu Zhang, Qiyong Gong, and Kui Luo. "Polysaccharide-based nanomedicines for cancer immunotherapy: A review." *Bioactive Materials* 6, no. 10 (2021): 3358–3382. https://doi.org/10.1016/j.bioactmat.2021.03.008

7 Toxicity Profiles of Nanoparticles Synthesized with Plants

Mansi Butola, Yogita Ale, Shilpa Rana, Vanshika Sautha, and Himanko Gogoi

7.1 INTRODUCTION

Two important concepts that are frequently used to categorize a wide range of research activities globally are nanoscience and nanotechnology. A subfield of nanotechnology, nanobiotechnology is a multidisciplinary field. The use of nanoparticles in biological systems is the topic of discussion here. Physical and chemical processes were employed to produce the nanoparticles. The employment of these complex and costly procedures results in the release of toxic material into the environment. In addition to high levels of radiation, they have also been using powerful reductants and stabilizing chemicals, which are dangerous to human health as well as the environment. Therefore, developing a different method to make nanoparticles will be crucial (Pirsaheb et al. 2024). This method uses materials science technology that is environmentally sustainable to create metallic nanoparticles in a sustainable manner. The suitability of green synthesis for large-scale manufacturing has aided in its advancement when compared to traditional methods. The nanoparticles are produced using environmentally benign methods at room temperature, with a pH of neutral, and at a reasonable cost. The reagents utilized are safe for usage, environmentally friendly, and non-toxic. These nanoparticles have a number of qualities, such as improved stability and appropriate size. They can be synthesized in a single step with a minimum amount of energy. The need for ecologically friendly nanoparticle synthesis has arisen from the extensive usage of metal-based nanoparticles across numerous industries, including agriculture, health, environmental research, and industry (Al-darwesh, Ibrahim, and Mohammed 2024).

7.1.1 Green Synthesis and Evaluation of Plant-Based Nanoparticles

The term "green chemistry" for "sustainable development" has been under thorough research in the last 10 years. Sustainable development implies a balance between present needs and the capacity of future generations to satisfy their own. Different chemistry-related industries are heavily reliant on sustainable development, as they are confronted with issues of pollution and the wanton use of natural resources. The three basic prerequisites for green nanoparticle synthesis are the use of a harmless

DOI: 10.1201/9781003559900-7

stabilizing agent, a green non-toxic reducing agent, and an environmentally friendly solvent. Nevertheless, physical, chemical, and biosynthetic methods are the most frequently used synthetic procedures to produce nanoparticles. Chemical processes are generally overly expensive due to the use of toxic and harmful materials, and they introduce various environmental issues. Green synthesis, however, is a biocompatible, environment-friendly, and safe method for the synthesis of nanoparticles for several applications, including biomedicine. Green synthesis has been accomplished by bacteria, plants, fungi, and algae. Yet various types of nanoparticles have been synthesized from plant biomass, such as leaves, fruits, roots, stems, and seeds. In addition, plant extracts may be used to create nanoparticles with accurate size, shape, and fabrication. Moreover, since the extract has multiple phytochemicals, it could potentially function as a natural reducing and/or stabilizing agent for the production of nanoparticles. It is generally known that in comparison to chemically synthesized nanoparticles, plant nanoparticles are less susceptible to inducing harmful effects on human beings. Moreover, they hold a lot of biological potential that can be applied in bioengineering, cosmetics, nanomedicine, food science and technology, agriculture, and protection of human health (Abada et al. 2024). Table 7.1 highlights several NP kinds according to their composition, form, preparation process, and functional characteristics along with examples and applications.

7.1.2 Overview of the Various Types of Plant-Derived Nanoparticles

Zinc oxide nanoparticles (ZnONPs) have attracted much interest in recent years due to their numerous possible uses in electronics, biology, optics, and cosmetics. Various experiments have been done on the biosynthesis and applications of ZnONPs by microorganisms, plants, and other living organisms. Their simple, safe, and cost-effective synthesis has been explored in many investigations. Numerous plant parts, including roots, flowers, seeds, and leaves, can be employed for the synthesis of ZnONPs. The nanoparticles possess a wide bandgap of 3.37 eV and a remarkable exciton binding energy of 60 me V, which render them a range of semiconducting characteristics. *Artemisia pallens* (Gomathi and Suhana 2021), *Cayratia pedate* (Jayachandran and Nair 2021), *Euphorbia hirta* (Ahmad and Kalra 2020), *Eucalyptus globules* (H. Ahmad et al. 2020), *Tecoma castanifolia* (Sharmila, Thirumarimurugan, and Muthukumaran 2019), *Azadirachta indica* (Noorjahan et al. 2024), *Solanum nigrum* (Ramesh, Anbuvannan, and Viruthagiri 2015), *Olea europea* (Awwad, Albiss, and Ahmad 2014), and *Azadirachta indica* (Oudhia, Kulkarni, and Sharma 2024) are among the numerous studies that have investigated the plant-assisted synthesis of zinc oxide NPs.

Plant-mediated synthesis of silver nanoparticles (AgNPs) is the most simple approach. A biological reducing agent and an aqueous solution of silver metal ions are required for the green synthesis of silver nanoparticles. The reduction and stabilization of Ag ions by a variety of biomolecules, such as polysaccharides, vitamins, proteins, amino acids, phenolics, saponins, alkaloids, and/or terpenes, is the most efficient and cost-effective technique for the preparation of AgNPs. Nearly all species possess the ability to generate AgNPs. There are many studies done on the silver nanoparticle plant-mediated production, including *Morinda citrifolia* L.

TABLE 7.1
Classification of Various Types of Nanoparticles

Classification Criteria	Categories	Examples	Applications	References
Based on composition	Metallic NPs	Silver, zinc oxide, gold	Antibacterial agent, targeted drug delivery	(Khan et al. 2022)
	Semiconducting NPs	Titanium dioxide, silicon, cadmium sulphide	UV protection, bioimaging and drug delivery, anti-cancer	(Mekuye and Abera 2023)
	Polymeric NPs	Poly(lactic-co-glycolic acid) (PLGA), chitosan polyethylene glycol (PEG)	Vaccine delivery, gene delivery, brain drug delivery	(Article et al. 2021)
	Metal oxide NPs	Iron oxide, copper oxide, magnesium oxide	Hyperthermia therapy, anti-fungal, gastroprotective agent	(Mekuye and Abera 2023)
	Composite NPs	Polymer-lipid hybrid (PLHNs), metal ceramic composite, metal-polymer composite	Controlled release, drug encapsulation, photothermal therapy	(Pacheco and Buzea 2021)
Based on shape	Spherical NPs	Quantum dots (spherical semiconductor), calcium phosphate spherical, carbon-based spherical	Bioimaging, bone regeneration, antioxidant therapy	(Joudeh and Linke 2022)
	Cylindrical NPs	Peptide-based nanofibers, hybrid cylindrical, cylindrical micelles	Controlled release, targeted therapy, tumor penetration	(Joudeh and Linke 2022)
	Tube-like NPs	Polymeric nanotubes, polymer nanotubes, boron nitride nanotubes	Tissue engineering, tissue repair, antioxidant	(Joudeh and Linke 2022)
	Sheet-like NPs	Graphene and graphene oxide (GO), polymer-based nanosheets, clay-based nanosheets	Photothermal therapy, wound healing	(Joudeh and Linke 2022)

Based on method of preparation	Top-down methods	Doxorubicin-loaded nanocrystals, carbamazepine NPs, curcumin-loaded NPs	Cancer therapy, improve bioavailability, anti-inflammatory	(Article et al. 2021)
	Bottom-up methods	Liposomes, nanocrystals, polymeric micelles	Drug encapsulation, improved bioavailability	(Ijaz et al. 2020)
Based on functional properties	Magnetic NPs	Iron oxide NPs (Fe_3O_4), magnetite NPs (Fe_3O_4), cobalt ferrite NPs ($CoFe_2O_4$)	MRI imaging, antibacterial activity, cancer therapy	(Al-Harbi and Abd-Elrahman 2024)
	Thermal NPs	Gold NPs (AuNPs), pluronic NPs, mesoporous silica NPs (MSNs)	Photothermal therapy, thermo-responsive drug delivery, thermal-controlled drug release	(Joudeh and Linke 2022)
	Catalytic NPs	Palladium NPs (PdNPs), rhodium NPs (RhNPs), copper NPs (CuNPs)	Hydrogenation, pharmaceutical synthesis	(Joudeh and Linke 2022)

(Morales-Lozoya et al. 2021), *Nymphae odorata* (Gudimalla et al. 2021), *Capparis zeylanica L* (Nilavukkarasi, Vijayakumar, and Prathip Kumar 2020), *Caesalpinia pulcherrima* (Moteriya and Chanda 2020), *Carya illinoinensis* (Javan bakht Dalir et al. 2020), *Mentha piperita* (Khatoon et al. 2018), and *Jatropha curcas* (Bar et al. 2009).

Due to their special qualities, which include significant therapeutic potential, low toxicity, superior biocompatibility, and simplicity in surface functionalization, gold nanoparticles (AuNPs) have garnered a lot of attention. Several chemical components of biogenic complexes function as reducing agents throughout the gold nanoparticle manufacturing process. Consequently, nanoparticles are created and the concentration of gold metal ions in the solution decreases. The reduction of metal ions and the encapsulation of gold nanoparticles in plant extracts depend on biomolecules such as proteins, phenols, and flavonoids, according to a number of studies. Numerous studies on the generation of gold nanoparticles by plants have been carried out including *Curcuma pseudomontana* (Muniyappan, Pandeeswaran, and Amalraj 2021), *Lawsoniainermis* (Kumari and Meena 2020), *Cinnamon* (ElMitwalli et al. 2020), *Croton Caudatus Geisel* (Vijaya Kumar, Mary Jelastin Kala, and Prakash 2019), *Tamarind* (Ankamwar, Chaudhary, and Sastry 2005), *Mentha*, *Ocimum*, *Eucalyptus* (Jha and Prasad 2012), *Murraya koenigii* (Philip et al. 2011), *Aegle marmelos* (Jha and Prasad 2011), and *Rosa hybrid* (Noruzi et al. 2011).

Because of its electrical band structure, huge specific surface area, stability, quantum efficiency, and chemical inertness, titanium dioxide nanoparticles (TiNPs) have attracted a lot of attention. Titanium nitride nanoparticles may reduce the toxicity of synthetic colors, medications, and wastewater treatment. Researchers are interested in the low cost, environmental friendliness, and reproducibility of using biological approaches to produce TiO_2 nanoparticles on a big scale. There is now a lot of research being done on the synthesis of TiO_2 nanoparticles using microorganisms (including bacteria and fungi), algae, plant materials, and enzymes. *Ledebouria revoluta* (Aswini, Murugesan, and Kannan 2021), *Pouteria campechiana* (Narayanan et al. 2021), *Syzygium cumini* (Sethy et al. 2020), *Mentha arvensis* (Ahmad, Jaiswal, and Soni 2020), *Azadirachta indica* (Thakur, Kumar, and Kumar 2019), *Psidium guajava* (Santhoshkumar et al. 2014), and *Calotropis gigantea* (Marimuthu et al. 2013) are among the numerous studies that have been conducted on the plant-assisted synthesis of titanium dioxide NPs.

7.1.3 Toxicity Profiles

7.1.3.1 Silver NPs

The capability of silver nanoparticles to eliminate bacteria, fungi, viruses, and parasites makes them suitable for numerous biomedical applications such as drug delivery, molecular imaging, biomedical sensing, and photodynamic cancer treatment. Toxicity effects of silver nanoparticles have been investigated despite the fact that they possess multiple benefits in medical uses. Silver nanoparticles can accumulate in various tissues and organs through multiple entry channels into the body, and they can also cross the blood-brain barrier to reach the brain. In comparison to other nanosized metals, silver nanoparticles are more toxic due to the generation of reactive oxygen species and the release of the enzyme lactate dehydrogenase within the body. The chemical and physical properties of silver nanoparticles, the environment in which they exist, the interaction dynamics, and other parameters influence their harmfulness. Mulenos et al. indicate that the toxicity of silver nanoparticles is quite moderate. Low concentrations of silver ions, that is, 6.9 μg/mL in toxic conditions and 0.2 μg/mL in anoxic conditions, are released as a consequence. Compared to natural concentrations, these are far lower. Ion production is responsible for both the onset and management of toxicity. They found comparable evidence for titania and copper (Mulenos et al. 2020).

7.1.3.2 Gold NPs

Because of its advantageous properties, which include durable optical capabilities, easy modification, easy manufacture, and customizable diameters, applications for gold nanoparticles in biomedicine are numerous. Gold nanoparticle toxicity is still a controversial topic. According to some research, these particles are biocompatible and show minimal toxicity, but other investigations have shown that their physicochemical properties make them highly poisonous and capable of inducing oxidative stress. As per Senut's study, human embryonic stem cells treated with gold nanoparticles coated with 1.5 nm mercaptosuccinic acid had no adhesion but rather became detached and spherical in shape. This, at as low as 0.1 μg/mL, caused the irreversible death of the cells. But up to a concentration of 10 μg/mL, other gold nanoparticles (4 and 14 nm) had no detectable effect on human embryonic stem cells (Senut et al. 2016). At below 0.2 or 2 μg/mL, Nelson et al. established that citrate-stabilized gold nanoparticles (10, 30, or 60 nm) do not yield free radicals or induce cytotoxicity, nor do they trigger definite, dose-dependent oxidative damage to DNA in calf thymus DNA or human liver cancer cells (Nelson et al. 2013).

7.1.3.3 Titanium Dioxide NPs

One of the top five nanoparticles in consumer products is titanium dioxide, which is produced at a quicker rate. Various in vitro and in vivo experiments have considered the cytotoxic and genotoxic potential of titanium oxide nanoparticles, but the toxicological findings are still ambiguous. Through apoptosis, inflammation, and oxidant formation, titanium dioxide nanoparticles have been shown in several studies to be cytotoxic, genotoxic, and oxidative in effect. However, some studies in the literature have reported negligible or no adverse effects.

Based on research conducted by Proquin et al., human colon cancer cells can be killed by food-grade titanium dioxide at concentrations that are less than nano- and micro-sized titanium dioxide particles. But when administered to human colon adenocarcinoma cell cultures in doses up to 100 μg/cm^2, food-grade titanium dioxide is not observed to have any adverse effects (Proquin et al. 2017). Under ultraviolet irradiation, Yin et al. evaluated the phototoxicity of titanium dioxide NPs of four diameters (less than 25 nm, 31 nm, less than 100 nm, and 325 nm) and two crystalline structures (rutile and anatase) in human skin keratinocytes. They discovered that when titanium dioxide nanoparticles (NPs) are irradiated using ultraviolet light, they generate reactive oxygen species (ROS) regardless of whether they are anatase or mixed anatase/rutile. Moreover, in comparison to larger particles, small titanium dioxide nanoparticles were more phototoxic (Yin et al. 2012).

7.1.3.4 Zinc Oxide NPs

Because of their exceptional resistance to oxidation and corrosion, photocatalytic capabilities, and superior absorption and shielding properties against ultraviolet radiation, zinc oxide nanoparticles have a wide range of applications, such as paints, sunscreens, solar cells, varistors, gas sensors, transparent conductive films, wave filters, UV detectors, and cosmetics. De Berardis et al. evaluated the toxicological effects and physicochemical properties of zinc oxide nanoparticles. Human colon cancer cells were exposed to zinc oxide nanoparticles (5 $\mu g/cm^2$) for a full day. The result was a marked decrease in cell viability, coupled with a rise in ROS production, which triggered apoptosis and IL-8 secretion. In addition, there were notable changes in morphology. After 24 hours of treatment, the cytotoxicity rose to almost 98% at higher concentrations (10, 20, and 40 $\mu g/cm^2$), with a cell survival rate of less than 5% (De Berardis et al. 2010). Xiao et al. demonstrated in both in vitro and in vivo investigations that zinc oxide nanoparticles caused oxidative stress, which was characterized by a rise in ROS and malondialdehyde levels and a fall in superoxide dismutase levels, leading to kidney injury (Xiao et al. 2016).

7.1.3.5 Platinum NPs

Although platinum nanoparticles have been observed to have negative side effects, they have demonstrated exceptional therapeutic efficacy against cancer cells. Citrate-coated platinum nanoparticles have been shown by Gatto et al. to be non-toxic to monocytes and immunologically compatible. Additionally, they observed that platinum nanoparticles reduced reactive oxygen species and acknowledged that they may control gene transcription without affecting the production of inflammatory cytokines (Gatto et al. 2018). In an in vitro investigation, Bendale et al. investigated how platinum nanoparticles destroy human lung adenocarcinoma, ovarian teratocarcinoma, pancreatic cancer cells, and healthy peripheral blood mononucleocyte cells. By causing ovarian teratocarcinoma cells to undergo apoptosis, the anticancer potential was assessed. At the highest dose, the researchers discovered that platinum nanoparticles (NPs) had no cytotoxic effect on normal cells but that they did have a deadly effect on cancer cell types. The research showed that via inducing apoptosis and cell cycle arrest, platinum nanoparticles had potent anticancer effects on the ovarian teratocarcinoma cell line (Bendale, Bendale, and Paul 2017).

7.1.3.6 Cobalt NPs

Energy storage devices, sensors, pigments, catalysts, and magnetic contrast agents are some of the uses of cobalt-based nanoparticles. In numerous cell types, cobalt or cobalt-containing nanoparticle exposure resulted in oxidative stress, DNA damage, morphological alterations, and inflammatory responses. Consistent with vast experimental evidence, the analysis of carcinogens ranked two cobalt-containing chemicals—cobalt sulfate and cobalt-tungsten carbide—as probable human carcinogens. Cobalt metal and cobalt (II) oxide nanoparticles, based on Cappellini et al., are capable of oxidative damage and DNA strand breaking. They also demonstrated that lung cells are capable of taking up three various cobalt-based nanoparticles (Cappellini et al. 2018). Cellular signaling pathways associated with the oxidative stress induced by cobalt oxide nanoparticles were explored by Chattopadhyay et al. It has been shown that uncoated cobalt oxide nanoparticles reduce cell viability extensively by the formation of reactive oxygen species, leading to activation of tumor necrosis factor alpha. Because it triggers caspase-8, which in turn triggers p38 MAPK and caspase-3, tumor necrosis factor alpha is essential for cellular apoptosis. Additionally, this study showed that human immune cells are negatively impacted by exposed cobalt oxide nanoparticles (Chattopadhyay et al. 2015).

7.1.3.7 Aluminium Oxide NPs

Based on the worldwide market for aluminum oxide nanoparticles, 20% of nanosized compounds are composed of aluminum. They have been used in various industries such as paints, plastics, coatings, textiles, fuel cells, solar energy, airbag propellants, and energetic materials, due to their superior mechanical strength and toughness. Exposure to aluminum oxide nanoparticles has been found to cause mitochondrial dysfunction, elevated oxidative stress, blood-brain barrier protein expression changes, and reduced cell viability, according to research (Kim, Baek, and Choi 2010). Srikanth et al. state that aluminum oxide nanoparticles induce membrane damage and cytotoxicity in Chinook salmon cells. The reasons for this damage may be oxidative stress, lipid peroxidation, and morphological alterations. Di Virgilio et al. identified dose-dependent cytotoxicity after exposing Chinese hamster ovary cells to titanium oxide and aluminum oxide nanoparticles for 24 hours. Doses as low as 5 μg/mL were identified. Because of the drastic increase in micronucleus frequencies at 0.5 and 1 μg/mL titanium oxide and 0.5–10 μg/mL aluminum oxide NPs, they also reported genotoxic effects (Di Virgilio et al. 2010).

7.2 CONCLUSION

Increased need for green chemistry and nanotechnology has driven the development of environmentally friendly and sustainable synthetic processes for the production of nanomaterials from microorganisms, plants, and other natural sources. Sustainable and environmentally friendly nanoparticle production has been the subject of research. Because of their eco-friendliness, accessibility, safety, and affordability, an extensive amount of research has been carried out to synthesize nanoparticles using plant extracts and to consider their potential in a number of applications. An interesting and futuristic area of nanotechnology with large environmental impacts that promotes long-term sustainability is using plants to form green nanoparticles. The environmental fate and toxicity of different types of nanoparticles have been the focal points of debate. While earlier studies have suggested the possible dangers of these nanoparticles, the pathogenesis of their toxicity remains unknown. Future studies need to give high priority to evaluating the safety of complex, multi-component, and functional nanomaterials. In addition, researchers must develop validated models that can assess how produced materials are released, transported, transformed, accumulated, and absorbed in the environment. The models must enable a comprehensive approach to solving the possible implications on the affected population by predicting the possible effects of artificial nanomaterials and nanoproducts and correlating the physical and chemical properties of nanomaterials with their behavior.

REFERENCES

Abada, Emad, Abdullah Mashraqi, Yosra Modafer, Mohamed A. Al Abboud, and A. El-Shabasy. 2024. "Review Green Synthesis of Silver Nanoparticles by Using Plant Extracts and Their Antimicrobial Activity." *Saudi Journal of Biological Sciences* 31 (1): 103877. https://doi.org/10.1016/J.SJBS.2023.103877.

Ahmad, Hilal, Krishnan Venugopal, Kalyanaraman Rajagopal, Savitha De Britto, Boregowda Nandini, Hosur Gnanaprakash Pushpalatha, Narasimhamurthy Konappa, Arakere C. Udayashankar, Nagaraja Geetha, and Sudisha Jogaiah. 2020. "Green Synthesis and Characterization of Zinc Oxide Nanoparticles Using Eucalyptus Globules and Their Fungicidal Ability Against Pathogenic Fungi of Apple Orchards." *Biomolecules* 10 (3): 425. https://doi.org/10.3390/BIOM10030425.

Ahmad, Waseem, and Divya Kalra. 2020. "Green Synthesis, Characterization and Anti-Microbial Activities of ZnO Nanoparticles Using Euphorbia Hirta Leaf Extract." *Journal of King Saud University—Science* 32 (4): 2358–64. https://doi.org/10.1016/J.JKSUS.2020.03.014.

Ahmad, Waseem, Krishna Kumar Jaiswal, and Shivani Soni. 2020. "Green Synthesis of Titanium Dioxide (TiO2) Nanoparticles by Using Mentha Arvensis Leaves Extract and Its Antimicrobial Properties." *Inorganic and Nano-Metal Chemistry* 50 (10): 1032–38. https://doi.org/10.1080/24701556.2020.1732419.

Al-darwesh, Mouhaned Y., Sattar S. Ibrahim, and Mohammed A. Mohammed. 2024. "A Review on Plant Extract Mediated Green Synthesis of Zinc Oxide Nanoparticles and Their Biomedical Applications." *Results in Chemistry* 7 (January): 101368. https://doi.org/10.1016/J.RECHEM.2024.101368.

Al-Harbi, Nuha, and Nabil K. Abd-Elrahman. 2024. "Physical Methods for Preparation of Nanomaterials, Their Characterization and Applications: A Review." *Journal of Umm Al-Qura University for Applied Sciences* (May): 1–22. https://doi.org/10.1007/S43994-024-00165-7/FIGURES/16.

Ankamwar, Balaprasad, Minakshi Chaudhary, and Murali Sastry. 2005. "Gold Nanotriangles Biologically Synthesized Using Tamarind Leaf Extract and Potential Application in Vapor Sensing." *Synthesis and Reactivity in Inorganic, Metal-Organic and Nano-Metal Chemistry* 35 (1): 19–26. https://doi.org/10.1081/SIM−200047527.

Article, Review, Nilesh Patil, Rajveer Bhaskar, Vishal Vyavhare, Rahul Dhadge, Vaishnavi Khaire, and Yogesh Patil. 2021. "Overview on Methods of Synthesis of Nanoparticles." *International Journal of Current Pharmaceutical Research* (March): 11–16. https://doi.org/10.22159/IJCPR.2021V13I2.41556.

Aswini, R., S. Murugesan, and Karthik Kannan. 2021. "Bio-Engineered TiO2 Nanoparticles Using Ledebouria Revoluta Extract: Larvicidal, Histopathological, Antibacterial and Anticancer Activity." *International Journal of Environmental Analytical Chemistry* 101 (15): 2926–36. https://doi.org/10.1080/03067319.2020.1718668.

Awwad, Akl M., Borhan Albiss, and Ahmad L. Ahmad. 2014. "Advanced Materials Letters Green Synthesis, Characterization and Optical Properties of Zinc Oxide Nanosheets Using Olea Europea Leaf Extract." *Research Article Advanced Materials Letters* 5 (9): 520–24. https://doi.org/10.5185/amlett.2014.5575.

Bar, Harekrishna, Dipak Kr Bhui, Gobinda P. Sahoo, Priyanka Sarkar, Sankar P. De, and Ajay Misra. 2009. "Green Synthesis of Silver Nanoparticles Using Latex of Jatropha Curcas." *Colloids and Surfaces A: Physicochemical and Engineering Aspects* 339 (1–3): 134–39. https://doi.org/10.1016/J.COLSURFA.2009.02.008.

Bendale, Yogesh, Vineeta Bendale, and Saili Paul. 2017. "Evaluation of Cytotoxic Activity of Platinum Nanoparticles against Normal and Cancer Cells and Its Anticancer Potential through Induction of Apoptosis." *Integrative Medicine Research* 6 (2): 141–48. https://doi.org/10.1016/J.IMR.2017.01.006.

Berardis, Barbara De, Gabriele Civitelli, Maria Condello, Pasquale Lista, Roberta Pozzi, Giuseppe Arancia, and Stefania Meschini. 2010. "Exposure to ZnO Nanoparticles Induces Oxidative Stress and Cytotoxicity in Human Colon Carcinoma Cells." *Toxicology and Applied Pharmacology* 246 (3): 116–27. https://doi.org/10.1016/J.TAAP.2010.04.012.

Cappellini, Francesca, Yolanda Hedberg, Sarah McCarrick, Jonas Hedberg, Remco Derr, Giel Hendriks, Inger Odnevall Wallinder, and Hanna L. Karlsson. 2018. "Mechanistic Insight into Reactivity and (Geno)Toxicity of Well-Characterized Nanoparticles of Cobalt Metal and Oxides." *Nanotoxicology* 12 (6): 602–20. https://doi.org/10.1080/17435390.2018.1470694.

Chattopadhyay, Sourav, Sandeep Kumar Dash, Satyajit Tripathy, Balaram Das, Debasis Mandal, Panchanan Pramanik, and Somenath Roy. 2015. "Toxicity of Cobalt Oxide Nanoparticles to Normal Cells; an In Vitro and In Vivo Study." *Chemico-Biological Interactions* 226 (January): 58–71. https://doi.org/10.1016/J.CBI.2014.11.016.

ElMitwalli, Omar S., Omar A. Barakat, Rabbani M. Daoud, Sultan Akhtar, and Fryad Z. Henari. 2020. "Green Synthesis of Gold Nanoparticles Using Cinnamon Bark Extract, Characterization, and Fluorescence Activity in Au/Eosin Y Assemblies." *Journal of Nanoparticle Research* 22 (10): 1–9. https://doi.org/10.1007/S11051-020-04983-8/FIGURES/7.

Gatto, Francesca, Mauro Moglianetti, Pier Paolo Pompa, and Giuseppe Bardi. 2018. "Platinum Nanoparticles Decrease Reactive Oxygen Species and Modulate Gene Expression without Alteration of Immune Responses in THP-1 Monocytes." *Nanomaterials* 8 (6): 392. https://doi.org/10.3390/NANO8060392.

Gomathi, R., and Harindran Suhana. 2021. "Green Synthesis, Characterization and Antimicrobial Activity of Zinc Oxide Nanoparticles Using Artemisia Pallens Plant Extract." *Inorganic and Nano-Metal Chemistry* 51 (12): 1663–72. https://doi.org/10.1080/24701556.2020.1852256.

Gudimalla, Apparao, Jiya Jose, Rajendran Jose Varghese, and Sabu Thomas. 2021. "Green Synthesis of Silver Nanoparticles Using Nymphae Odorata Extract Incorporated Films and Antimicrobial Activity." *Journal of Polymers and the Environment* 29 (5): 1412–23. https://doi.org/10.1007/S10924-020-01959-6/METRICS.

Ijaz, Irfan, Ezaz Gilani, Ammara Nazir, and Aysha Bukhari. 2020. "Detail Review on Chemical, Physical and Green Synthesis, Classification, Characterizations and Applications of Nanoparticles." *Green Chemistry Letters and Reviews* 13 (3): 59–81. https://doi.org/10.1080/17518253.2020.1802517/ASSET/2171C3FE-8071-413C-AEED-7756197F1B3B/ASSETS/IMAGES/TGCL_A_1802517_F0017_OC.JPG.

Javan bakht Dalir, Sahar, Hoorieh Djahaniani, Farzaneh Nabati, and Malak Hekmati. 2020. "Characterization and the Evaluation of Antimicrobial Activities of Silver Nanoparticles Biosynthesized from Carya Illinoinensis Leaf Extract." *Heliyon* 6 (3): e03624. https://doi.org/10.1016/J.HELIYON.2020.E03624.

Jayachandran, Ashwini, Aswathy T.R., and Achuthsankar S. Nair. 2021. "Green Synthesis and Characterization of Zinc Oxide Nanoparticles Using Cayratia Pedata Leaf Extract." *Biochemistry and Biophysics Reports* 26 (July): 100995. https://doi.org/10.1016/J.BBREP.2021.100995.

Jha, Anal K., and K. Prasad. 2011. "Biosynthesis of Gold Nanoparticles Using Bael (Aegle Marmelos) Leaf: Mythology Meets Technology." *International Journal of Green Nanotechnology* 3 (2): 92–97. https://doi.org/10.1080/19430892.2011.574560.

Jha, Anal K., and K. Prasad. 2012. "Biosynthesis of Gold Nanoparticles Using Common Aromatic Plants." *International Journal of Green Nanotechnology* 4 (3): 219–24. https://doi.org/10.1080/19430892.2012.706070.

Joudeh, Nadeem, and Dirk Linke. 2022. "Nanoparticle Classification, Physicochemical Properties, Characterization, and Applications: A Comprehensive Review for Biologists." *Journal of Nanobiotechnology* 20 (1). https://doi.org/10.1186/S12951-022-01477-8.

Khan, Yousaf, Haleema Sadia, Syed Zeeshan Ali Shah, Muhammad Naeem Khan, Amjad Ali Shah, Naimat Ullah, Muhammad Farhat Ullah, et al. 2022. "Classification, Synthetic, and Characterization Approaches to Nanoparticles, and Their Applications in Various Fields of Nanotechnology: A Review." *Catalysts* 12 (11): 1386. https://doi.org/10.3390/CATAL12111386.

Khatoon, Aisha, Farheen Khan, Nabeel Ahmad, Sibhghatulla Shaikh, Syed Mohd Danish Rizvi, Shazi Shakil, Mohammad H. Al-Qahtani, et al. 2018. "Silver Nanoparticles from Leaf Extract of Mentha Piperita: Eco-Friendly Synthesis and Effect on Acetylcholinesterase Activity." *Life Sciences* 209 (September): 430–34. https://doi.org/10.1016/J.LFS.2018.08.046.

Kim, In Sun, Min Baek, and Soo Jin Choi. 2010. "Comparative Cytotoxicity of Al2O3, CeO2, TiO2 and ZnO Nanoparticles to Human Lung Cells." *Journal of Nanoscience and Nanotechnology* 10 (5): 3453–58. https://doi.org/10.1166/JNN.2010.2340.

Kumari, Priyanka, and Abha Meena. 2020. "Green Synthesis of Gold Nanoparticles from Lawsoniainermis and Its Catalytic Activities Following the Langmuir-Hinshelwood Mechanism." *Colloids and Surfaces A: Physicochemical and Engineering Aspects* 606 (December): 125447. https://doi.org/10.1016/J.COLSURFA.2020.125447.

Marimuthu, Sampath, Abdul Abdul Rahuman, Chidambaram Jayaseelan, Arivarasan Vishnu Kirthi, Thirunavukkarasu Santhoshkumar, Kanayairam Velayutham, Asokan Bagavan, et al. 2013. "Acaricidal Activity of Synthesized Titanium Dioxide Nanoparticles Using Calotropis Gigantea against Rhipicephalus Microplus and Haemaphysalis Bispinosa." *Asian Pacific Journal of Tropical Medicine* 6 (9): 682–88. https://doi.org/10.1016/S1995-7645(13)60118–2.

Mekuye, Bawoke, and Birhanu Abera. 2023. "Nanomaterials: An Overview of Synthesis, Classification, Characterization, and Applications." *Nano Select* 4 (8): 486–501. https://doi.org/10.1002/NANO.202300038.

Morales-Lozoya, Violeta, Heriberto Espinoza-Gómez, Lucía Z. Flores-López, Erika Lis Sotelo-Barrera, Alfredo Núñez-Rivera, Rubén Darío Cadena-Nava, Gabriel Alonso-Nuñez, and Ignacio A. Rivero. 2021. "Study of the Effect of the Different Parts of Morinda Citrifolia L. (Noni) on the Green Synthesis of Silver Nanoparticles and Their Antibacterial Activity." *Applied Surface Science* 537 (January): 147855. https://doi.org/10.1016/J.APSUSC.2020.147855.

Moteriya, Pooja, and Sumitra Chanda. 2020. "Green Synthesis of Silver Nanoparticles from Caesalpinia Pulcherrima Leaf Extract and Evaluation of Their Antimicrobial, Cytotoxic and Genotoxic Potential (3-in-1 System)." *Journal of Inorganic and Organometallic Polymers and Materials* 30 (10): 3920–32. https://doi.org/10.1007/S10904-020-01532-7/METRICS.

Mulenos, Marina R., Jiaqi Liu, Henry Lujan, Binglin Guo, Eric Lichtfouse, Virender K. Sharma, and Christie M. Sayes. 2020. "Copper, Silver, and Titania Nanoparticles Do Not Release Ions under Anoxic Conditions and Release Only Minute Ion Levels under Oxic Conditions in Water: Evidence for the Low Toxicity of Nanoparticles." *Environmental Chemistry Letters* 18 (4): 1319–28. https://doi.org/10.1007/S10311-020-00985-Z/METRICS.

Muniyappan, N., M. Pandeeswaran, and Augustine Amalraj. 2021. "Green Synthesis of Gold Nanoparticles Using Curcuma Pseudomontana Isolated Curcumin: Its Characterization, Antimicrobial, Antioxidant and Anti-Inflammatory Activities." *Environmental Chemistry and Ecotoxicology* 3 (January): 117–24. https://doi.org/10.1016/J.ENCECO.2021.01.002.

Narayanan, Mathiyazhagan, P. Gokila Devi, Devarajan Natarajan, Sabariswaran Kandasamy, Kesavan Devarayan, Mishal Alsehli, Ashraf Elfasakhany, and Arivalagan Pugazhendhi. 2021. "Green Synthesis and Characterization of Titanium Dioxide Nanoparticles Using Leaf Extract of Pouteria Campechiana and Larvicidal and Pupicidal Activity on Aedes Aegypti." *Environmental Research* 200 (September). https://doi.org/10.1016/J.ENVRES.2021.111333.

Nelson, Bryant C., Elijah J. Petersen, Bryce J. Marquis, Donald H. Atha, John T. Elliott, Danielle Cleveland, Stephanie S. Watson, et al. 2013. "NIST Gold Nanoparticle Reference Materials Do Not Induce Oxidative DNA Damage." *Nanotoxicology* 7 (1): 21–29. https://doi.org/10.3109/17435390.2011.626537.

Nilavukkarasi, M., S. Vijayakumar, and S. Prathip Kumar. 2020. "Biological Synthesis and Characterization of Silver Nanoparticles with Capparis Zeylanica L. Leaf Extract for Potent Antimicrobial and Anti Proliferation Efficiency." *Materials Science for Energy Technologies* 3 (January): 371–76. https://doi.org/10.1016/J.MSET.2020.02.008.

Noorjahan, C. M., S. K. Jasmine Shahina, T. Deepika, and Summera Rafiq. 2024. "Green Synthesis and Characterization of Zinc Oxide Nanoparticles from Neem (Azadirachta Indicia)." Accessed December 27. www.ijsetr.com.

Noruzi, Masumeh, Davood Zare, Kamyar Khoshnevisan, and Daryoush Davoodi. 2011. "Rapid Green Synthesis of Gold Nanoparticles Using Rosa Hybrida Petal Extract at Room Temperature." *Spectrochimica Acta. Part A, Molecular and Biomolecular Spectroscopy* 79 (5): 1461–65. https://doi.org/10.1016/J.SAA.2011.05.001.

Oudhia, Anjali, Pragya Kulkarni, and Savita Sharma. 2024. "Research Paper Green Synthesis of ZnO Nanotubes for Bioapplications." *International Journal of Advanced Engineering Research*. Accessed December 27.

Pacheco, Ivan, and Cristina Buzea. 2021. "Nanomaterials and Nanocomposites: Classification and Toxicity." *Handbook of Nanomaterials and Nanocomposites for Energy and Environmental Applications: Volumes 1–4* 1 (January): 3–39. https://doi.org/10.1007/978-3-030-36268-3_1.

Philip, Daizy, C. Unni, S. Aswathy Aromal, and V. K. Vidhu. 2011. "Murraya Koenigii Leaf-Assisted Rapid Green Synthesis of Silver and Gold Nanoparticles." *Spectrochimica Acta. Part A, Molecular and Biomolecular Spectroscopy* 78 (2): 899–904. https://doi.org/10.1016/J.SAA.2010.12.060.

Pirsaheb, Meghdad, Tahereh Gholami, Hooman Seifi, Elmuez A. Dawi, Esraa Ahmed Said, Abdul Hameed M. Hamoody, Usama S. Altimari, and Masoud Salavati-Niasari. 2024. "Green Synthesis of Nanomaterials by Using Plant Extracts as Reducing and Capping Agents." *Environmental Science and Pollution Research* 31 (17): 24768–87. https://doi.org/10.1007/S11356-024-32983-X/METRICS.

Proquin, Héloïse, Carolina Rodríguez-Ibarra, Carolyn G.J. Moonen, Ismael M. Urrutia Ortega, Jacob J. Briedé, Theo M. de Kok, Henk van Loveren, and Yolanda I. Chirino. 2017. "Titanium Dioxide Food Additive (E171) Induces ROS Formation and Genotoxicity: Contribution of Micro and Nano-Sized Fractions." *Mutagenesis* 32 (1): 139–49. https://doi.org/10.1093/MUTAGE/GEW051.

Ramesh, M., M. Anbuvannan, and G. Viruthagiri. 2015. "Green Synthesis of ZnO Nanoparticles Using Solanum Nigrum Leaf Extract and Their Antibacterial Activity." *Spectrochimica Acta. Part A, Molecular and Biomolecular Spectroscopy* 136 Pt B (PB): 864–70. https://doi.org/10.1016/J.SAA.2014.09.105.

Santhoshkumar, Thirunavukkarasu, Abdul Abdul Rahuman, Chidambaram Jayaseelan, Govindasamy Rajakumar, Sampath Marimuthu, Arivarasan Vishnu Kirthi, Kanayairam Velayutham, John Thomas, Jayachandran Venkatesan, and Se Kwon Kim. 2014. "Green Synthesis of Titanium Dioxide Nanoparticles Using Psidium Guajava Extract and Its Antibacterial and Antioxidant Properties." *Asian Pacific Journal of Tropical Medicine* 7 (12): 968–76. https://doi.org/10.1016/S1995-7645(14)60171-1.

Senut, Marie Claude, Yanhua Zhang, Fangchao Liu, Arko Sen, Douglas M. Ruden, and Guangzhao Mao. 2016. "Size-Dependent Toxicity of Gold Nanoparticles on Human Embryonic Stem Cells and Their Neural Derivatives." *Small* 12 (5): 631–46. https://doi.org/10.1002/SMLL.201502346.

Sethy, Naresh Kumar, Zeenat Arif, Pradeep Kumar Mishra, and Pradeep Kumar. 2020. "Green Synthesis of TiO2 Nanoparticles from Syzygium Cumini Extract for Photo-Catalytic Removal of Lead (Pb) in Explosive Industrial Wastewater." *Green Processing and Synthesis* 9 (1): 171–81. https://doi.org/10.1515/GPS-2020-0018/MACHINEREADABLECITATION/RIS.

Sharmila, Govindasamy, Marimuthu Thirumarimurugan, and Chandrasekaran Muthukumaran. 2019. "Green Synthesis of ZnO Nanoparticles Using Tecoma Castanifolia Leaf Extract: Characterization and Evaluation of Its Antioxidant, Bactericidal and Anticancer Activities." *Microchemical Journal* 145 (March): 578–87. https://doi.org/10.1016/J.MICROC.2018.11.022.

Thakur, B. K., A. Kumar, and D. Kumar. 2019. "Green Synthesis of Titanium Dioxide Nanoparticles Using Azadirachta Indica Leaf Extract and Evaluation of Their Antibacterial Activity." *South African Journal of Botany* 124 (August): 223–27. https://doi.org/10.1016/J.SAJB.2019.05.024.

Vijaya Kumar, P., S. Mary Jelastin Kala, and K. S. Prakash. 2019. "Green Synthesis of Gold Nanoparticles Using Croton Caudatus Geisel Leaf Extract and Their Biological Studies." *Materials Letters* 236 (February): 19–22. https://doi.org/10.1016/J.MATLET.2018.10.025.

Virgilio, A. L. Di, M. Reigosa, P. M. Arnal, and M. Fernández Lorenzo de Mele. 2010. "Comparative Study of the Cytotoxic and Genotoxic Effects of Titanium Oxide and Aluminium Oxide Nanoparticles in Chinese Hamster Ovary (CHO-K1) Cells." *Journal of Hazardous Materials* 177 (1–3): 711–18. https://doi.org/10.1016/J.JHAZMAT.2009.12.089.

Xiao, Lu, Chunhua Liu, Xiaoniao Chen, and Zhuo Yang. 2016. "Zinc Oxide Nanoparticles Induce Renal Toxicity through Reactive Oxygen Species." *Food and Chemical Toxicology* 90 (April): 76–83. https://doi.org/10.1016/J.FCT.2016.02.002.

Yin, Jun Jie, Jun Liu, Marilyn Ehrenshaft, Joan E. Roberts, Peter P. Fu, Ronald P. Mason, and Baozhong Zhao. 2012. "Phototoxicity of Nano Titanium Dioxides in HaCaT Keratinocytes—Generation of Reactive Oxygen Species and Cell Damage." *Toxicology and Applied Pharmacology* 263 (1): 81–88. https://doi.org/10.1016/J.TAAP.2012.06.001.

8 Plant-Based Carbon Dots and Their Bio-Medical Applications

Shashi Pandey, Neeraj Mohan Gupta, Akanksha Upadhyay, Kundan Singh Rawat, Rashmi Priya, and Shachi Mishra

8.1 INTRODUCTION

Research in the realm of nanotechnology occupies a pivotal position within the landscape of scientific advancement, providing innovative resolutions to complex issues in the domains of healthcare, electronics, energy production, and environmental science. The Nobel Prize-winning identification of nanoscale semiconductor quantum dots (QDs) has fundamentally transformed numerous disciplines, including but not limited to optical display technology, medical imaging techniques, solar energy harnessing, and quantum computational systems (The Noble Prize in Chemistry 2023). However, despite many merits, QDs still have issues with efficiency, toxicity, and instability. Traditional QDs are often made from materials like cadmium or lead which possess toxic effects and thus are a threat to environment and health. Additionally, their stability is limited, as exposure to light, heat, or moisture can degrade their performance (Wang et al. 2019b; Zuo et al. 2015). As a promising alternative to semiconductor QDs, carbon dots (CDs), a zero-dimensional quasi-spherical carbon-based nanomaterial, have gained immense popularity. Initially, these entities were fortuitously identified by Xu et al. in the year 2004 during the process of purifying single-walled carbon nanotubes via preparative electrophoresis (Xu et al. 2004). In the year 2006, Sun and their team effectively produced surface-passivated nanoscale carbon nanoparticles that displayed heightened fluorescence emission, thereby introducing the term "carbon dots" to refer to these materials (Sun et al. 2006). However, low quantum yield of these surface passivated CDs and complicated preparation process limited their development. In the year 2013, Yang and his associates employed citric acid and ethylenediamine as foundational materials to produce remarkably luminous polymer-like CDs using a simple hydrothermal technique (Zhu et al. 2013). This development catalyzed a substantial increase in scholarly exploration focused on carbon dots, attributed to their straightforward synthesis, elevated quantum yield (QY), minimal toxicity, and significant resilience to photobleaching (Hailing et al. 2020). Since then, countless essential progressions have been made within the sphere of carbon dot exploration, which includes multicolor

DOI: 10.1201/9781003559900-8

and near-infrared (NIR) emissions, two-photon and multiphoton photoluminescence, chirality, phosphorescence at room temperature (RTP), and thermally activated delayed fluorescence (TADF), alongside an abundance of practical applications (Qiu et al. 2025; Zhai et al. 2021; Han et al. 2018, 2022; Wu et al. 2022; Yuan et al. 2023; Liu et al. 2020a, 2020b).

In a broad and encompassing sense, the term "carbon dots" denotes nanoscale luminous carbon-based materials that are fundamentally composed of both sp^2 and sp^3 hybridized carbon configurations, along with various oxygen- and nitrogen-containing functional groups and possess at least one dimension measuring smaller than 10 nanometers and exhibit intrinsic fluorescence characteristics as a defining trait. Right now, the categorization of carbon dots has been outlined into different individual categories, incorporating but not solely confined to graphene quantum dots (GQDs), carbon quantum dots (CQDs), carbon nanodots (CNDs), and carbonized polymer dots (CPDs), categorized based on their unique formation techniques, the specific micro- or nanostructural qualities they showcase, together with the varying properties they reveal in multiple environments. This categorization serves not only as a means of organizing these materials for research and application purposes but also highlights the complexity and versatility inherent in the study of nanoscale carbon materials, reflecting the ongoing advancements in materials science and nanotechnology that continue to emerge in contemporary scientific discourse and investigation (Xia et al. 2019). GQDs are disc-shaped nanoparticles mainly produced by exfoliation of larger graphitized carbon materials and consist of single or multiple-layer graphite structures with chemical groups on the surface/edge or within the interlayer defect. Their photoluminescence (PL) is mainly determined by conjugated π-domains and is greatly affected by the surface or edge structure. GQDs exhibit quantum confinement effect (Li et al. 2013; Gao et al. 2017). The CQDs are quasi-spherical nanoparticles with a multi-layered crystalline graphitic core structure and chemical groups on the surface. They show intrinsic state luminescence and quantum confinement effect (Naik et al. 2022). CNDs, on the other hand, are amorphous, quasi-spherical nanoparticles having high carbonization degree but no obvious crystal lattice structure. Carbon nanodots (CNDs) predominantly exhibit excited states that resemble molecular configurations, and the phenomenon of their photoluminescence can primarily be ascribed to the presence of defect or surface states, as well as subdomain states situated within the intricate graphite carbon core that constitutes their structural foundation. Meanwhile, carbonized polymer dots (CPDs) represent complex hybrid structures that combine polymer and carbon elements, characterized by an abundance of surface-functionalized groups and polymer chains intertwined with carbon cores that may either be entirely carbonized or exhibit a paracrystalline nature, featuring minuscule carbon clusters enveloped by polymer frameworks or an extensively dehydrated, intricately crosslinked, and closely-knit polymer structure that provides additional stability and functionality. The origins of photoluminescence observed in CPDs can be principally traced back to molecular states, alongside the effects of crosslinked enhanced emission (CEE), which collectively contribute to the enhanced optical characteristics that these materials exhibit under specific excitatory conditions. Hence, it is vital to comprehend the complicated interactions between structural attributes and electronic states in both CNDs and CPDs, given that these interactions substantially impact their photoluminescent characteristics and potential roles in multiple sectors of advanced materials science and nanotechnology (Han et al. 2018).

Utilization of biomass including plant-derived or animal-derived or small biomolecules precursors for the synthesis of fluorescent CDs has garnered significant attention, since it is abundant, heterogeneous, organic and biodegradable (Rajesh et al. 2022; Wang et al. 2023). Also, this leads to biomass valorization converting low value biomass waste to useful materials. Further, utilization of plant parts as green precursors for CDs synthesis helps in satisfying the pressing need for large-scale production and thus creation of sustainable applications (Zulfajri et al. 2020). Plant parts including root, stem, leaf, fruit, flower, and seed contain various carbohydrates, proteins, amino acids, and other biomolecules that provide sufficient elements for surface functionality of green CDs preparation, and thus use of separate reactants for doping, surface passivation, or post modification is not required. Several plant-derived resources like neem leaves (Vijeata et al. 2023), banana petioles (Korram et al. 2023), orange juice (Sahu et al. 2012), beet root juice (Singh et al. 2018), green tea leaves (Shi et al. 2017), coffee (Chem et al. 2012), garlic (Zhao et al. 2015), ginger (Li et al. 2014), and bio wastes such as watermelon peels (Zhou et al. 2012), lemon peels (Tyagi et al. 2016), tangerine seeds (Wang et al. 2020), etc. have been used as carbon sources in developing fluorescent CDs and studied for biomedical applications. This chapter summarizes recent advances in the synthetic routes, property tunability, and biomedical applications of plant-derived green CDs along with the challenges associated and future perspectives.

8.2 SYNTHESIS OF CARBON DOTS DERIVED FROM PLANT SOURCES

In recent years, a long-standing development objective has been to find inexpensive, environmentally friendly, nontoxic carbon source precursors for the large-scale production of CDs. Plant parts are widely used for CDs synthesis because they provide a sustainable, eco-friendly, and cost-effective source of carbon. They are rich in organic compounds such as cellulose, lignin, carbohydrates, and polyphenols and thus serve as excellent precursors. The diverse chemical compositions of different plant materials allow for tailoring the optical and structural properties of CDs, making them suitable for applications in bioimaging, sensing, catalysis, and environmental remediation (Tejwan et al. 2019; Sharma et al. 2017). Their intrinsic functional groups, like hydroxyl, carboxyl, and amino groups, enhance the surface properties of CDs, improving their water solubility, fluorescence, and biocompatibility. Moreover, plant parts are abundant, renewable, and often derived from agricultural waste, aligning with green chemistry principles and reducing production costs. This approach not only promotes sustainable nanotechnology but also adds value to otherwise discarded biological resources (Rajesh et al. 2022).

These advanced carbon dots are derived from multiple sources of plant materials and are crafted via sustainable synthesis practices that seek to lessen any adverse consequences on the ecosystem, thus responding to current environmental concerns. The array of techniques available for carbon dot synthesis is extensive, with each

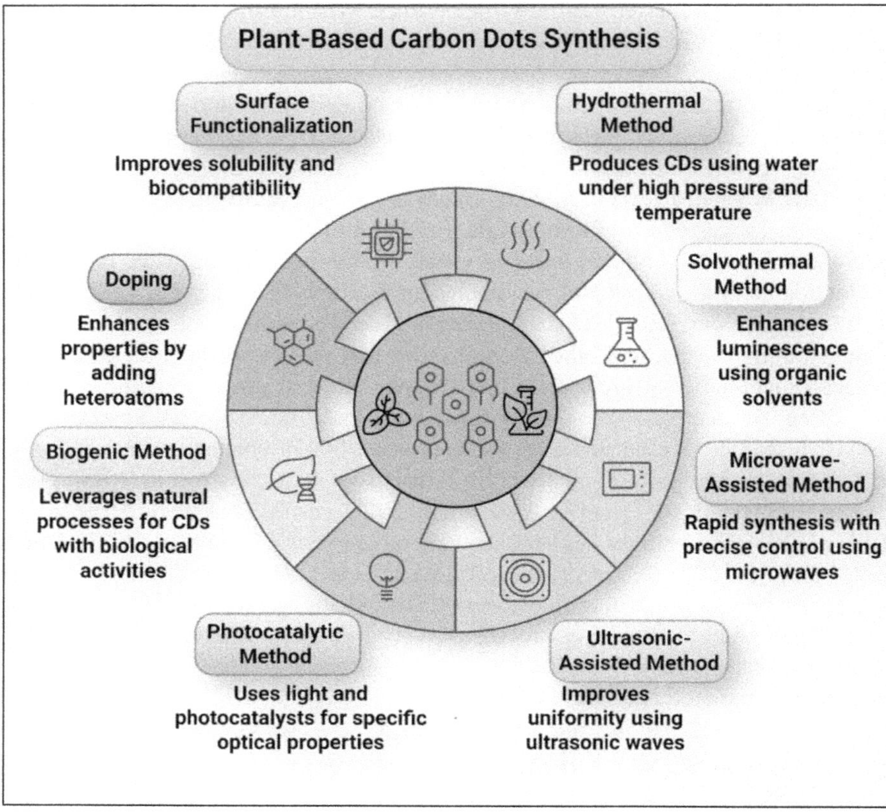

FIGURE 8.1 Different synthetic approaches for plant-based carbon dots (CDs).

method bringing its own unique benefits concerning efficiency, the extent of control over the final material properties, and their general utility across several research domains and industries. Figure 8.1 illustrates the different synthetic approaches for plant-based CDs.

8.2.1 Hydrothermal and Solvothermal Methods

In hydrothermal scenarios, plant-sourced substances experience substantial heat and pressure in a solution of water, consequently yielding CDs that exhibit notable features. This methodology is widely regarded as the predominant approach for the synthesis of plant-derived CDs due to its straightforward and effective pathway for CD formation, utilizing water as the solvent. Reactions generally happen in a sealed chamber (Teflon-lined autoclave) at heat levels from 120°C to 250°C and under high pressure for durations of two to 14 hours. Employing this methodology grants the ability to modify morphology, surface attributes, and crystalline structure through variations in pressure, temperature, solvent, and reaction time. The process encompasses phases of carbonization, dehydration, passivation, and polymerization of the precursors (Meng et al. 2019; Mansuriya and Altintas 2021). Conversely, the technique known as solvothermal synthesis engages organic solvents under analogous high-pressure and high-temperature circumstances to fabricate CDs. A number of researchers have also utilized oil baths, sand baths, or direct heating under reflux conditions to achieve CDs through thermal means alone. With the solvothermal strategy, one can achieve superior regulation of the reaction setting, factoring in characteristics like polarity, viscosity, and reactivity (Bhartiya et al. 2016). Furthermore, the solvents that are part of this technique significantly contribute to the functionalization of CDs (Fang et al. 2023).

8.2.2 Microwave-Assisted and Ultrasonic-Assisted Methods

Employing microwave-assisted synthesis integrates microwave exposure particularly for forming carbon dots. Through the utilization of this innovative approach, one can achieve the production of CDs characterized by a remarkably uniform size distribution, which is critical for ensuring consistent performance in various applications. Moreover, this advanced method facilitates the rapid synthesis of CDs, accomplishing this task within a remarkably short time frame that typically ranges from as brief as 30 seconds to a maximum of 30 minutes, thereby demonstrating a significant time efficiency when juxtaposed with traditional heating methodologies. Remarkably, this strategy reveals a substantial compatibility with an assortment of precursors, both from nature and those made artificially, which are applied in the generation of CDs, enhancing its potential and relevance in diverse scientific fields. The mechanism of microwave irradiation not only promotes reactions free from solvents but also allows for the conduction of reactions mediated by water or other solvents, all while maintaining a precise control over the heating process, which is vital for optimizing the synthesis conditions (Rajesh et al. 2022; Wareing et al. 2021).

In a different approach, synthesis aided by ultrasonic waves relies on the application of ultrasonic frequencies to help and quicken the development of carbon dots, thus taking advantage of the underlying mechanics of sound waves to boost the reaction rates. This specific methodology markedly augments the kinetics of the chemical reactions involved, resulting in a substantial enhancement of the overall velocity of the synthesis process, which constitutes an essential attribute for researchers endeavoring to refine their experimental conditions with efficacy. Moreover, the incorporation of ultrasonic waves in this synthesis methodology facilitates the production of homogeneous carbon dots, thereby ensuring that the resultant materials exhibit uniform properties, a crucial factor for their prospective applications across a multitude of scientific and technological fields. When evaluating the broader context, the synthesis strategies that employ microwave and ultrasonic assistance illustrate groundbreaking advancements in carbon dot fabrication, each proffering distinct advantages that contribute to the burgeoning research and practical applications within the realm of nanotechnology (Usman and Cheng 2024).

8.2.3 Photocatalytic and Biogenic Synthesis

Employing a photocatalytic synthesis technique involves leveraging light to promote the generation of carbon dots, typically alongside an appropriate photocatalyst that improves the process's effectiveness. One of the primary advantages of this method lies in its capacity to produce carbon dots with specific and tailored optical properties that can be beneficial for various applications (Usman and Cheng 2024). On the other hand, biogenic synthesis represents an innovative methodology that capitalizes on living organisms, including plants and microorganisms, to generate carbon dots. This eco-conscious strategy leverages organic biological processes, culminating in the formation of carbon dots that inherently carry biological activities, like antimicrobial and anticancer traits, thereby significantly widening their potential uses in biomedicine (Sahu and Sahoo 2024).

8.2.4 Doping and Surface Functionalization

One prominent and highly significant methodology currently employed within the scientific domain to enhance the optical and electronic properties of carbon dots involves the strategic doping of these materials with heteroatoms, such as nitrogen and sulfur, which are known to impart distinct advantages (Krishnaiah et al. 2025). Additionally, adopting surface engineering strategies, like the direct incorporation of different phytochemicals, may substantially improve the solubility and biocompatibility of carbon dots, thus making them notably more fitting for bioimaging applications and a wide range of other pertinent and burgeoning sectors of nanotechnology and biomedical exploration (Annisa et al. 2024).

Although it is evident that carbon dots derived from botanical sources present numerous benefits and opportunities for refinement, the intricate phytochemical composition of these plants complicates achieving batch-to-batch uniformity, consequently impacting their scalability, biocompatibility, and practical applications (Nekoueian et al. 2019). To address this challenge, scholars advocate for the meticulous optimization of reaction parameters alongside post-synthesis modifications involving the incorporation of heteroatoms (doping) and surface passivation techniques (Rajesh et al. 2022). Additionally, it is imperative to consider the phytochemical constituents found in plant biomass sourced from varying regions, batches, or origins (Rajesh et al. 2022). Numerous components of plants, such as leaves, fruits, flowers, stems, roots, and seeds, along with wasted peels, have undergone research as forerunners to the development of carbon dots, which are afterward applied in medical contexts. Table 8.1 provides several instances of plant-derived carbon dots, detailing their synthesis methods, along with their physical and optical characteristics relevant for biomedical applications since 2019.

8.3 PROPERTIES OF PLANT-DERIVED CDS

Plant-derived CDs exhibit distinctive physical and optical characteristics that render them exceptionally valuable across diverse applications, especially biomedical applications (Figure 8.2). They are highly biocompatible and non-toxic, owing to

TABLE 8.1
Plant-Derived CDs and Their Applications

Plant Source	Synthesis Method	Size (nm)	Fluorescence Color	Quantum Yield	Biomedical Application	Reference
Date palm leaves	Hydrothermal	5 ~ 15 nm	Blue	–	Cell imaging	(Athinarayanan et al. 2022)
Aloe barbadensis miller	Microwave-assisted	3.2 nm	Blue	0.31	Live-cell bio-imaging studies	(Malavika et al. 2021)
Water hyacinth	Hydrothermal	6.2 nm	Blue	–	Bioimaging and anticancer activity	(Paul and Kurian 2020)
Mexican mint extract	Microwave assisted	2.43 nm	Blue	0.17	Fe^{3+} detection (LOD 53μM) and bioimaging (MCF-7 cells)	(Architha et. al. 2021)
Grape seeds	Hydrothermal	8.9 nm	Green	0.275	Detecting Cu(II) (LOD 0.048 mg L^{-1}) and AA (LOD and 0.036 mg L^{-1}), bioimaging	(Li et al. 2021)
Abelmoschus manihot flower	Hydrothermal	9 nm	Blue	30.8	Bioimaging (CDs were explored for cytotoxicity in three cell lines and cellular imaging in Hela cells.)	(Wang et al. 2019a)
Muskmelon	Acid oxidation	~3.5, ~4.3, and ~5.8 nm	Blue, green, yellow	7.07, 26.9, and 14.3%	Detect Hg^{2+} ions in biological and water samples. The fabricated multicolor-emissive CMCDs possess the cell (Cunninghamella elegans, Aspergillus flavus, and Rhizoctonia solani) imaging property, suggesting the biocompatible nature for multicolor imaging of various cells.	(Desai et al. 2019)
Agarose waste	Hydrothermal	2–6	Blue	62	Aspartic acid sensing	(Chauhan et al. 2020)
Allium fistulosum	Hydrothermal	4.22	Blue	10.48	Bioimaging	(Wei et al. 2019)
Artemisia argyi Folium Carbonisata	Pyrolysis	6–10 nm	Violet-blue	0.91	Biomedical (frostbite treatment)	(Kong et al. 2020)
Azadirachta indica	Acidic oxidation	4–10 nm		–	Photothermal therapy	(Meena et al. 2019)

(Continued)

TABLE 8.1 (*Continued*)
Plant-Derived CDs and Their Applications

Plant Source	Synthesis Method	Size (nm)	Fluorescence Color	Quantum Yield	Biomedical Application	Reference
Banana peel	Hydrothermal	5		20	Bioimaging	(Atchudan et al. 2020)
Bean pods, onion	Hydrothermal	6		5.55	Bioimaging	(Lu et al. 2019)
Betel leaf	Hydrothermal	3–6		12	Sensor (Fe^{3+})	(Kamilah et al. 2021)
Black sesame	Hydrothermal	2.5		7.15	Sensor (Fe^{3+})	(Zhang et al. 2019)
Catharanthus Roseus	Hydrothermal	5		28.2	Bioimaging	(Arumugham et al. 2020)
Citrus aurantium	Hydrothermal pyrolysis	1–2		19.9	Cell imaging	(Fatahi et al. 2019)
Citrus fruit peel	Sand bath assisted	4.6		–	Bioimaging, free radical scavenging	(Gudimella et al. 2021)
Cranberry seed	Hydrothermal	3.96		10.8	Sensor (Fe3+)	(Zulfajri et al. 2020)
Crown daisy leaf waste	Hydrothermal	5–10		–	Sensor (Cu2+)	(Wang et al. 2020)
Custard apple peel	Microwave	3–8		–	Bioimaging	(Babu et al. 2021)
Dunaliella salina	Hydrothermal	2–8		8	Sensor (Hg2+, Cr4+)	(Singh et al. 2019)
Fennel seeds	Hydrothermal pyrolysis	3.9		9.5	Biosensor	(Permatasari et al. 2021)
Fenugreek seeds	Microwave plasma enhanced decomposition	–		4.96	Fluorescent protein crystal	(Dager et al. 2020)
Ginkgo biloba	Hydrothermal	4.11		21.7	Electrocatalysis	(Tuna et al. 2020)
Grass	Reflux method	20		-	Detection of Cu2+ ions	(Khalid et al. 2022)
Jackfruit	Hydrothermal	5.4		5.2	Sensing acetone vapors	(Thongsai et al. 2019a)

Latex of *Ficus benghalensis*	Hydrothermal	3.7	41.2	Sensor (Tyrosine), Schottky barrier diode	(Yadav et al. 2021)
Lawsonia inermis (Henna)	Hydrothermal	3–7	28.7	Biocompatible probe for methotrexate	(Shahshahanipour et al. 2019)
Lemon and onion juice	Carbonization	4.23–8.22	23.6	Determination of riboflavin	(Monte-Filho et al. 2019)
Lemon juice	Hydrothermal	4.6	28	Bioimaging	(Ding et al. 2019)
Lemon juice	Hydrothermal	1–6	31	Bioimaging, sensor (Hg2+)	(Tadesse et al. 2020)
Lotus seed pod	Hydrothermal	2.1	1.9	Sensor (Fe3+)	(Xie et al. 2021)
Marigold	Hydrothermal	3.2	7.84	Sensor (Fe 3+)	(Zhang et al. 2019)
Millet	Hydrothermal	6–10	–	Bioimaging	(Quang and Ha 2019)
Mulberry leaves	Hydrothermal	2–4	–	Bioimaging, drug delivery	(Shao et al. 2020)
Onion	Hydrothermal	9	28	Bioimaging	(Debabrata et al. 2021)
Orange peel	Microwave	4.2	16.2	Assay E. coli	(Hu et al. 2021)
Osmanthus leaves	Hydrothermal	18	13.70	HASE analysis	(Liu et al. 2019)
Pomegranate peel, watermelon peel	Microwave	1–5	–	Biomedical applications	(Muktha et al. 2020)
Prosopis juliflora leaves	Pyrolysis	5.8	5	Sensor (Hg^{2+}, Chimet drug)	(Pourreza and Ghomi 2019)
Purple perilla	Hydrothermal	<5	9.01	Bioimaging	(Zhao et al. 2015)
Quinoa	Hydrothermal		14	In vitro imaging	(Xu et al. 2020)
Quinoa saponin	Hydrothermal	2.25	22.2	Sensor (Co^{2+})	(Zhou et al. 2021)
Red cabbage	Hydrothermal	3	8.3	Antioxidant, bio labeling	(Sharma et al. 2020)
Rise husk	Hydrothermal	4–5	3	Detecting alcohol vapors	(Thongsai et al. 2019a)
Sugarcane industrial waste	Hydrothermal	0.75–2.75	17.98	Bioimaging	(Pandiyan et al. 2020)

(Continued)

TABLE 8.1 (*Continued*)
Plant-Derived CDs and Their Applications

Plant Source	Synthesis Method	Size (nm)	Fluorescence Color	Quantum Yield	Biomedical Application	Reference
Taro peels	Chemical oxidation	–		26.20	Fluoride ion detection	(Boruah et al. 2020)
Taxus leaves	Solvothermal (acetone)	2.7		59	Bioimaging	(Liu et al. 2020a)
Tomato	Chemical oxidation	–		2.76–12.7	HeLa cell imaging	(Kailasa et al. 2019)
Trapa bispinosa peel	Thermal oxidation	5–10		1.2	Bio compatible against MDCK cells	(Nair et al. 2020)
Tulsi	Hydrothermal	5–7		–	Forward osmosis	(Doshi and Mungray 2020)
Vigna radiata seeds	Hydrothermal Carbonization	<10		58	Sensor (Fe^{3+}), cancer therapy	(Kaur et al. 2022)
Wheat straw and bamboo residues	Hydrothermal	3.5		13	Bioimaging	(Huang et al. 2019)
Zapota fruit	Acidic oxidation, carbonization	1.9–4.5	Blue, green, yellow	5.2–7.9	Bacterial/fungal bio-imaging	(Bhamore et al. 2019)
Watermelon juice/Orange juice/Lemon juice/Cantaloupe juice/Red plum juice/Green plum juice/Carrot juice/Red pitaya juice/White pitaya juice	Hydrothermal	1.6–5.6		13–25	ROS sensing, in-vivo Zebrafish imaging	(Zhu et al. 2022)

Turmeric	Hydrothermal		Green		Anti-bacterial activity	(Roy et al. 2022)
Beetroot	Hydrothermal	<5.0		11.6	Anticancer/scavenging ROS	(Smrithi et al. 2021)
Curcumin	Powder carbonization	4.8		–	Anti-viral	(Chen et al. 2022)
Calotropis gigantea (Crown Flower) leaf extract	Microwave assisted	2.7 to 10.4 nm	Blue	4.24	Bio-imaging of bacteria, fungi, and plant cell	(Sharma et al. 2022)
Tangerine seeds	Hydrothermal	1.276 nm	Blue	39	Sensing ketorolac tromethamine (KETO)	(Rizk et al. 2024)
Salvadora persica root	Hydrothermal	–	Green	19	Sensing Congo Red (CR) in living cells and zebrafish	(Durrani et al. 2022)
Scutellaria baicalensis	Hydrothermal	2.67	Blue	–	ROS scavenging, antioxidant activity	(Wang et al. 2022)
Typha angustata Bory (Patera)	Hydrothermal	2–7	Blue	83	Sensing Hg2+ and cysteine	(Samota et al. 2022)
Wet pomace oil waste	Hydrothermal	2-4.4	Blue	21	Cytotoxic effects in L929 and HeLa cell lines	(Sousa et al. 2022)
Cinchona Pubescens Vahl extract	Hydrothermal	5	Blue	–	Drug nanocarriers for carboplatin (CBP) delivery	(González-Reyna et al. 2023)
Polyalthia longifolia	Hydrothermal	3.3	Red	22	Cd^{2+} sensing (LOD 2.4 nm)	(Kolaprath et al. 2023)
Vitex negundo leaves	Hydrothermal	6.5 nm	Blue	–	pH sensor and fluorescent ink	(Korah and Mathew 2023)
Crescentia cujete fruit waste	Hydrothermal	–	Green	–	Selective and sensitive detection of Fe^{3+} ions (LOD 0.257 μM)	(Elizabeth et al. 2023)
Tagetes patula flowers	Hydrothermal	5.15 nm	Blue	29.88%.	Fe3+ sensor, antifungal activity	(Patra et al. 2023)
Cinnamon	Hydrothermal	–	Blue	–	Cr^{6+} and Mn^{7+} metal ion sensor	(Kolekar et al. 2024)
Brewery spent grain	Microwave irradiation	5.3 nm	Blue	14%	Cu^{2+}, Fe^{3+}, Zn^{2+} and Al^{2+} sensor	(Nkeumaleu et al. 2022)
Curcumin	Hydrothermal	2.92 nm	Green	-	Antibacterial activity	(Miao et al. 2025)

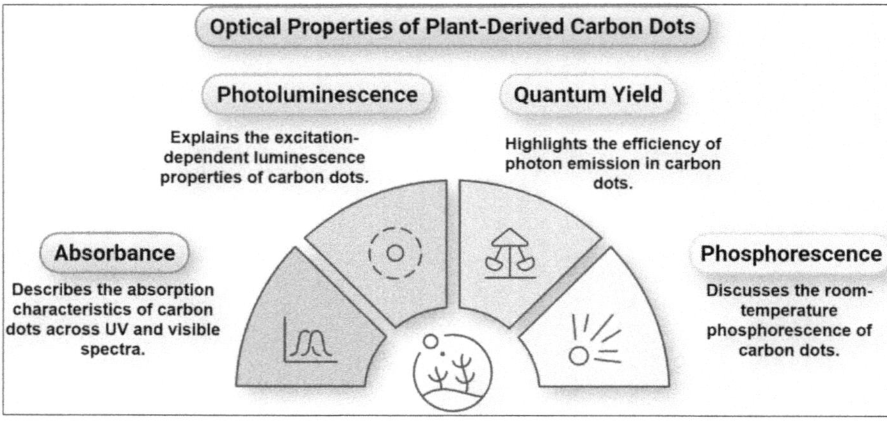

FIGURE 8.2 Optical properties of plant-derived CDs.

their natural origin and eco-friendly synthesis process, which minimizes the use of harmful chemicals. These CDs possess strong and tunable photoluminescence, making them excellent candidates for bioimaging and diagnostic purposes. Their surfaces are rich in functional groups such as hydroxyl (–OH), carboxyl (–COOH), and amino ($-NH_2$) groups, which enhances water solubility and enables functionalization with biomolecules, drugs, or targeting agents (Nekoueian et al. 2019). Their substantial hydrophilicity guarantees suitability within aqueous biological contexts. Plant-derived carbon dots are distinguished by their heightened fluorescence intensity and exceptional quantum yields, which render them particularly advantageous for imaging applications. Moreover, they demonstrate notable photostability, allowing them to withstand prolonged exposure to irradiation, thus establishing them as reliable candidates for extended imaging endeavors (Rajesh et al. 2022). Furthermore, carbon dots sourced from plants demonstrate both antioxidant and antimicrobial qualities, consequently improving their utility in therapy. Furthermore, their diminutive size and substantial surface area promote efficient cellular uptake and interaction with biomolecules (Kumar et al. 2023). These multifaceted attributes position plant-derived carbon dots as a promising resource for various applications in bioimaging, drug delivery, biosensing, and an array of therapeutic interventions.

8.3.1 Absorbance

The significance of absorbance is paramount in the comprehensive characterization of CDs, which are increasingly utilized in a myriad of applications spanning various scientific and industrial domains. The carbon dots synthesized through diverse methodologies exhibit unique and distinctive absorption spectra that reflect their structural and compositional disparities. Overall, these carbon dots are distinguished for their impressive absorbance properties in the ultraviolet section, mainly between 200 and 400 nanometers, with this absorbance gradually moving into the visible spectrum ranging from 400 to 700 nanometers. Typically, it is noted that entities classified as CDs demonstrate a marked degree of absorbance within the ultraviolet segment of the electromagnetic spectrum, specifically within the wavelength interval of 200 to 400 nanometers, while concurrently revealing a gradual yet significant extension into the visible light spectrum, which encompasses wavelengths from 400 to 700 nanometers. Crucially, the absorption peak that is distinctly recognized at a wavelength of 230 nanometers can be closely linked to the π–π^* electronic transition that pertains to carbon-carbon double bonds (C=C), while the comparatively faint shoulder peak, located roughly at 300 nanometers, is mainly tied to the n–π^* electronic transition of functional groups like carbonyl (C=O) or imine (C=N) bonds. A myriad of factors, including the various types and quantities of surface functional groups, the dimensions of the π-conjugated regions, and the alterations in oxygen and nitrogen content within the carbon core structure, significantly impact the absorbance characteristics of carbon dots. In endeavors to optimize or alter the absorption capacities of carbon quantum dots (CQDs), numerous methodologies are employed by experts in the field. For instance, surface engineering strategies can be effectively implemented to finely tune both the absorption and emission spectra of these quantum dots, while the integration of heteroatoms via doping techniques can facilitate modifications in the absorption properties by influencing the energy levels associated with the π–π^* transitions and introducing a range of surface defects. In a noteworthy scientific investigation conducted by Wei and colleagues (Wei et al. 2019), they proficiently synthesized carbon dots that were co-doped with nitrogen and sulfur through hydrothermal treatment of *Allium fistulosum*, commonly referred to as green onion, resulting in a novel variant of carbon dots (Af-CDs) exhibiting enhanced properties owing to the strategic incorporation of these heteroatoms. The synthesized CDs demonstrated remarkable stability in photoluminescence across varying pH levels, elevated NaCl concentrations, and in the presence of diverse metal ions. They were adeptly used for multihued imaging within the cytoplasm of MCF-7 and K562 cell lines, indicating their prospective roles in sensing and biotechnological areas.

8.3.2 Photoluminescence

Highlighting the diverse attributes of CDs, their PL is of special interest, demonstrating excitation-dependent luminescence that spans ultraviolet, visible, and NIR light. This excitation-dependent attribute is integral to the functional application of CDs, as it facilitates the modulation of their optical, electrical, and chemical properties through the regulation of the excitation state, thereby unveiling a broad spectrum of functionalities and potential applications. The alteration of the excitation wavelength permits a precise enhancement of PL characteristics, leveraging the principles inherent in quantum confinement. As a result, a singular CD can produce multiple emission spectra through the variation of the excitation wavelength, while its chemical composition and dimensional attributes remain invariant. Usually, CDs present photoluminescence in the blue and green spectral areas; however, it's achievable to stretch this emission toward the red and NIR spectra by optimizing the reaction circumstances and selecting fitting carbon precursors. Furthermore, the PL spectrum can be modulated through techniques such as surface modification, careful precursor selection, and diverse synthesis methodologies (Wu et al. 2022; Liu et al. 2020a; Xia et al. 2019). This unique optical property makes CDs promising for in vivo bioimaging, offering enhanced photon penetration through tissue, reduced background autofluorescence, and lower photon-induced toxicity, particularly in the NIR region. Liu et al. (Liu et al. 2020a) synthesized CDs with an exceptionally narrow FWHM of 20 nm from taxus leaves using a solvothermal method and purification via silica gel column chromatography. The remarkable deep red-emissive CDs displayed a quantum yield (QY) of 59% with excitation at 413 nm and a QY of 31% when interacting with the deep red fluorescence domain. Comprehensive investigations indicated that the CDs exhibited distinctive polymer-like properties alongside two-photon capabilities. Their utility in deep tissue bioimaging employing both one-photon and two-photon modalities was explored in Eca-109 cells (carcinoma cells) and C2C12 cells (myoblast cells). In vivo experiments conducted on nude mouse models illustrated the swift elimination of CDs through the renal and hepatobiliary pathways, thereby underscoring their potential in biomedical applications. Ding et al. (Ding et al. 2019) utilized solvothermal approach to prepare NIR emitting CDs from pulp-free lemon juice as carbon precursor and formamide solution as solvent. The NIR-CDs were safely applied for in vitro cell imaging in HeLa cells and for in vivo bioimaging in nude mouse model subcutaneously as well as intravenously.

8.3.3 Quantum Yield

QY constitutes a fundamental metric for CDs, particularly in the context of photodevice applications. It quantifies the proportion of photons emitted relative to those absorbed by the CDs (Liu et al. 2020b; Naik et al. 2022; Meng et al. 2019). Significant scholarly attention has been directed towards augmenting the QY of CDs, as it is vital for their operational efficacy. This improvement may be accomplished via surface modification and heteroatom doping. By modifying the surface, one can transform groups that pull electrons into those that give them, ensuring the carbon nanostructure remains intact. Concurrently, heteroatom doping modifies the bandgap and electron density attributes, thereby facilitating an increase in the QY of CDs (Xia et al. 2019).

8.3.4 Phosphorescence

Room-temperature phosphorescence represents a captivating characteristic of CDs, arising from intersystem transitions between the lowest excited singlet state and a triplet state, succeeded by radiative transitions from the triplet state back to the ground state. The initial process fosters intersystem crossing by amplifying spin-orbit coupling, typically accomplished through the integration of transition metals. The subsequent process mitigates non-radiative transitions by constraining molecular rotation and vibration. Both mechanisms are pivotal for realizing phosphorescence at ambient temperature in organic substances. This phenomenon can be manifested in CDs with extensively cross-linked frameworks incorporating non-conjugated groups such as C=O and C=N, which demonstrate pronounced spin-orbit coupling. Furthermore, doping with heteroatoms (e.g., N, P) enhances the $n-\pi^*$ transitions of C=O and C=N bonds, thereby further promoting phosphorescence at ambient temperature (Cheng et al. 2022).

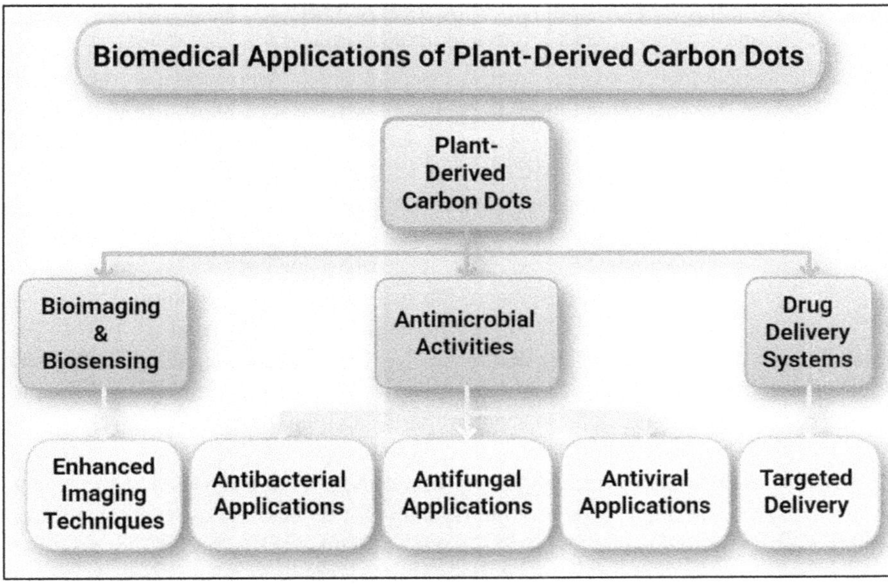

FIGURE 8.3 Biomedical applications of plant-derived carbon dots (CDs).

8.4 BIOMEDICAL APPLICATIONS OF PLANT-DERIVED CDS

CDs obtained from plant origins have progressively taken center stage as exceptionally promising substances that show a vast range of advantageous characteristics, which is particularly beneficial for various uses in the biomedical arena, owing to their distinctive traits such as natural biocompatibility, superior water solubility, and noteworthy photoluminescent capabilities. The spectrum of biomedical applications for plant-derived carbon dots is impressively diverse, encompassing critical functions that range from advanced bioimaging techniques and highly sensitive biosensing methods to innovative drug delivery systems and effective antimicrobial activities (Figure 8.3). Pivotal biomedical applications of plant-based CDs are discussed next.

8.4.1 Bioimaging and Biosensing

Bioimaging techniques are crucial in both fundamental research and clinical settings, offering non-invasive detailed monitoring of biological processes to study cell morphology and physiological functions. Imaging agents should possess essential properties like water solubility and non-toxicity to biological systems, along with favorable photophysical features such as large Stokes shifts, high quantum yields, resistance to photobleaching, prolonged lifetimes, and excitation/emission properties compatible with commercially available equipment (Hickey et al. 2021). Conventionally, chemically synthesized CDs and organic dyes have been extensively utilized in bioimaging; however, their high toxicity poses significant risks to living organisms. In contrast, plant-derived CDs are regarded as a superior alternative for biomedical imaging due to their minimal harmful effects, eco-friendliness, excellent fluorescence intensity, and robust photostability. Additionally, the carbon framework of CDs can be easily modified in multiple ways, allowing for the attachment of various recognition elements for targeted biological and cellular applications (Jelinek 2017). An impressive count of 122 distinct carbon dot species obtained from a wide selection of plant origins has been expertly utilized in imaging practices for several cell lines, including HeLa, HepG2, MCF-7, MDA-MB 468, MDA-MB-231, HaCaT, MC3T3, SiHa, HT-29, PC3, K-562, Hep3B, MG-63, and others, showcasing the adaptability and utility of these nanomaterials sourced from plants in pioneering cellular imaging approaches across different biological research contexts. In most cases, these plant-derived CDs stained the cytoplasmic region, however the difference in luminance property was observed based on different precursors (Kumar et al. 2023). The most plausible explanation for this phenomenon may be attributed to the diverse array of phytochemicals that can be found within the complex structure of plant biomass, which inherently contribute in a substantial manner to the overall fluorescent properties exhibited by CDs. The research by Zhao and colleagues highlights how crucial these compounds are, indicating that their varying interactions and levels can greatly affect the optical properties of the created carbon dots, thus broadening our comprehension of their possible uses across different scientific disciplines (Zhao et al. 2015). Paul et al. (Paul and Kurian 2020) synthesized CDs using water hyacinth as a precursor through a one-step hydrothermal method. These CDs exhibited uniform size, an amorphous graphitic structure, excellent water solubility, strong excitation-dependent light emission, and remarkable photostability. They emitted blue fluorescence in Klebsiella cells, effectively penetrating the cells, with fluorescence cell sorting revealing a 41.78% internalization rate. The attributes delineated herein underscore the remarkable potential of these materials to function as highly efficient fluorescent probes specifically designed for the purpose of imaging Klebsiella, a significant bacterial pathogen. In a related investigation, Samran Durrani and colleagues conducted a study wherein they successfully synthesized CDs from the roots of the Salvadora persica plant, which were subsequently employed for the quantitative detection of Congo red (CR) in various samples, including fish tissues and wastewater samples (Durrani et al. 2022). The resulting carbon dots exhibited exceptional sensitivity and selectivity in the detection of Congo red through the mechanism of fluorescence quenching, thereby marking a significant milestone as it represents the inaugural instance of probing Congo red within living cells and zebrafish specimens. In parallel, Samota and her research team undertook the synthesis of nitrogen- and sulfur-doped carbon dots (N, S-CDs), achieving an impressive quantum yield of 83% through the application of a hydrothermal synthesis method utilizing Typha angustata Bory powder in conjunction with 0.5 grams of thiourea as a precursor (Samota et al. 2022). This innovative approach not only emphasizes the versatility of carbon dots in various applications but also significantly contributes to the advancement of methodologies aimed at enhancing the detection capabilities of fluorescent probes in biological systems. These sensors enabled fluorescence-based "on-off" detection of mercury ions and "off-on" detection of cysteine.

8.4.2 Anti-Microbial Activity

Carbon dots from plant origins have gained notable recognition as highly promising antimicrobial compounds in the medical sector and reveal extensive antimicrobial capabilities that proficiently address a multitude of pathogens, such as bacteria, fungi, and viruses thus establishing their potential as solutions for the urgent and widespread challenge of antimicrobial resistance that has escalated into a significant global health issue.

8.4.2.1 Antibacterial Activity

Plant-based CDs display broad spectrum antibacterial activity. Miao et al. (Miao et al. 2025) synthesized novel curcumin-derived carbon dots Cur-CDs using curcumin and ethylenediamine that exhibited significant antibacterial activity against a wide range of bacteria, including both Gram-positive and Gram-negative strains and notably encompassing drug-resistant variants such as methicillin-resistant *Staphylococcus aureus* (MRSA). These carbon dots manifested their impacts by undermining the structural integrity of bacterial cells, along with the creation of reactive oxygen species (ROS), recognized for their lethality against bacterial entities. Further, employing carbon dots has also been recognized as a particularly effective method to stop biofilm formation on different medical devices and implants, which are often related to lingering and recurring infections. Their remarkable ability to penetrate established biofilms and disrupt microbial communication mechanisms, as well as adhesion processes, significantly contributes to their value in mitigating biofilm-related infections (Priyadarshini et al. 2024).

8.4.2.2 Antifungal Activity

Recent studies have identified plant-derived carbon dots as being effective antifungal agents, with the capability to inhibit the growth of opportunistic fungi that have developed resistance to conventional antifungal therapies. Moreover, their unique properties enable them to function as fluorescent nano-trackers, thus facilitating bioimaging and monitoring of fungal infections, which is pivotal for timely and effective treatment interventions (Sturabotti et al. 2024). The antifungal efficacy exhibited by carbon dots can be largely attributed to their ability to disrupt the integrity of fungal cell membranes, coupled with their interference in critical cellular processes, ultimately leading to cell death and the resolution of fungal infections (Sturabotti et al. 2024). Patra et al. (2023) prepared CDs from tagetes patula flowers using the green hydrothermal approach and investigated their antifungal activity on the mycelial growth of *Magnaporthe oryzae* (a plant pathogenic fungus). Further, fluorescence quenching behavior of the obtained CDs towards Fe^{3+} ions with a limit of detection of 0.32 μM was observed.

8.4.2.3 Antiviral Activity

Recent advancements in carbon nanomaterials highlight their significant promise in fighting and mitigating the threats that viral infections present to public wellness. The intrinsic fluorescent characteristics inherent to carbon dots markedly augment their applicability in the bioimaging processes associated with the visualization and identification of viral pathogens, thus providing invaluable assistance to healthcare professionals in the accurate diagnosis and subsequent treatment of viral infections, which is fundamentally essential for the efficient management of patients and the overall improvement of health outcomes. In this way, employing these sophisticated materials not only boosts our comprehension of viral activity but also opens up new avenues for inventive treatment options that could notably shift the path of viral diseases (Rai et al. 2024).

In an important development within the complex and swiftly advancing sector of nanotechnology, paired with thorough research into new antimicrobial options, Chen and his colleagues (Chen et al. 2022) have carefully crafted Curcumin-derived carbon dots, generally known as Cur-CQDs, through an intentional process that involves gentle pyrolysis-driven polymerization and later carbonization, a method that has not only yielded considerably improved water solubility and lessened cytotoxicity but also shown remarkable antiviral strength against the Japanese encephalitis virus (JEV) infection, thereby underscoring the possible applications of such nanomaterials in treatment.

While the prospect of employing plant-derived carbon dots as antimicrobial agents is notably promising, it is imperative to recognize the myriad challenges and limitations associated with their successful integration into healthcare settings. The assessment of the long-term safety and efficacy of these pioneering carbon dots within authentic clinical environments necessitates meticulous and comprehensive scrutiny, and there is an urgent demand for more expansive research focused on enhancing their synthesis and functionalization for specific medical applications. Moreover, it is essential to understand the potential ecological repercussions of carbon dots to promote their sustainable advancement and judicious use within the healthcare domain, thereby ensuring that they yield beneficial outcomes for both public health and environmental integrity.

8.4.3 Drug Delivery

Carbon dots display noteworthy drug delivery applications in antimicrobial, anticancer, and neurodegenerative therapies. Conventional chemotherapeutic drugs frequently come with significant limitations, such as high costs as well as harmful side effects on healthy and active tissues. This has driven the demand for affordable therapeutic anticancer agents derived from natural sources with minimal adverse effects. Due to their non-toxic nature, plant-derived CDs are increasingly explored for use in targeted gene and drug delivery systems. CDs interact with drug molecules through the π-π interactions, electrostatic forces, and chemical bonds, effectively binding the drug to the matrix. Despite their small size, CDs possess a substantial surface area that facilitates the encapsulation and loading of drug molecules. Further, the green carbon dots, created from diverse natural origins such as fruits, vegetables, and various therapeutic plants provide a range of benefits particularly in terms of greater stability and safety profiles that are essential in drug delivery context (Archana and Mohanan 2024). Shao et al. (Shao et al. 2020) synthesized carbon dots from mulberry leaves residue for intracellular fluorescence imaging of human hepatocellular carcinoma cells (HepG2) and investigated its drug delivery application for the loading of Lycorine, an anti-cancer alkaloid extracted from bulbs of *Lycoris radiata* L.

8.5 CONCLUSION

Plant-based CDs possess notable potential for biomedical applications owing to their distinctive attributes, including adjustable photoluminescence and effective radiation absorption for therapeutic use. The utilization of sources originating from botanical entities for the synthesis of CDs is particularly attractive due to their abundant availability, inherent safety for human health, and associated environmental advantages. This methodology is consistent with the principles of green chemistry and fosters a circular economy by reducing ecological footprints. However, maintaining uniform control over the dimensions, morphology, and surface characteristics of plant-derived CDs remains a significant challenge. The application of sophisticated characterization methodologies and systematic optimization of processes is essential for addressing these obstacles and enhancing scalability. The variability of phytochemical constituents across diverse plant sources complicates the consistency from batch to batch, thereby affecting the scalability and biocompatibility of CDs. To alleviate these challenges, it is prudent to optimize reaction parameters and post-synthetic modification techniques, such as doping and surface passivation. CDs from plant origins reveal impressive antioxidant and antimicrobial abilities, positioning them as highly suitable for various therapeutic roles, encompassing drug delivery, bioimaging, and biosensing. The environmental ramifications of CDs necessitate thorough evaluation to ensure their sustainable development and responsible application in healthcare, ultimately benefiting both public health and ecological sustainability. Future investigative efforts must aim at perfecting the creation and adaptation of CDs tailored for specialized medical functions, which would ensure their lasting safety and utility in therapeutic contexts. Overall, plant-sourced carbon dots indicate a highly promising research avenue with the capability to innovate biomedical practices. Nonetheless, it is crucial to address the prevailing challenges through innovative strategies and sustainable practices to enable their effective incorporation into healthcare solutions.

REFERENCES

Annisa, W.D., Permatasari, F.A., Iskandar, F. and Rachmawati, H., 2024. Functionalized phytochemicals-embedded carbon dots derived from medicinal plant for bioimaging application. *ACS Applied Bio Materials*, 7(1). https://doi.org/10.1021/acsabm.3c00575.

Archana, P. and Mohanan, J., 2024. An overview of herbal-carbon dots. *Preprints*. https://doi.org/10.20944/preprints202406.0120.v1.

Architha, N. et al., 2021. Microwave-assisted green synthesis of fluorescent carbon quantum dots from Mexican mint extract for Fe3+ detection and bio-imaging applications. *Environmental Research*, 199, 111263. https://doi.org/10.1016/j.envres.2021.111263.

Arumugham, T. et al., 2020. A sustainable synthesis of green carbon quantum dot (CQD) from *Catharanthus roseus* (white flowering plant) leaves and investigation of its dual fluorescence responsive behavior in multi-ion detection and biological applications. *Sustainable Materials and Technologies*, 23, e00138. https://doi.org/10.1016/j.susmat.2019.e00138.

Atchudan, R. et al., 2020. Sustainable synthesis of carbon quantum dots from banana peel waste using hydrothermal process for in vivo bioimaging. *Physica E: Low-dimensional systems and Nanostructures*, 126, 114417. https://doi.org/10.1016/j.physe.2020.114417.

Athinarayanan, J., Periasamy, V.S. and Alshatwi, A.A., 2022. Unveiling the biocompatible properties of date palm tree nanoparticles. *ACS Omega*, 7(23), 19270–19279. https://doi.org/10.1021/acsomega.2c00753.

Babu, J. et al., 2021. Photoluminescence carbon nano dots for the conductivity based optical sensing of dopamine and bioimaging applications. *Optical Materials*, 117, 111120. https://doi.org/10.1016/j.optmat.2021.111120.

Bhamore, J.R., Jha, S., Park, T.J. and Kailasa, S.K., 2019. Green synthesis of multi-color emissive carbon dots from *Manilkara zapota* fruits for bioimaging of bacterial and fungal cells. *Journal of Photochemistry and Photobiology B: Biology*, 191, 150–155. https://doi.org/10.1016/j.jphotobiol.2018.12.023.

Bhartiya, P., Singh, A., Kumar, H., Jain, T., Singh, B.K. and Dutta, P.K., 2016. Carbon dots: chemistry, properties and applications. *Journal of the Indian Chemical Society*, 93(7), 759–766.

Boruah, A., Saikia, M., Das, T., Goswamee, R.L. and Saikia, B.K., 2020. Blue-emitting fluorescent carbon quantum dots from waste biomass sources and their application in fluoride ion detection in water. *Journal of Photochemistry and Photobiology B: Biology*, 209, 111940. https://doi.org/10.1016/j.jphotobiol.2020.111940.

Chauhan, P., Saini, J., Chaudhary, S. and Bhasin, K.K., 2020. Sustainable synthesis of carbon dots from agarose waste and prospective application in sensing of L-aspartic acid. *Materials Research Bulletin*, 111113. https://doi.org/10.1016/j.materresbull.2020.111113.

Chem, G. et al., 2012. Synthesis and analytical applications of photoluminescent carbon nanodots. *Green Chemistry*, 14, 917–920. https://doi.org/10.1039/C2GC16451E.

Chen, H.H. et al., 2022. Development of antiviral carbon quantum dots that target the Japanese encephalitis virus envelope protein. *The Journal of Biological Chemistry*, 298(6), 101957. https://doi.org/10.1016/j.jbc.2022.101957.

Cheng, M., Cao, L., Guo, H., Dong, W. and Li, L., 2022. Green synthesis of phosphorescent carbon dots for anticounterfeiting and information encryption. *Sensors*, 22(8), 2944. https://doi.org/10.3390/s22082944.

Dager, A., Baliyan, A., Kurosu, S. et al., 2020. Ultrafast synthesis of carbon quantum dots from fenugreek seeds using microwave plasma enhanced decomposition: application of C-QDs to grow fluorescent protein crystals. *Scientific Reports*, 10, 12333. https://doi.org/10.1038/s41598-020-69264-9.

Debabrata, G.D., Payel, M., Dipanjan, G. and Devdut, B., 2021. Carbon quantum dots prepared from onion extract as fluorescence turn-on probes for selective estimation of Zn^{2+} in blood plasma. *Colloids and Surfaces A: Physicochemical and Engineering Aspects*, 611, 0927–7757. https://www.sciencedirect.com/science/article/pii/S0927775720313741.

Desai, M.L. et al., 2019. Acid oxidation of muskmelon fruit for the fabrication of carbon dots with specific emission colors for recognition of Hg2+ ions and cell imaging. *ACS Omega*, 4(21), 19332–19340. https://doi.org/10.1021/acsomega.9b02730.

Ding, H., Zhou, X., Qin, B., Zhou, Z. and Zhao, Y., 2019. Highly fluorescent near-infrared emitting carbon dots derived from lemon juice and its bioimaging application. *Journal of Luminescence*, 211, 298–304. https://doi.org/10.1016/j.jlumin.2019.03.064.

Doshi, K. and Mungray, A.A., 2020. Bio-route synthesis of carbon quantum dots from tulsi leaves and its application as a draw solution in forward osmosis. *Journal of Environmental Chemical Engineering*, 8(5), 104174. https://doi.org/10.1016/j.jece.2020.104174.

Durrani, S. et al., 2022. Plant-derived Ca, N, S-Doped carbon dots for fast universal cell imaging and intracellular Congo red detection. *Analytica Chimica Acta*, 1202, 339672. https://doi.org/10.1016/j.aca.2022.339672.

Elizabeth, A.T. et al., 2023. Green synthesis of value-added nitrogen doped carbon quantum dots from *Crescentia cujete* fruit waste for selective sensing of Fe3+ ions in aqueous medium. *Inorganic Chemistry Communications*, 110427. https://doi.org/10.1016/j.inoche.2023.110427.

Fang, M., Wang, B., Qu, X., Li, S. and Huang, J., 2023. State-of-the-art of biomass-derived carbon dots: preparation, properties, and applications. *Chinese Chemical Letters*, 35(1), 108423. https://doi.org/10.1016/j.cclet.2023.108423.

Fatahi, Z. et al., 2019. Physicochemical and cytotoxicity analysis of green synthesis carbon dots for cell imaging. *EXCLI Journal*, 18, 454–466. https://doi.org/10.17179/excli2019-1465.

Gao, T. et al., 2017. Red, yellow, and blue luminescence by graphene quantum dots: syntheses, mechanism, and cellular imaging. *ACS Applied Materials & Interfaces*, 9, 24846–24856. https://doi.org/10.1021/acsami.7b05569.

González-Reyna, M.A. et al., 2023. Green nanoarchitectonics of carbon quantum dots from *Cinchona pubescens* Vahl as targeted and controlled drug cancer nanocarrier. *Biomaterials Advances*, 213561. https://doi.org/10.1016/j.bioadv.2023.213561.

Gudimella, K.K., Appidi, T., Wu, H.F., Battula, V., Jogdand, A., Rengan, A.K. and Gedda, G., 2021. Sand bath assisted green synthesis of carbon dots from citrus fruit peels for free radical scavenging and cell imaging. *Colloids and Surfaces B: Biointerfaces*, 197, 111362. https://doi.org/10.1016/j.colsurfb.2020.111362.

Han, M. et al., 2018. Recent progress on the photocatalysis of carbon dots: classification, mechanism and applications. *Nano Today*, 19, 201–218. https://doi.org/10.1016/j.nantod.2018.02.008.

Han, T. et al., 2022. Near-infrared carbonized polymer dots for NIR-II bioimaging. *Advanced Science*, 9, 2203474. https://doi.org/10.1002/advs.202203474.

Hailing, Y. et al., 2020. Doxorubicin-loaded fluorescent carbon dots with PEI passivation as a drug delivery system for cancer therapy. *Nanoscale*, 12(33), 17222–17237. https://doi.org/10.1039/d0nr01236j.

Hickey, S.M. et al., 2021. Fluorescence microscopy—an outline of hardware, biological handling, and fluorophore considerations. *Cells*, 11(1), 35. https://doi.org/10.3390/cells11010035.

Hu, X., Li, Y., Xu, Y., Gan, Z., Zou, X., Shi, J., Huang, X., Li, Z. and Li, Y., 2021. Green one-step synthesis of carbon quantum dots from orange peel for fluorescent detection of *Escherichia coli* in milk. *Food Chemistry*, 339, 127775. https://doi.org/10.1016/j.foodchem.2020.127775.

Huang, C. et al., 2019. Synthesis of carbon quantum dot nanoparticles derived from byproducts in bio-refinery process for cell imaging and in vivo bioimaging. *Nanomaterials*, 9(3), 387. https://doi.org/10.3390/nano9030387.

Jelinek, R., 2017. Carbon nanostructures. In: R. Jelinek, ed. *Carbon Quantum Dots*. Berlin: Springer, p. 61.

Kailasa, S.K. et al., 2019. Tuning of carbon dots emission color for sensing of Fe^{3+} ion and bioimaging applications. *Materials Science and Engineering: C*, 98, 834–842. https://doi.org/10.1016/j.msec.2019.01.002.

Kamilah, N. et al., 2021. A review on multifunctional carbon-dots synthesized from biomass waste: design/fabrication, characterization and applications. *Frontiers in Energy Research*, 9, 1–22. https://doi.org/10.3389/fenrg.2021.626549.

Kaur, N. et al., 2022. Photoactivatable carbon dots as a label-free fluorescent probe for picric acid detection and light-induced bacterial inactivation. *Journal of Photochemistry and Photobiology B: Biology*, 229, 112412. https://doi.org/10.1016/j.jphotobiol.2022.112412.

Khalid, H., Khan, M.F., Ahmad, B., Ismail, M., Zahid, M. and Ismail, A., 2022. Carbon dots from natural sources for biomedical applications. *Plasma Processes and Polymers*, 39(9), 0934–0866. https://doi.org/10.1002/ppsc.202200017.

Kolaprath, M.K.A., Benny, L. and Varghese, A., 2023. A facile, green synthesis of carbon quantum dots from *Polyalthia longifolia* and its application for the selective detection of cadmium. *Dyes and Pigments*, 210, 111048. https://doi.org/10.1016/j.dyepig.2022.111048.

Kolekar, A.G. et al., 2024. Green, facial zinc doped hydrothermal synthesis of cinnamon derived fluorescent carbon dots (Zn-Cn-CDs) for highly selective and sensitive Cr6+ and Mn7+ metal ion sensing application. *Spectrochimica Acta Part A: Molecular and Biomolecular Spectroscopy*, 304, 123413. https://doi.org/10.1016/j.saa.2023.123413.

Korah, B.K. and Mathew, B., 2023. Sustainable carbon quantum dots from *Vitex negundo* leaves as a synergistic nanoplatform for triple object sensing and anticounterfeiting applications. *Materials Today Sustainability*, 21, 100273. https://doi.org/10.1016/j.mtsust.2022.100273.

Kong, H. et al., 2020. Carbon dots from *Artemisiae Argyi Folium Carbonisata*: strengthening the anti-frostbite ability. *Artificial Cells, Nanomedicine, and Biotechnology*, 49(1), 11–19. https://doi.org/10.1080/21691401.2020.1862134.

Korram, J., et al., 2023. Biomass-derived carbon dots as nanoprobes for smartphone – paper-based assay of iron and bioimaging application. *ACS Omega*, 8, 31410–31418. https://doi.org/10.1021/acsomega.3c03969.

Krishnaiah, P., Atchudan, R., Gangadaran, P. et al., 2025. A sustainable synthesis and applications of biomass waste-derived tunable fluorescent carbon dots: in vitro and in vivo fluorescent imaging. *Journal of Photochemistry and Photobiology A: Chemistry*, 458, 115944. https://doi.org/10.1016/j.jphotochem.2024.115944.

Kumar, A., Kumar, D., Saikia, M., Devi, R. and Subramania, A., 2023. A review on plant derived carbon quantum dots for bio-imaging. *Materials Advances*, 4(18), 3951–3966. https://doi.org/10.1039/D3MA00254C.

Li, C. L. et al., 2014. Carbon dots prepared from ginger exhibiting efficient inhibition of human hepatocellular carcinoma cells. *The Journal of Materials Chemistry B*, 2, 4564–4571. https://doi.org/10.1039/C4TB00216D.

Li, J. et al., 2021. A facile green and one-pot synthesis of grape seed-derived carbon quantum dots as a fluorescence probe for Cu(II) and ascorbic acid. *RSC Advances*, 11, 34107–34116. http://dx.doi.org/10.1039/D1RA05656E.

Li, L. et al., 2013. Focusing on luminescent graphene quantum dots: current status and future perspectives. *Nanoscale*, 5, 4015–4403. https://doi.org/10.1039/C3NR33849E.

Liu, J. et al., 2020a. Deep red emissive carbonized polymer dots with unprecedented narrow full width at half maximum. *Advanced Materials*, 32, 1906641. https://doi.org/10.1002/adma.201906641.

Liu, J., Li, R. and Yang, B., 2020b. Carbon dots: a new type of carbon-based nanomaterial with wide applications. *ACS Central Science*, 6, 2179–2195. https://doi.org/10.1021/acscentsci.0c01306.

Liu, X., Wang, T., Lu, Y., Wang, W., Zhou, Z. and Yan, Y., 2019. Constructing carbon dots and CdTe quantum dots multi-functional composites for ultrasensitive sensing and rapid degrading ciprofloxacin. *Sensors and Actuators B: Chemical*, 289, 242. https://doi.org/10.1016/j.snb.2019.03.094.

Lu, H. et al., 2019. Biomass-derived sulfur, nitrogen co-doped carbon dots for colorimetric and fluorescent dual mode detection of silver(I) and cell imaging. *ACS Omega*, 4, 21500–21508. https://doi.org/10.1021/acsomega.9b03198.

Malavika, J.P. et al., 2021. A sustainable green synthesis of functionalized biocompatible carbon quantum dots from *Aloe barbadensis Miller* and its multifunctional applications. *Environmental Research*, 200, 111414. https://doi.org/10.1016/j.envres.2021.111414.

Mansuriya, B.D. and Altintas, Z., 2021. Carbon dots: classification, properties, synthesis, characterization, and applications in health care—an updated review (2018–2021). *Nanomaterials*, 11(10), 2525. https://doi.org/10.3390/nano11102525.

Meena, R. et al., 2019. Fluorescent carbon dots driven from ayurvedic medicinal plants for cancer cell imaging and phototherapy. *Heliyon*, 5(9), e02483. https://doi.org/10.1016/j.heliyon.2019.e02483.

Meng, W. et al., 2019. Biomass-derived carbon dots and their applications. *Energy & Environmental Materials*, 2, 172–192. https://doi.org/10.1002/eem2.12038.

Miao, H. et al., 2025. Highly efficient and broad-spectrum antibacterial carbon dots combat antibiotic resistance. *Talanta*, 281, 126926. https://doi.org/10.1016/j.talanta.2024.126926.

Monte-Filho, S.S., Andrade, S.I.E., Lima, M.B. and Araújo, M.C.U., 2019. Synthesis of highly fluorescent carbon dots from lemon and onion juices for determination of riboflavin in multivitamin/mineral supplements. *Journal of Pharmaceutical Analysis*, 9(3), 209–216. https://doi.org/10.1016/j.jpha.2019.02.003.

Muktha, H., Sharath, R., Kottam, N., Smrithi, S.P., Samrat, K. and Ankitha, P., 2020. Green synthesis of carbon dots and evaluation of its pharmacological activities. *BioNanoScience*, 10, 731–744. https://doi.org/10.1007/s12668-020-00741-1.

Naik, K., Chaudhary, S., Ye, L. and Parmar, A.S., 2022. A strategic review on carbon quantum dots for cancer-diagnostics and treatment. *Frontiers in Bioengineering and Biotechnology*, 10, 1–21. https://doi.org/10.3389/fbioe.2022.882100.

Nair, A., Haponiuk, J.T., Thomas, S. and Gopi, S., 2020. Natural carbon-based quantum dots and their applications in drug delivery: a review. *Biomedicine & Pharmacotherapy*, 132, 110834. https://doi.org/10.1016/j.biopha.2020.110834.

Nekoueian, K. et al., 2019. Carbon-based quantum particles: an electroanalytical and biomedical perspective. *Chemical Society Reviews*, 48, 4281–4316. https://doi.org/10.1039/C8CS00445E.

Nkeumaleu, A.T. et al., 2022. Brewery spent grain derived carbon dots for metal sensing. *RSC Advances*, 12(19), 11621–11627. https://doi.org/10.1039/D2RA00048B.

The Noble Prize in Chemistry 2023. NobelPrize.org. Nobel Prize Outreach AB, 2025. Available at: https://www.nobelprize.org/prizes/chemistry/2023/summary/ (Accessed: 19 January 2025).

Pandiyan, S. et al., 2020. Biocompatible carbon quantum dots derived from sugarcane industrial wastes for effective nonlinear optical behavior and antimicrobial activity applications. *ACS Omega*, 5(47), 30363–30372. https://doi.org/10.1021/acsomega.0c03290.

Patra, S. et al., 2023. One-step green synthesis of in-situ functionalized carbon quantum dots from *Tagetes patula* flowers: applications as a fluorescent probe for detecting Fe3+ ions and as an antifungal agent. *Journal of Photochemistry and Photobiology A: Chemistry*, 442, 114779. https://doi.org/10.1016/j.jphotochem.2023.114779.

Paul, A. and Kurian, M., 2020. N-doped photoluminescent carbon dots from water hyacinth for tumour detection. *Materials Today: Proceedings*, 25(2), 213–217. https://doi.org/10.1016/j.matpr.2020.01.038.

Permatasari, F.A., Irham, M.A. and Bisri, S.Z., 2021. Carbon-based quantum dots for supercapacitors: recent advances and future challenges. *Nanomaterials*, 11(1), 91. https://doi.org/10.3390/nano11010091.

Pourreza, N. and Ghomi, M., 2019. Green synthesized carbon quantum dots from *Prosopis juliflora* leaves as a dual off-on fluorescence probe for sensing mercury (II) and chemet drug. *Materials Science and Engineering: C*, 98, 887–896. https://doi.org/10.1016/j.msec.2018.12.141.

Priyadarshini, E. et al., 2024. Biofilm inhibition on medical devices and implants using carbon dots: an updated review. *ACS Applied Bio Materials*, 7(5), 2604–2619. https://doi.org/10.1021/acsabm.4c00024.

Qiu, S. et al., 2025. Advanced room-temperature cured encapsulant film for crystalline silicon solar modules: enhancing efficiency with luminescent down-shifting flame retardancy and UV resistance. *Materials Horizons*, [online]. https://doi.org/10.1039/D4MH01668H.

Quang, N.K. and Ha, C.T.C., 2019. Low-cost synthesis of carbon nanodots from millets for bioimaging. *MRS Advances*, 4, 249. https://doi.org/10.1557/adv.2019.

Rai, M., Ingle, A.P., Törős, G. and Prokisch, J., 2024. Assessing the efficacy of carbon nanodots derived from curcumin on infectious diseases. *Expert Review of Anti-Infective Therapy*, 22(12), 1107–1121. https://doi.org/10.1080/14787210.2024.2409401.

Rajesh, G.A. et al., 2022. Carbon dots from natural sources for biomedical applications. *Particle & Particle Systems Characterization*, 39, Article ID 2200017. https://doi.org/10.1002/ppsc.202200017.

Rizk, M., Ramzy, E. and Toubar, S., 2024. Bioinspired carbon dots-based fluorescent sensor for the selective determination of a potent anti-inflammatory drug in the presence of its photodegradation products. *ACS Omega*, 9(25), 27517–27527. https://doi.org/10.1021/acsomega.4c02757.

Roy, S., Ezati, P. and Rhim, J.W., 2022. Preparation of turmeric-derived sulfur-functionalized carbon dots: antibacterial and antioxidant activity. *Journal of Materials Science*, 57, 2941–2952. https://doi.org/10.1007/s10853-021-06804-2.

Sahu, S., Behera, B., T.K. Maiti and S. Mohapatra et al., 2012. Simple one-step synthesis of highly luminescent carbon dots from orange juice: application as excellent bio-imaging agents. *Chemical Communications*, 48, 8835–8837. https://doi.org/10.1039/C2CC33796G.

Sahu, V. and Sahoo, S.K., 2024. Biogenic synthesis of carbon dots with inbuilt biological activity. *Next Nanotechnology*, 5, 100034. https://doi.org/10.1016/j.nxnano.2023.100034.

Samota, S. et al., 2022. Carbon dot nanosensors for ultra-low level, rapid assay of mercury ions synthesized from an aquatic weed, *Typha angustata* Bory (Patera). *Diamond and Related Materials*, 130, 109433. https://doi.org/10.1016/j.diamond.2022.109433

Shao, Y., Zhu, C., Fu, Z., Lin, K., Wang, Y., Chang, Y., Han, L., Yu, H. and Tian, F., 2020. Green synthesis of multifunctional fluorescent carbon dots from mulberry leaves for simultaneous intracellular imaging and drug delivery. *Journal of Nanoparticle Research*, 22, 229. https://doi.org/10.1007/s11051-020-04917-4.

Shahshahanipour, M., Rezaei, B., Ensafi, A.A. and Etemadifar, Z., 2019. An ancient plant for the synthesis of a novel carbon dot and its applications as an antibacterial agent and probe for sensing of an anti-cancer drug. *Materials Science & Engineering C*, 98, 826–833. https://doi.org/10.1016/j.msec.2019.01.041.

Sharma, N., Das, G.S. and Yun, K., 2020. Green synthesis of multipurpose carbon quantum dots from red cabbage and estimation of their antioxidant potential and bio-labeling activity. *Applied Microbiology and Biotechnology*, 104(16), 7187–7200. https://doi.org/10.1007/s00253-020-10726-5.

Sharma, N., Sharma, I. and Bera, M.K., 2022. Microwave-assisted green synthesis of carbon quantum dots derived from *Calotropis gigantea* as a fluorescent probe for bioimaging. *Journal of Fluorescence*, 32(3), 1039–1049. https://doi.org/10.1007/s10895-022-02923-4.

Sharma, V., Tiwari, P. and Mobin, S.M., 2017. Sustainable carbon-dots: recent advances in green carbon dots for sensing and bioimaging. *Journal of Materials Chemistry B*, 5, 8904–8924. https://doi.org/10.1039/C7TB02484C.

Shi, L. et al., 2017. Green-fluorescent nitrogen-doped carbon nanodots for biological imaging and paper-based sensing. *Analytical Methods*, 9, 2197–2204. https://doi.org/10.1039/C7AY00163K.

Singh, A.K., Singh, V.K., Singh, M., Singh, P., Khadim, S.R., Singh, U., Koch, B., Hasan, S.H. and Asthana, R.K., 2019. One pot hydrothermal synthesis of fluorescent NP-carbon dots derived from *Dunaliella salina* biomass. *Journal of Photochemistry and Photobiology A: Chemistry*, 376, 63–71. https://doi.org/10.1016/j.jphotochem.2019.02.023.

Singh, V. et al., 2018. Biocompatible fluorescent carbon quantum dots prepared from beetroot extract for in vivo live imaging in *C. elegans* and BALB/c mice. *Journal of Materials Chemistry B*, [online]. https://doi.org/10.1039/C8TB00503F.

Smrithi, S.P. et al., 2021. Carbon dots derived from *Beta vulgaris*: evaluation of its potential as antioxidant and anticancer agent. *Nanotechnology*, 33(4). https://doi.org/10.1088/1361-6528/ac30f1.

Sousa, D.A. et al., 2022. Luminescent carbon dots from wet olive pomace: structural insights, photophysical properties and cytotoxicity. *Molecules*, 27, 6768. https://doi.org/10.3390/molecules27196768.

Sturabotti, E., Camilli, A., Leonelli, F. and Vetica, F., 2024. Carbon dots as bioactive antifungal nanomaterials. *ChemMedChem*, 19(23), e202400463. https://doi.org/10.1002/cmdc.202400463.

Sun, Y.P. et al., 2006. Quantum-sized carbon dots for bright and colorful photoluminescence. *Journal of the American Chemical Society*, 128(24), 7756–7757. https://doi.org/10.1021/ja062677d.

Swami, S., 2024. Captivating nano sensors for mercury detection: a promising approach for monitoring of toxic mercury in environmental samples. *RSC Advances*, 18907–18941. https://doi.org/10.1039/d4ra02787f.

Tadesse, A., Hagos, M., RamaDevi, D., Basavaiah, K. and Belachew, N., 2020. Fluorescent nitrogen-doped carbon quantum dots derived from citrus lemon juice: green synthesis, mercury(II) ion sensing, and live cell imaging. *ACS Omega*, 5(8), 3889–3898. https://doi.org/10.1021/acsomega.9b03175.

Tejwan, N., Saha, S.K. and Das, J., 2019. Multifaceted applications of green carbon dots synthesized from renewable sources. *Advances in Colloid and Interface Science*, Article 102046. https://doi.org/10.1016/j.cis.2019.102046.

Thongsai, N., Jaiyong, P., Kladsomboon, S. and In, I., 2019a. Utilization of carbon dots from jackfruit for real-time sensing of acetone vapor. *Applied Surface Science*, 487, 1233–1244. https://doi.org/10.1016/j.apsusc.2019.04.269.

Thongsai, N., Tanawannapong, N., Praneerad, J., Kladsomboon, S., Jaiyong, P. and Paoprasert, P., 2019b. Real-time detection of alcohol vapors and volatile organic compounds via optical electronic nose using carbon dots from rice husk. *Colloids and Surfaces A*, 560, 278–287. https://doi.org/10.1016/j.colsurfa.2018.09.077.

Tuna, M., Yanalak, G., Arslan, G. and Hatay, I., 2020. Green preparation of carbon quantum dots using *Ginkgo biloba* to sensitize TiO_2. *Materials Science in Semiconductor Processing*, 109, 104945. https://doi.org/10.1016/j.mssp.2020.104945.

Tyagi, A., et al. 2016. Green synthesis of carbon quantum dots from lemon peel waste : applications in sensing and photocatalysis. *RSC Advances*, 6, 72423–72432. https://doi.org/10.1039/c6ra10488f.

Usman, M. and Cheng, S., 2024. Recent trends and advancements in green synthesis of biomass-derived carbon dots. *Eng*, 5, 2223–2263. https://doi.org/10.3390/eng5030116.

Vijeata, A. et al., 2023. Biogenic synthesis of highly fluorescent carbon dots using *Azadirachta indica* leaves: an eco-friendly approach with enhanced photocatalytic degradation efficiency towards Malachite green. *Chemosphere*, 341, 139946. https://doi.org/10.1016/j.chemosphere.2023.139946.

Wang, C. et al., 2023. An overview of the potential of food-based carbon dots for biomedical applications. *International Journal of Molecular Sciences*, 24, 16579. https://doi.org/10.3390/ijms242316579.

Wang, H. et al., 2022. Inhibition of UV-B stress in lettuce through enzyme-like *Scutellaria baicalensis* carbon dots. *Ecotoxicology and Environmental Safety*, 246, 114177. https://doi.org/10.1016/j.ecoenv.2022.114177.

Wang, M. et al., 2019a. Green synthesis of carbon dots using the flowers of *Osmanthus fragrans* (Thunb.) Lour. as precursors: application in Fe3+ and ascorbic acid determination and cell imaging. *Analytical and Bioanalytical Chemistry*, 411(12), 2715–2727. https://doi.org/10.1007/s00216-019-01712-6.

Wang, X., Feng, Y., Dong, P. and Huang, J., 2019b. A mini review on carbon quantum dots: preparation, properties, and electrocatalytic application. *Frontiers in Chemistry*, 7, 1–9. https://doi.org/10.3389/fchem.2019.00671.

Wang, X.Y. et al., 2020. Crown daisy leaf waste-derived carbon dots: a simple and green fluorescent probe for copper ion. Surface and Interface Analysis, 52, 148–155. https://doi.org/10.1002/sia.6733.

Wang, Y., Wang, S., Xie, T. and Cao, J., 2020. Activated carbon derived from waste tangerine seed for the high-performance adsorption of carbamate pesticides from water and plant. *Bioresource Technology*, 316, 123929. https://doi.org/10.1016/j.biortech.2020.123929.

Wareing, T.C., Gentile, P. and Phan, A.N., 2021. Biomass-based carbon dots: current development and future perspectives. *ACS Nano*, 15, 15471–15493. https://doi.org/10.1021/acsnano.1c03886.

Wei, Z. et al., 2019. Green synthesis of nitrogen and sulfur co-doped carbon dots from *Allium fistulosum* for cell imaging. *New Journal of Chemistry*, 43(2), 718–723. https://doi.org/10.1039/C8NJ05783D.

Wu, H., Lu, S. and Yang, B., 2022. Carbon-dot-enhanced electrocatalytic hydrogen evolution. *Accounts of Materials Research*, 3, 319. https://doi.org/10.1021/accountsmr.1c00194.

Xia, C., Zhu, S., Feng, T., Yang, M. and Yang, B., 2019. Evolution and synthesis of carbon dots: from carbon dots to carbonized polymer dots. *Advanced Science*, [online]. https://doi.org/10.1002/advs.201901316.

Xie, Z., Sun, J., Zhou, Z. and Xia, S., 2021. Lotus seedpod-based carbon quantum dots: preparation, characterization and application for Fe(III) detection. *Materials Technology*, 55, 141. https://doi.org/10.17222/mit.2020.160.

Xu, D., Li, Y., Li, N., Lei, F., Liu, J., Shi, Y., Yin, L. and Zhang, L., 2020. A facile synthesis of CDs from quinoa for nanosensors and bio-imaging. *Nano Express*, 1(2), 020001. httpss://doi.org/10.1088/2632-959X/ab9f53.

Xu, X. et al., 2004. Electrophoretic analysis and purification of fluorescent single-walled carbon nanotube fragments. *Journal of the American Chemical Society*, 126(40), 12736–12737. https://doi.org/10.1021/ja040082h.

Yadav, P., Upadhyay, R., Kumar, D., Bano, D., Chandra, S., Jit, S. and Hasan, S., 2021. Synthesis of green fluorescent carbon quantum dots from *Ficus benghalensis* latex. *New Journal of Chemistry*, 45, 12549. https://doi.org/10.1039/D1NJ01655E.

Yuan, T. et al., 2023. Carbon quantum dots with near-unity quantum yield bandgap emission for electroluminescent light-emitting diodes. *Angewandte Chemie International Edition*, 62. https://doi.org/10.1002/anie.202218568.

Zhai, Y. et al., 2021. Carbon dots as new building blocks for electrochemical energy storage and electrocatalysis. *Advanced Energy Materials*, [online]. https://doi.org/10.1002/aenm.202103426.

Zhang, Y.P., Ma, J.M., Yang, Y.S., Ru, J.X., Liu, X.Y., Ma, Y. and Guo, H.C., 2019. Synthesis of nitrogen-doped graphene quantum dots (N-GQDs) from marigold. *Spectrochimica Acta Part A*, 217, 60. https://doi.org/10.1016/j.saa.2019.03.044.

Zhao, S. et al., 2015. Green synthesis of bifunctional fluorescent carbon dots from garlic for cellular imaging and free radicals scavenging. *ACS Applied Materials & Interfaces*, 7, 1944–8244. https://doi.org/10.1021/acsami.5b03228h.

Zhou, C., Wu, S., Qi, S., Song, W. and Sun, C., 2021. Facile and high-yield synthesis of N-doped carbon quantum dots from biomass quinoa saponin. *Journal of Analytical Methods in Chemistry*, 2021, 9732364. https://doi.org/10.1155/2021/9732364.

Zhou, J., Sheng, Z., Han, H., Zou, M. and Li, C., 2012. Facile synthesis of fluorescent carbon dots using watermelon peel as a carbon source. *Materials Letters*, 66(1), 222–224. https://doi.org/10.1016/j.matlet.2011.08.081.

Zhu, S. et al., 2013. Highly photoluminescent carbon dots for multicolor patterning, sensors, and bioimaging. *Angewandte Chemie International Edition*, 52(14), 3953–3957. https://doi.org/10.1002/anie.201300519.

Zhu, W. et al., 2022. Screening of multifunctional fruit carbon dots for fluorescent labeling and sensing in living immune cells and zebrafishes. *Mikrochimica Acta*, 189(6), 223. https://doi.org/10.1007/s00604-022-05318-1.

Zulfajri, M. et al., 2020. Plant part-derived carbon dots for biosensing. *Biosensors*, 10, 68. https://doi.org/10.3390/bios10060068.

Zuo, J., Jiang, T., Zhao, X., Xiong, X., Xiao, S. and Zhu, Z., 2015. Preparation and application of fluorescent carbon dots. *Journal of Nanomaterials*, 2015, Article ID 787862. https://doi.org/10.1155/2015/787862.

9 Phyto-Assisted Biogenic Nanoparticles of d^8 (Nickel, Palladium, and Platinum) System

An Advancement in Medicinal Chemistry

Sayantani Paul, Atif Aziz Chowdhury, and Nilendu Basak

9.1 INTRODUCTION

Metal nanoparticles (NPs) have been extensively popularized since the late 20th century because of their diverse properties. Various sectors such as cosmetics industries, biofuel production, food processing, medicinal treatment, and tissue engineering require the bio-compatibility of NPs and nanocomposites (Singh et al., 2021). The rising need for environmentally sustainable and economically viable NP synthesis has motivated researchers to investigate green nanotechnology approaches. Other approaches such as conventional chemical methodologies, despite being efficient, often use harmful precursors, produce large waste byproducts, and pose major ecological and human health risks. The green synthetic techniques are taken into account due to their minimized environmental effects, cost-effectiveness, eco-friendliness, and biocompatibility nature. The shape and size of NPs are controlled through their synthesis process which can impact their functionality to a substantial degree (Bogireddy et al., 2021; Dinesh et al., 2017; Naseer et al., 2020). Scientists can use natural materials and technologies to synthesize NPs not only well suited to the demands of particular applications but also with low environmental burdens, facilitating sustainable development in a wide range of industries. These advancements in the production of NPs introduce engaging opportunities for novel applications in medicinal sectors that involve drug delivery, image processing, also medical regeneration (Abdullah et al., 2023; Khan et al., 2014; Tahir et al., 2016a). These NPs have potential applications within biomedical disciplines including cancer therapy, anti-diabetic, anti-fungal, and antibacterial activities (Adams et al., 2014; Armijo et al., 2020). These techniques possibly lessen toxicity from conventional medicinal

DOI: 10.1201/9781003559900-9

systems while improving the effectiveness of therapies. Researchers are using the natural capacities of these organisms to create more effective and safer treatment options.

The production and study of metal NPs are becoming increasingly focused on non-toxic d^8 metals, such as platinum (Pt), palladium (Pd), and nickel (Ni) NPs. The d^8 metals are particularly interesting due to their exceptional physicochemical properties and potential biomedical applications (Alaqarbeh et al., 2023). Although Ni-NPs are the least studied among the other NPs, the size and form of Pt and Pd NPs can be modified through various biosynthetic techniques. Additionally, d^8 metal NPs produced from plant extracts had special therapeutic uses; they could be modified by the use of diverse sources of phytochemicals, such as proteins, alcohols, sugars, amides, alkaloids, phenols, terpenoids, quinines, and flavonoids (Naseer et al., 2020). Although Pt has excellent refractory properties, it was not widely used until the 16th century (Armijo et al., 2020). These metal NPs are very useful in drug delivery, sensing, and as a good catalyst, due to their inert nature (Ghosh, 2019). Moreover, the diverse use of NPs can be found in the industrial and medicinal domains of dentistry, microwave technology, electroplating, and cancer treatment because of their special qualities, which allow for improved performance in chemical reactions. However, the cytotoxic, neurotoxic, and nephrotoxic effects of such NPs can be offset by their sustainable generation (Dinesh et al., 2017). Bacteria, fungi, and plants are examples of bio-factories that are essential to the production of such NPs. Their antibacterial and anticancer properties make them an interesting topic for further research. These bio-factories aid in the development of new medicinal agents in addition to offering a more environmentally friendly method of synthesizing NPs (Armijo et al., 2020; Dinesh et al., 2017; Ghosh, 2019). Further investigation in this field may lead to advancements in targeted drug delivery techniques, which could ultimately improve patient outcomes for a range of medical treatments. This study presents a brief overview of the phyto-derived Pt, Pd, and Ni-NPs, along with their implementation in the medicinal fields.

9.2 SYNTHESIS OF PHYTO-ASSISTED NPS

Many researchers have documented various synthetic techniques to produce d^8 metal NPs. These are usually categorized into two broad approaches:

Top-down method

The top-down method includes the synthesis of nanomaterials through the breaking of bulk metallic particles. The major benefit of this process is the development of NPs with homogeneous shape and size (Soundarrajan et al., 2012). Sputtering, chemical etching, thermal breakdown, and lithography are among the main physical processes used in this approach (Mafuné and Kondow, 2004; Park et al., 2010). These methods require high mechanical pressure, energy, evaporation, and condensation. The primary drawbacks of this method are high expenditure, radiation exposure, high temperature, and limited production (Figure 9.1).

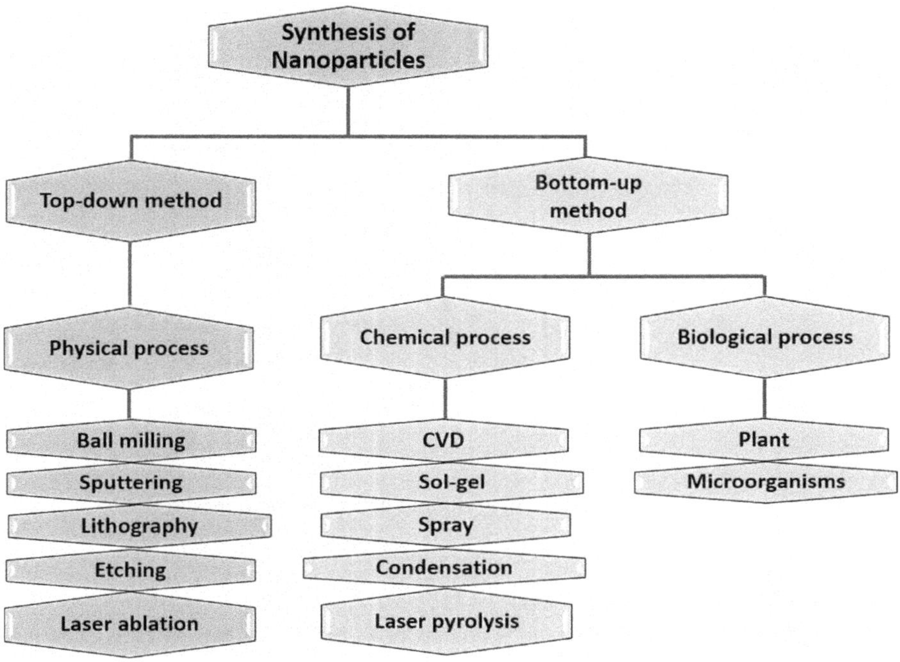

FIGURE 9.1 Bottom-up/top-down tactics for nanoparticle synthesis.

Bottom-up method

The bottom-up strategy initiates with individual atoms or molecules. Atoms are arranged in layers to form crystal planes, which are then stacked to create NPs (Nichols et al., 2006). This method includes both chemical and biological processes. Sol-gel technique, laser pyrolysis, plasma spraying, and chemical vapour deposition (CVD) are some examples of the chemical process. The biological method comprises phyto and microbe-based synthetic techniques (Figure 9.2). The synthesis of highly functionalized NPs with controlled shape and size could be monitored in this bottom-up method, which is the primary advantage of this technique. However, low-scale production and implementation of hazardous stabilizing agents are some major drawbacks of this method. This chapter is mainly focused on the plant-based synthesis of Ni/Pd/Pt-NPs.

9.2.1 Synthesis of Pt-NPs

The Pt-NPs could be synthesized utilizing plant extracts with biological derivatives. Synthesis of Pt-NPs from plant extract, i.e., leaf, seed, bark, and even whole plant extract, has been widely investigated to date.

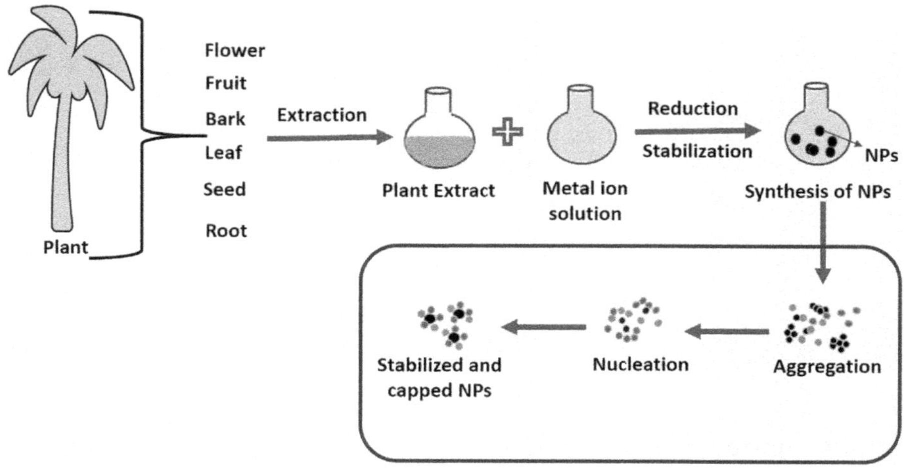

FIGURE 9.2 Schematic diagram of phyto-derived nanoparticle synthesis.

9.2.1.1 Pt-NPs from Leaf Extract

Many studies have documented the synthesis of Pt-NPs from leaf extracts. The primary ingredients of leaf extract include nitrogen, soluble polymeric polysaccharides, organic acids, and minerals. These substances not only serve as reducing agents but also as capping and stabilizing agents (Naseer et al., 2020). Kumar et al. (2019) established the fabrication of Pt-NPs from *Xanthium strumarium* leaf extract with a particle size of 22 nm, whereas neem leaf extract was utilized to synthesize Pt-NPs of sizes ranging from 5 to 50 nm (Thirumurugan et al., 2016). The leaf extract from *Fumariae herba* was used to synthesize the Pt-NPs with a size of 30 nm (Dobrucka, 2019). Soundarrajan et al. (2012) reported Pt-NPs with an average size of 23 nm using *Ocimum sanctum* leaf extract and $H_2PtCl_6 \cdot 6H_2O$ as the precursor. Rod-shaped crystalline Pt-NPs were synthesized using the leaf extract of *Anacardium occidentale* (Sheny et al., 2013). *Costus speciosus* leaf extract assisted with TiO_2 was utilized for the synthesis of spherical-shaped, 10–50 nm sized NPs (Surya et al., 2019). Moreover, synthesis of the NPs from *Bacopa monnieri* (Nellore et al., 2013), *Mentha piperita* (Yang et al., 2017), *Quercus glauca* (Karthik et al., 2016), *Lantana camara* (Mavukkandy et al., 2016), *Jatropa gossypifolia* and *J. glandulifera* (Jeyapaul et al., 2018) leaf extract were also investigated. Table 9.1 lists a few Pt-NPs synthesized using leaf extracts with their corresponding particle sizes.

9.2.1.2 Pt-NPs from Root/Bark/Fruit/Flower/Whole Plant Extract

Other plant components, including the stem, roots, bark, flowers, fruits, and even the peels and bio-derivatives (such as extracts, gums, etc.) have also been used to synthesize Pt-NPs. These plant parts contain proteins, enzymes, flavonoids, polyphenols, cannabinoids, terpenoids, glycosides, sugars, alcohols, aldehydes, amines, carbonyls, and other biomolecules (Roy et al., 2022). In general, the biomolecular components aid in the reduction and stabilization of Pt salts, allowing for the formation of

TABLE 9.1
Synthesis of Pt-NPs from Leaf Extract

Plant Species	General or Local Name	Particle Size (nm) and Shape	Synthesis Process and Parameter: Time	Temperature	pH	Extract Concentration	Activity and Property	References
X. strumarium	Cocklebur	22; cubic/ rectangular	1 h	100 °C	–	–	Antibacterial and antifungal	(Kumar et al., 2019)
A. indica	Neem	5–50; spherical	1 h	100 °C	–	–	–	(Thirumurugan et al., 2016)
F. herba	Fumitory	10–30; hexagonal and pentagonal	4 h	50 °C	–	–	Catalytic (MB and CV)	(Dobrucka, 2019)
D. kaki	Japanese persimmom	2–20; spherical	150 min	95 °C	–	20%	–	(Song et al., 2010)
O. sanctum	Tulsi	23; rectangular and triangular	1 h	100 °C		5%	–	(Soundarrajan et al., 2012)
B. monnieri	Water hyssop	5–20; spherical	2 h	80 °C		50 g	Antioxidant and neuro rescue (zebrafish)	(Nellore et al., 2013)
A. occidentale	Cashew	Crystalline and irregular rod	RT	–	6.0–9.0	50 mg	Catalytic (4-nitrophenol)	(Sheny et al., 2013)
E. crassipes	Water hyacinth	3.74–73.3; spherical	1 h	90 °C	–	10%	–	(Anyik and Oluwafemi, 2017)
M. piperita	Peppermint	54.3; spherical	2 h	60 °C	–	20 %	Cytotoxicity (colon cell; HCT-116)	(Yang et al., 2017)
Q. Glauca	Ring-cupped oak	5–15; spherical	30 min	80 °C	–	40 %	Electrocatalytic	(Karthik et al., 2016)
L. Camara	Lantana	35; spherical	8 min	95 °C	–	–	–	(Mavukkandy et al., 2016)
J. gossypifolia	Bellyache bush	20; different morphology	20 min	30 °C		10 mL	Antibacterial	(Jeyapaul et al., 2018)

(Continued)

TABLE 9.1 (*Continued*)
Synthesis of Pt-NPs from Leaf Extract

Plant Species	General or Local Name	Particle Size (nm) and Shape	Synthesis Process and Parameter				Activity and Property	References
			Time	Temperature	pH	Extract Concentration		
Maytenus royleanus	Prickly spike thorn	5; spherical	3 h	90 °C		10 mL	Anticancer (lung cancer cells A549), photocatalytic (MB)	(Ullah et al., 2017)
Barleria prionitis	Porcupine flower	1–2; irregular	5 h	100 °C	–	5%	Anticancer (breast adenocarcinoma cell MCF-7)	(Rokade et al., 2017)
Piper betle	Betel	2.1	10 min	Sunlight	–	–	Toxicity (*Daphnia magna*)	(Rajasekharreddy and Rani, 2014)
Withania somnifera	Ashwagandha	12; spherical	3 h	60 °C	–	30 g	Anti-diabetic	(Li et al., 2017)
Antigonon leptopus	Coral vine	5–190; spherical	–	95 °C	–	–	–	(Ganaie et al., 2018)
Mimosa pudica	Lajvanti	~1.4; spherical	15 min	RT	–	2 mL	–	(Henam et al., 2018)
Tragia involucrata	Indian stinging nettle	10; spherical and cubes	1 h	50 °C		10 mL	Antioxidant, antibacterial, and anticancer (cervical cancer HeLa cells)	(Selvi et al., 2020)

MB = Methylene blue; **CV** = Crystal violet; **RT** = Room temperature corresponding metal NPs. Stable Pt-NP dispersions have also been made from plant seeds, which are rich in cellulose, minerals, proteins, and carbohydrates. Woody vascular plants' bark, comprising about 10–20% of their weight, is composed of tannins, suberin, biopolymers, lignin, suberin, and polysaccharides (Ganaie et al., 2018; Selvi et al., 2020). Additionally, the effectiveness of these compounds for the creation of NPs has been studied. Table 9.2 provides a summary of the synthesis of Pt-NPs from plant extract and the associated particle sizes.

9.2.2 Synthesis of Pd-NPs

Plant-assisted Pd-NPs are synthesized using biomolecules such as proteins, polyols, flavones, flavonoids, and tannins (Kardel et al., 2013). The advanced X-ray diffraction (XRD) study reveals that plant-based Pd-NPs belong to the face-centered cubic lattice with a series of aggregated oval, sphere-shaped particles that range in size from 10–35 nm (Costantino et al., 2015). This synthetic technique requires high temperatures and pressures with comparatively larger production yield due to the greater stability and substantial abundance of phytochemicals (Shaik et al., 2017; Siddiqi and Husen, 2016). Different plant components that are used in the fabrication of Pd-NPs are listed in Table 9.3.

9.2.3 Synthesis of Ni-NPs

There is not much research in the literature that discusses the synthesis of Ni-NPs using plant extract. Abdullah et al. reported the fabrication of Ni-NPs from *Zingiber officinale*, i.e., ginger (Abdullah et al., 2023). This plant extract contains bioactive phytochemicals, gingerols and shogaols, which operate as both reducing agents and stabilizers for the synthesis of Ni-NPs of ~ 75 nm of particle size (Abdullah et al., 2023). Aqueous leaf extract of native *Aegle marmelos Correa* was used as a reducing, stabilizing, and capping agent for the phyto-synthesis of Ni-NPs. These phyto-fabricated triangular-shaped Ni-NPs belong to face-centered cubic lattices with particle sizes ranging from 80–100 nm (Angajala et al., 2014). *Ocimum sanctum* leaf extract was utilized to synthesize Ni-NPs of particle size 58 nm (Rameshthangam and Chitra, 2018).

9.3 MEDICINAL APPLICATION OF D^8-NPS

9.3.1 Anticancer Activity

The d^8-NPs, especially Pt and Pd-NPs, have the potential to revolutionize the treatment of cancer. Despite this, Ni-NPs' anticancer potential has not yet been investigated. Platinum-based anticancer drugs, like cisplatin, are widely used because of their cytotoxicity and antibacterial qualities. However, these medications have several adverse effects, such as peripheral nerve damage, anaemia, kidney damage, fatigue, nausea, and alopecia (Ghosh, 2019; Zhang et al., 2022). Several important anticancer herbs have been identified to date, e.g., *Azadirachta indica*, *Cannabis sativa*, *Phoenix dactylifera*, *Zingiber officinale*, etc. These plant extracts-mediated synthesized

TABLE 9.2
Synthesis of Pt-NPs from Root/Bark/Fruit/Flower/Whole Plant extract

				Synthesis Process and Parameter					
Plant Species	General or Local Name	Plant Part Used in Synthesis	Particle Size (nm) and Shape	Time	Temperature	pH	Extract Concentration	Activity and Property	References
N. sativa L.	Black cumin	Seed	1–6; spherical	2 d	75 °C	–	15 mL	Antibacterial and cytotoxicity (HeLa and breast cancer cell MDA-MB-231)	(Aygun et al., 2020)
C. sinensis	Orange	Fruit peel	~2.2; spherical	–	80 °C	2 and 7	9:1 ratio (precursor: extract)	Catalytic (p-nitrophenol)	(Castro et al., 2015)
Coffea arabica	Coffee	Seed	2; spherical with 2D superstructures	35 min	85 °C	6	0.6 g	Catalytic (4-nitrophenol and MB)	(Bogireddy et al., 2021)
Taraxacum laevigatum	Dandelion	Whole plant	2–7; spherical	10 min	90 °C	–	10 mL	Antibacterial	(Tahir et al., 2017)
Punica granatum	Pomegranate	Fruits peel	~20; spherical	24 h	RT	–	220 mL	Cytotoxicity (human breast adenocarcinoma cell line MCF-7)	(Şahin et al., 2018)
Terminalia chebula	Haritaki	Fruits	3.26; spherical	–	100 °C	–	10 mL	–	(Kumar et al., 2013)

Prunus yedoensis	Yoshino cherry	Gum	10–50; spherical (regular and irregular)	30 min	–	8.0	20 mL	Antifungal	(Velmurugan et al., 2016)
Cochlospermum gossypium	Sonali Simul or Yellow Silk Cotton	Gum	2.4; spherical	15 min	120 °C	8.0	10 mL	–	(Vinod et al., 2011)
Phoenix dactylifera	Date	Fruit	~2; spherical	7 h	100 °C	–	1–6 mL	Hepatoprotective	(Al-Radadi and Adam, 2020)
Punica granatum	Pomegranate	Peel	16–23; spherical	30 min	90 °C	–	100 mL	Catalytic (3-nitrophenol)	(Dauthal and Mukhopadhyay, 2015)
Heterotheca inuloides	Mexican arnica	Whole plant	7.1; quasi-spherical	1 h	–	–	975 mg	Catalytic (2-butyne-1,4-diol)	(Gama-Lara et al., 2019)

MB = Methylene blue

TABLE 9.3
Synthesis of Pd-NPs from Plant Extract

Plant Species	General or Local Name	Plant Part Used in Synthesis	Particle Size (nm) and Shape	Synthesis Process and Parameter				Activity and Property	References
				Time	Temperature	pH	Extract Concentration		
Pulicaria glutinosa	False fleabane	Whole plant	~20–25	2 h	90 °C		5 mL	–	(Khan et al., 2014)
Phyllanthus emblica	Amla or Indian gooseberry	Seed	30; spherical	3 h	60 °C		20 mL	Antibacterial, Hemolytic, toxicity (*Artemia salina*) cytotoxicity (HeLa cell)	(Dinesh et al., 2017)
Phoenix dactylifera	Date palm	Leaves	2–5; spherical	10 min	60 °C	–	10 mL	Antibacterial, antioxidant, Hemolytic, catalytic (4-NP)	(Tahir et al., 2016a)
Prunus x yedoensis	Yoshino cherry	Leaves	50–150; spherical	30 min	80 °C		5 mL	Antibacterial	(Manikandan et al., 2016)
Origanum vulgare L.	Oregano	Aerial part	2–20; spherical	120 min	90 °C		1 mL	Catalytic	(Shaik et al., 2017)
Cinnamom zeylanicum	Cinnamon	Bark	15–20; spherical	72 h	30 °C	<5	5 mL	–	(Sathishkumar et al., 2009)
Glycine max	Soybean	Leaves	~15; spherical	–	–	–	10 mL	–	(Petla et al., 2011)
Moringa oleifera	Drumstick tree	Peel	~27; spherical	2 min	65 °C (micro-waved-400 W)	–	1 g	Antibacterial and Hemolytic	(Surendra et al., 2016)
Melia azedarach	Chinaberry	Leaves	10–20; spherical	20 min	100 °C	–	1 mL	Antibacterial, antifungal, and larvicidal	(Bhakyaraj et al., 2017)
Filicium decipiens	Fern	Leaves	2–22; spherical	4 d	RT		10 mL	Antibacterial	(Sharmila et al., 2017)

Sapium sebiferum	Chinese tallow plant	Leaves	2–5; spherical	–	60 °C		10 mL	Antibacterial, Hemolytic, and photocatalytic (MB)	(Tahir et al., 2016b)
Citrullus lanatus	Watermelon	Rinds	~96.4; spherical	24 h	30 °C	6	10 mL	–	(Lakshmipathy et al., 2015)
Euphorbia thymifolia L.	Gulf Sandmat	Leaves	20–30; spherical	20 min	80 °C		15 mL	–	(Nasrollahzadeh et al., 2015)
Hibiscus sabdariffa L.	Hibiscus	Flower	5–8; spherical	5 h	100 °C	–	10 g	Catalytic	(Hekmati et al., 2017)
Camellia sinensis	Green tea	Leaves	7–10; spherical	1 h	100 °C		10 mL	Catalytic	(Veisi et al., 2017)
Poplar	Cottonwood	Leaves	2.2–6.8; spherical	12 h	25 °C		30 mL	Catalytic	(Liu and Bai, 2017)
Salvadora persica L.	Miswak	Root	2.2–15; spherical	2 h	90 °C		2.5 mL	Catalytic	(Khan et al., 2017)
Rosa canina	Dog rose	Fruit	7–10	2 h	100 °C		10 mL	Catalytic	(Veisi et al., 2016)

MB = Methylene blue

Pt-NPs are among the most promising candidates that are effective against a variety of cancer types as they contain essential oils, acids, alkaloids, and phytoncides. Additionally, Pt-NPs have a stronger anticancer impact when mixed with certain herbal ingredients, which removes the harmful effects of this metal (Alshatwi et al., 2015). For example, Pt-NPs fabricated from Ajwa and Barni Dates were used over HCT-116, MCF-7, and HePG-2 cancer cells. The efficacy of this was also compared with the commercial anticancer agent Doxorubicin HCl (Thirumurugan et al., 2016). Pt-NPs inhibited the proliferation of SiHa cells very efficiently and induced nuclear fragmentation and chromatin condensation (Naseer et al., 2020). Pt-NPs were also employed in cervical cancer treatment (Karthik et al., 2016). Because of their extremely low cytotoxicity, green-assisted Pt-NPs are highly suitable for use in gene transfer and drug discovery technologies (Nellore et al., 2013). Despite the lack of proper understanding of the underlying mechanism of the anticancer properties of Pt-NPs, a literature study by Alshatwi et al. (2015) revealed some aspects as the main mechanism for their anticancer activity. This includes: (a) the formation of reactive oxygen species (ROS), (b) disruption of the cell cycle, (c) DNA fragmentation, (d) disfunction of glutathione and mitochondria, (e) fabrication of several proinflammatory cytokines, and (f) elevation of multiple enzymes including lactate dehydrogenase and superoxide dismutase (Alshatwi et al., 2015).

Gorgizadeh et al. (2019) reported the anti-tumor activity of Ni-based nanocomposites applicable to melanoma tumors in the mouse model (Gorgizadeh et al., 2019). Furthermore, it was examined how well Ni-NPs and the anticancer medication Verbascoside (VB) worked to induce apoptosis in K562 cancer cells (Tailor et al., 2023). Balbín et al. (2014) first reported the cytotoxicity and anticancer activity of Pd-NPs against five different cancer cell lines such as A549, DLD-1, A253, A2780, and 8505C. The Pd-NPs derived from *Bauhinia variegates* prevent the proliferation of MCF-7 cells and showed an IC_{50} value of 41.37 g/mL (Vaghela et al., 2019). The underlying mechanism of anticancer activity of the Pd-NPs mainly involves the interactions of NPs with cellular proteins and DNA. The generation of a high level of lactate dehydrogenase, free radicals, and cellular damage hampers cell activities, while free radicals break down lipids and damage the protein chains and DNA molecules (Mubarak Ali et al., 2023).

9.3.2 Anti-Diabetic Activity

Pt-NPs have shown promise in the management of diabetes mellitus, offering potential advantages over conventional screening and therapeutic approaches. Streptozotocin-induced diabetic rats showed a significant drop in glucose levels following injection of Pt-NPs (Bag et al., 2023). Furthermore, Pt-NPs made from *P. salicifolium* function as anti-diabetic drugs primarily by blocking glucosidase (Hosny et al., 2022). The relevant process could be explained by the immediate connection between the phytoconstituents around Pt-NPs, the formation of excessive ROS, and/or unbalanced antioxidant defence systems (Hosny et al., 2022). Pt-NPs reduced fasting blood glucose levels, enhanced the genetic expression of antioxidant enzymes such as hemoxoigenase, glutathione peroxidase (GPx), and catalase (CAT), and improved the impaired ability to tolerate sugar in obese insulin-refractory type 2 diabetic KK-Ay mice (Shirahata et al., 2011).

9.3.3 Anti-Fungal Activity

Phyto-derived Pt-NPs have demonstrated significant antifungal activity against *A. parasiticus*, *A. flavus* (Godugu and Beedu, 2018), *Fusarium oxysporum*, *Colletotrichum acutatum*, and *Cladosporium fulvum* (Velmurugan et al., 2016). Additionally, Pt-NPs synthesized from *X. strumarium* are effective against *C. tropicalis*, *Candida albicans*, *Aspergillus flaves*, *C. parapsilosis*, and *A. niger* (Kumar et al., 2019). The antifungal activity of Pt nanocomposites is attributed to the induction of oxidative stress, leading to the production of ROS. Consequently, it disrupts the fungal mycelial membrane and shape, which eventually results in cell destruction and inhibition of fungal proliferation (Godugu and Beedu, 2018). In addition, Pt-NPs exhibited antifungal activity against *A. parasiticus* and *A. flavusfungi.* A literature survey showed that Pd-NPs acted as an antifungal agent against *Candida albicans* as evident from disk diffusion analysis (Sandhya et al., 2024), though standard ketoconazole was slightly more effective against this pathogen.

9.3.4 Antibacterial Activity

Applications of NPs as antimicrobial agents might result in the development of novel drugs to fight multi-drug resistant ESKAPE pathogens. Moreover, d^8-NPs, i.e., Pt, Pd, and Ni-NPs are well-known as antibacterial agents. Pt-NPs showed antibacterial activity against *V. parahaemolyticus*, as evident from the disk diffusion method (inhibition zone: 18 mm) (Palanisamy et al., 2023). Furthermore, Pd nanocrystals with sizes ranging from 1 nm exhibit strong antibacterial efficacy against both Gram-positive and Gram-negative bacteria, though these particles were more effective towards *E. coli* (Adams et al., 2014). Pd-NPs demonstrated a superior antibacterial effect against *S. aureus*, *E. coli*, *S. typhimurium*, and *B. subtilis* (Sandhya et al., 2024). Their efficacy is attributed to their oxidase and peroxidase activities. Specifically, the shape of Pd-NPs plays a crucial role in their antibacterial performance, and octahedron-shaped Pd-NPs have better antibacterial action. However, the cubes with facets are more successful in eliminating Gram-positive bacteria (Mubarak Ali et al., 2023). NiO-NPs extracted from *Dioscorea herbal* extract with nickel sulphate showed antibacterial activity against *S. aureus* and *K. pneumonia*, though in-depth research on the antimicrobial effects of NiO and its nanoform remains limited (Moulana Kareem et al., 2023).

The size, shape, and surface area influence the antibacterial activity of NPs. The underlying mechanism of antibacterial activities relies on these characteristics. Bacterial cell wall and cell membrane disruption are induced by the contact with the NPs. Mainly, when NPs interact with the cell membrane, they change the shape and increase permeability, metabolic disturbances, and mutagenesis of cell components leading to cell death (Palanisamy et al., 2023). Furthermore, the advantage of the use of phyto-derived NPs over chemically synthesized NPs is the presence

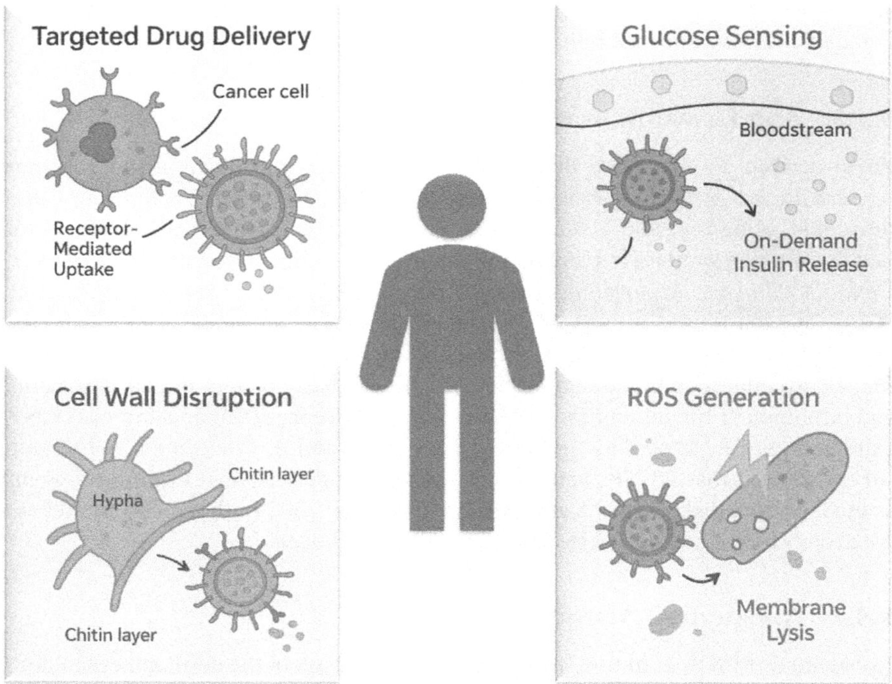

FIGURE 9.3 Various medicinal applications of phyto-derived nanoparticles.

of phytochemicals, which enhance the antibacterial activity of the NPs against Gram-positive and Gram-negative pathogens (Sandhya et al., 2024). An overview of the use of the phyto-derived NPs is demonstrated in Figure 9.3.

9.4 FUTURE ASPECT AND CONCLUSION

In medicinal chemistry, nanotechnology has emerged as a transformative solution. Biogenic NPs have created a novel path that includes applications in drug administration, microbial control, and the management of non-communicable diseases. Metal NPs, especially phyto-derived d^8-NPs, have special physicochemical characteristics that can be modified using eco-friendly methods, increasing their usefulness in such biological applications. The use of natural bio-factories like plants, fungi, and bacteria not only reduces environmental and health risks but also enables precise control over NP shape and size. The d^8 metals, especially Pt and Pd, show significant promise due to their inert nature and versatility in medical and industrial fields. Furthermore, the labour-intensive and time-consuming process of microbial culturing is avoided in the phyto-mediated production of biogenic NPs. On top of that, plant extracts improve biocompatibility by reducing metal ions, stabilizing the resultant NPs, and removing biosafety hazards related to endotoxins or pathogenic microorganisms.

Despite continuous progress in NP-based approaches in medicinal chemistry, there are still several challenges that need to be addressed in some sectors specifically. Their mechanism in the management of pathogens, especially the MDR pathogens, and their long-term impacts lack proper understanding. Furthermore, the exposure of NPs in the environment may disrupt ecosystems and beneficial microbiota, requiring studies on ecological impact and microbial balance. Therefore, evaluating long-term biocompatibility and toxicity is also crucial to ensure human and environmental safety. To increase effectiveness and reduce side effects, physicochemical characteristics such as size, shape, charge, and coating must be optimized. Targeted delivery is still difficult and necessitates sophisticated technologies for site-specific activity and controlled release. To handle the safety, disposal, and categorization of NPs as possible biohazards, regulatory frameworks must change. Inhibiting horizontal gene transfer is another way that NPs can be made to lessen the transmission of resistance genes. Additionally, surface alterations can decrease immunological detection, increasing circulation duration and enhancing therapeutic results. For antimicrobial NPs to be responsibly incorporated into clinical and environmental applications, these scientific and regulatory issues must be resolved. Despite having a lower carbon footprint than microbial alternatives, differences in plant metabolites necessitate standardized processes and sustainable procurement techniques. Despite being less studied, Ni-NPs have potential uses in the future. Utilizing biological resources to advance green nanotechnology helps create safer, more effective therapeutic agents and opens the door for novel techniques in targeted medicine.

9.5 DECLARATION

The authors declare that AI tools (Grammarly, Quillbot, and ChatGPT) were used in the writing process only to improve readability and language checking, not for generating text. After using the tools authors have reviewed and revised the content to maintain the accuracy.

REFERENCES

Abdullah, Hussain, T., Faisal, S., Rizwan, M., Almostafa, M.M., Younis, N.S., Yahya, G., 2023. Zingiber officinale rhizome extracts mediated Ni nanoparticles and its promising biomedical and environmental applications. *BMC Complementary Medicine and Therapies* 23, 1–18. https://doi.org/10.1186/s12906-023-04182-7

Adams, C.P., Walker, K.A., Obare, S.O., Docherty, K.M., 2014. Size-dependent antimicrobial effects of novel palladium nanoparticles. *PLoS ONE* 9, e85981. https://doi.org/10.1371/journal.pone.0085981

Alaqarbeh, M., Adil, S.F., Ghrear, T., Khan, M., Bouachrine, M., Al-Warthan, A., 2023. Recent progress in the application of palladium nanoparticles: A review. *Catalysts* 13, 1343. https://doi.org/10.3390/catal13101343

Al-Radadi, N.S., Adam, S.I.Y., 2020. Green biosynthesis of Pt-nanoparticles from Anbara fruits: Toxic and protective effects on CCl4 induced hepatotoxicity in Wister rats. *Arabian Journal of Chemistry* 13, 4386–4403. https://doi.org/10.1016/j.arabjc.2019.08.008

Alshatwi, A.A., Athinarayanan, J., Vaiyapuri Subbarayan, P., 2015. Green synthesis of platinum nanoparticles that induce cell death and G2/M-phase cell cycle arrest in human cervical cancer cells. *Journal of Materials Science: Materials in Medicine* 26, 7. https://doi.org/10.1007/s10856-014-5330-1

Angajala, G., Ramya, R., Subashini, R., 2014. In-vitro anti-inflammatory and mosquito larvicidal efficacy of nickel nanoparticles phytofabricated from aqueous leaf extracts of Aegle marmelos Correa. *Acta Tropica* 135, 19–26. https://doi.org/10.1016/j.actatropica.2014.03.012

Anyik, J.L., Oluwafemi, O.S., 2017. Plant-mediated synthesis of platinum nanoparticles using water hyacinth as an efficient biomatrix source—An eco-friendly development. *Materials Letters* 196, 141–144. https://doi.org/10.1016/j.matlet.2017.03.047

Armijo, L.M., Wawrzyniec, S.J., Kopciuch, M., Brandt, Y.I., Rivera, A.C., Withers, N.J., Cook, N.C., Huber, D.L., Monson, T.C., Smyth, H.D.C., Osiński, M., 2020. Antibacterial activity of iron oxide, iron nitride, and tobramycin conjugated nanoparticles against Pseudomonas aeruginosa biofilms. *Journal of Nanobiotechnology* 18, 35. https://doi.org/10.1186/s12951-020-0588-6

Aygun, A., Gülbagca, F., Ozer, L.Y., Ustaoglu, B., Altunoglu, Y.C., Baloglu, M.C., Atalar, M.N., Alma, M.H., Sen, F., 2020. Biogenic platinum nanoparticles using black cumin seed and their potential usage as antimicrobial and anticancer agent. *Journal of Pharmaceutical and Biomedical Analysis* 179, 112961. https://doi.org/10.1016/j.jpba.2019.112961

Bag, J., Mukherjee, S., Tripathy, M., Mohanty, R., Shendha, P.K., Hota, G., Mishra, M., 2023. Platinum as a novel nanoparticle for wound healing model in Drosophila melanogaster. *Journal of Cluster Science* 34, 1087–1098. https://doi.org/10.1007/s10876-022-02292-9

Balbín, A., Gaballo, F., Ceballos-Torres, J., Prashar, S., Fajardo, M., Kaluđerović, G.N., Gómez-Ruiz, S., 2014. Dual application of Pd nanoparticles supported on mesoporous silica SBA-15 and MSU-2: Supported catalysts for C–C coupling reactions and cytotoxic agents against human cancer cell lines. *RSC Advances* 4, 54775–54787. https://doi.org/10.1039/C4RA11759J

Bhakyaraj, K., Kumaraguru, S., Gopinath, K., Sabitha, V., Kaleeswarran, P.R., Karthika, V., Sudha, A., Muthukumaran, U., Jayakumar, K., Mohan, S., Arumugam, A., 2017. Eco-friendly synthesis of palladium nanoparticles using Melia azedarach leaf extract and their evaluation for antimicrobial and larvicidal activities. *Journal of Cluster Science* 28, 463–476. https://doi.org/10.1007/s10876-016-1114-8

Bogireddy, N.K.R., Pal, U., Kumar, M.K., Domínguez, J.M., Gomez, L.M., Agarwal, V., 2021. Green fabrication of 2D platinum superstructures and their high catalytic activity for mitigation of organic pollutants. *Catalysis Today* 360, 185–193. https://doi.org/10.1016/j.cattod.2019.06.044

Castro, L., Blázquez, M.L., González, F., Muñoz, J.Á., Ballester, A., 2015. Biosynthesis of silver and platinum nanoparticles using orange peel extract: Characterisation and applications. *IET Nanobiotechnology* 9, 252–258. https://doi.org/10.1049/iet-nbt.2014.0063

Costantino, F., Vivani, R., Bastianini, M., Ortolani, L., Piermatti, O., Nocchetti, M., Vaccaro, L., 2015. Accessing stable zirconium carboxy-aminophosphonate nanosheets as support for highly active Pd nanoparticles. *Communications Chemistry* 51, 15990–15993. https://doi.org/10.1039/C5CC06292F

Dauthal, P., Mukhopadhyay, M., 2015. Biofabrication, characterization, and possible bio-reduction mechanism of platinum nanoparticles mediated by agro-industrial waste and their catalytic activity. *Journal of Industrial and Engineering Chemistry* 22, 185–191. https://doi.org/10.1016/j.jiec.2014.07.009

Dinesh, M., Roopan, S.M., Selvaraj, C.I., Arunachalam, P., 2017. Phyllanthus emblica seed extract mediated synthesis of PdNPs against antibacterial, heamolytic and cytotoxic studies. *Journal of Photochemistry and Photobiology B: Biology* 167, 64–71. https://doi.org/10.1016/j.jphotobiol.2016.12.012

Dobrucka, R., 2019. Biofabrication of platinum nanoparticles using Fumariae herba extract and their catalytic properties. *Saudi Journal of Biological Sciences* 26, 31–37. https://doi.org/10.1016/j.sjbs.2016.11.012

Gama-Lara, S.A., Natividad, R., Vilchis-Nestor, A.R., López-Castañares, R., García-Orozco, I., Gonzalez-Pedroza, M.G., Morales-Luckie, R.A., 2019. Ultra-small platinum nanoparticles with high catalytic selectivity synthesized by an eco-friendly method supported on natural hydroxyapatite. *Catalysis Letters* 149, 3447–3453. https://doi.org/10.1007/s10562-019-02919-z

Ganaie, S.U., Abbasi, T., Abbasi, S.A., 2018. Biomimetic synthesis of platinum nanoparticles utilizing a terrestrial weed *Antigonon leptopus*. *Particulate Science and Technology* 36, 681–688. https://doi.org/10.1080/02726351.2017.1292336

Ghosh, S., 2019. Cisplatin: The first metal based anticancer drug. *Bioorganic Chemistry* 88, 102925. https://doi.org/10.1016/j.bioorg.2019.102925

Godugu, D., Beedu, S.R., 2018. Biopolymer-mediated synthesis and characterisation of platinum nanocomposite and its anti-fungal activity against *A. parasiticus* and *A. flavus*. *Micro & Nano Letters* 13, 1491–1496. https://doi.org/10.1049/mnl.2018.5097

Gorgizadeh, M., Azarpira, N., Lotfi, M., Daneshvar, F., Salehi, F., Sattarahmady, N., 2019. Sonodynamic cancer therapy by a nickel ferrite/carbon nanocomposite on melanoma tumor: In vitro and in vivo studies. *Photodiagnosis and Photodynamic Therapy* 27, 27–33. https://doi.org/10.1016/j.pdpdt.2019.05.023

Hekmati, M., Bonyasi, F., Javaheri, H., Hemmati, S., 2017. Green synthesis of palladium nanoparticles using *Hibiscus sabdariffa* L. flower extract: Heterogeneous and reusable nanocatalyst in Suzuki coupling reactions. *Applied Organom Chemis* 31, e3757. https://doi.org/10.1002/aoc.3757

Henam, P.S., Heikham, F.D., Henam, S.D., 2018. Sustainable synthesis of ultrasmall biogenic platinum nanoparticles for selective aqueous phase conversion of glucose and effective hydrogen peroxide decomposition. *Industrial & Engineering Chemistry Research* 57, 5190–5194. https://doi.org/10.1021/acs.iecr.7b05347

Hosny, M., Fawzy, M., El-Fakharany, E.M., Omer, A.M., El-Monaem, E.M.A., Khalifa, R.E., Eltaweil, A.S., 2022. Biogenic synthesis, characterization, antimicrobial, antioxidant, antidiabetic, and catalytic applications of platinum nanoparticles synthesized from Polygonum salicifolium leaves. *Journal of Environmental Chemical Engineering* 10, 106806. https://doi.org/10.1016/j.jece.2021.106806

Jeyapaul, U., Kala, M.J., Bosco, A.J., Piruthiviraj, P., Easuraja, M., 2018. An eco-friendly approach for synthesis of platinum nanoparticles using leaf extracts of Jatropa gossypifolia and Jatropa glandulifera and its antibacterial activity. *Oriental Journal of Chemistry* 34, 783–790.

Kardel, M., Taube, F., Schulz, H., Schütze, W., Gierus, M., 2013. Different approaches to evaluate tannin content and structure of selected plant extracts—review and new aspects. *Journal of Applied Botany and Food Quality* 86. https://doi.org/10.5073/JABFQ.2013.086.021

Karthik, R., Sasikumar, R., Chen, S.-M., Govindasamy, M., Kumar, J.V., Muthuraj, V., 2016. Green synthesis of platinum nanoparticles using extract and its electrochemical oxidation of hydrazine in water samples. *International Journal of Electrochemical Science* 11, 8245–8255. https://doi.org/10.20964/2016.10.62

Khan, Mujeeb, Albalawi, G.H., Shaik, M.R., Khan, Merajuddin, Adil, S.F., Kuniyil, M., Alkhathlan, H.Z., Al-Warthan, A., Siddiqui, M.R.H., 2017. *Miswak* mediated green synthesized palladium nanoparticles as effective catalysts for the Suzuki coupling reactions in aqueous media. *Journal of Saudi Chemical Society* 21, 450–457. https://doi.org/10.1016/j.jscs.2016.03.008

Khan, Mujeeb, Khan, Merajuddin, Kuniyil, M., Adil, S.F., Al-Warthan, A., Alkhathlan, H.Z., Tremel, W., Tahir, M.N., Siddiqui, M.R.H., 2014. Biogenic synthesis of palladium nanoparticles using Pulicaria glutinosa extract and their catalytic activity towards the Suzuki coupling reaction. *Dalton Transactions* 43, 9026–9031. https://doi.org/10.1039/C3DT53554A

Kumar, K.M., Mandal, B.K., Tammina, S.K., 2013. Green synthesis of nano platinum using naturally occurring polyphenols. *RSC Advances* 3, 4033–4039. https://doi.org/10.1039/C3RA22959A

Kumar, P.V., Jelastin Kala, S.M., Prakash, K.S., 2019. Green synthesis derived Pt-nanoparticles using Xanthium strumarium leaf extract and their biological studies. *Journal of Environmental Chemical Engineering* 7, 103146. https://doi.org/10.1016/j.jece.2019.103146

Lakshmipathy, R., Palakshi Reddy, B., Sarada, N.C., Chidambaram, K., Khadeer Pasha, Sk., 2015. Watermelon rind-mediated green synthesis of noble palladium nanoparticles: catalytic application. *Applied Nanoscience* 5, 223–228. https://doi.org/10.1007/s13204-014-0309-2

Li, Y., Zhang, J., Gu, J., Chen, S., 2017. Biosynthesis of polyphenol-stabilised nanoparticles and assessment of anti-diabetic activity. *Journal of Photochemistry and Photobiology B: Biology* 169, 96–100. https://doi.org/10.1016/j.jphotobiol.2017.02.017

Liu, G., Bai, X., 2017. Biosynthesis of palladium nanoparticles using Poplar leaf extract and its application in Suzuki coupling reaction. *IET Nanobiotechnology* 11, 310–316. https://doi.org/10.1049/iet-nbt.2016.0117

Mafuné, F., Kondow, T., 2004. Selective laser fabrication of small nanoparticles and nano-networks in solution by irradiation of UV pulsed laser onto platinum nanoparticles. *Chemical Physics Letters* 383, 343–347. https://doi.org/10.1016/j.cplett.2003.10.149

Manikandan, V., Velmurugan, P., Park, J.-H., Lovanh, N., Seo, S.-K., Jayanthi, P., Park, Y.-J., Cho, M., Oh, B.-T., 2016. Synthesis and antimicrobial activity of palladium nanoparticles from *Prunus × yedoensis* leaf extract. *Materials Letters* 185, 335–338. https://doi.org/10.1016/j.matlet.2016.08.120

Mavukkandy, M.O., Chakraborty, S., Abbasi, T., Abbasi, S.A., 2016. A clean-green synthesis of platinum nanoparticles utilizing a pernicious weed lantana (*Lantana Camara*). *American Journal of Engineering and Applied Sciences* 9, 84–90. https://doi.org/10.3844/ajeassp.2016.84.90

Moulana Kareem, Md., Hari Babu, Mr., Vijaya Lakshmi, G., 2023. Anticancer, antibacterial, antioxidant, and photo-catalytic activities of eco-friendly synthesized Ni nanoparticles. *Inorganic Chemistry Communications* 148, 110274. https://doi.org/10.1016/j.inoche.2022.110274

Mubarak Ali, D., Kim, H., Venkatesh, P.S., Kim, J.-W., Lee, S.-Y., 2023. A systemic review on the synthesis, characterization, and applications of palladium nanoparticles in biomedicine. *Applied Biochemistry and Biotechnology* 195, 3699–3718. https://doi.org/10.1007/s12010-022-03840-9

Naseer, A., Ali, A., Ali, S., Mahmood, A., Kusuma, H.S., Nazir, A., Yaseen, M., Khan, M.I., Ghaffar, A., Abbas, M., Iqbal, M., 2020. Biogenic and eco-benign synthesis of platinum nanoparticles (Pt NPs) using plants aqueous extracts and biological derivatives: environmental, biological and catalytic applications. *Journal of Materials Research and Technology* 9, 9093–9107. https://doi.org/10.1016/j.jmrt.2020.06.013

Nasrollahzadeh, M., Sajadi, S.M., Honarmand, E., Maham, M., 2015. Preparation of palladium nanoparticles using Euphorbia thymifolia L. leaf extract and evaluation of catalytic activity in the ligand-free Stille and Hiyama cross-coupling reactions in water. *New Journal of Chemistry* 39, 4745–4752. https://doi.org/10.1039/C5NJ00244C

Nellore, J., Pauline, C., Amarnath, K., 2013. *Bacopa monnieri* phytochemicals mediated synthesis of platinum nanoparticles and its neurorescue effect on 1-methyl 4-phenyl 1,2,3,6 tetrahydropyridine-induced experimental parkinsonism in Zebrafish. *Journal of Neurodegenerative Diseases* 2013, 1–8. https://doi.org/10.1155/2013/972391

Nichols, W.T., Sasaki, T., Koshizaki, N., 2006. Laser ablation of a platinum target in water. I. Ablation mechanisms. *Journal of Applied Physics* 100, 114911. https://doi.org/10.1063/1.2390640

Palanisamy, S., Anjali, R., Jeneeta, S., Mohandoss, S., Keerthana, D., Shin, I.-S., You, S., Prabhu, N.M., 2023. An effective bio-inspired synthesis of platinum nanoparticles using Caulerpa sertularioides and investigating their antibacterial and antioxidant activities. *Bioprocess and Biosystems Engineering* 46, 105–118. https://doi.org/10.1007/s00449-022-02816-7

Park, D.K., Lee, S.J., Lee, J.-H., Choi, M.Y., Han, S.W., 2010. Effect of polymeric stabilizers on the catalytic activity of Pt nanoparticles synthesized by laser ablation. *Chemical Physics Letters* 484, 254–257. https://doi.org/10.1016/j.cplett.2009.11.031

Petla, R.K., Vivekanandhan, S., Misra, M., Mohanty, A.K., Satyanarayana, N., 2011. Soybean (Glycine Max) leaf extract based green synthesis of palladium nanoparticles. *Journal of Biomaterials and Nanobiotechnology* 3, 14–19. https://doi.org/10.4236/jbnb.2012.31003

Rajasekharreddy, P., Rani, P.U., 2014. Biosynthesis and characterization of Pd and Pt nanoparticles using Piper betle L. plant in a photoreduction method. *Journal of Cluster Science* 25, 1377–1388. https://doi.org/10.1007/s10876-014-0715-3

Rameshthangam, P., Chitra, J.P., 2018. Synergistic anticancer effect of green synthesized nickel nanoparticles and quercetin extracted from Ocimum sanctum leaf extract. *Journal of Materials Science & Technology* 34, 508–522. https://doi.org/10.1016/j.jmst.2017.01.004

Rokade, S., Joshi, K., Mahajan, K., Tomar, G., Dubal, D., Parihar, vijay singh, kitture, R., Bellare, J., Ghosh, S., 2017. Novel anticancer platinum and palladium nanoparticles from Barleria prionitis. *Global Journal of Nanomedicine* 2. https://doi.org/10.19080/GJN.2017.02.555600

Roy, A., Khan, A., Ahmad, I., Alghamdi, S., Rajab, B.S., Babalghith, A.O., Alshahrani, M.Y., Islam, S., Islam, Md.R., 2022. Flavonoids a bioactive compound from medicinal plants and its therapeutic applications. *Biomed Research International* 2022, 5445291. https://doi.org/10.1155/2022/5445291

Şahin, B., Aygün, A., Gündüz, H., Şahin, K., Demir, E., Akocak, S., Şen, F., 2018. Cytotoxic effects of platinum nanoparticles obtained from pomegranate extract by the green synthesis method on the MCF-7 cell line. *Colloids and Surfaces B: Biointerfaces* 163, 119–124. https://doi.org/10.1016/j.colsurfb.2017.12.042

Sandhya, K., Bhagavanth Reddy, G., Ayodhya, D., Venkatesh, B., Kondaiah, S., Noorjahan, M., Yadagiri Swamy, P., Girija Mangatayaru, K., 2024. Ultrasound assisted biogenic synthesis of Palladium nanoparticles using Cocculus hirsutus leaf extract: antimicrobial, anti-inflammatory and catalytic activities. *Journal of Molecular Structure* 1306, 137848. https://doi.org/10.1016/j.molstruc.2024.137848

Sathishkumar, M., Sneha, K., Kwak, I.S., Mao, J., Tripathy, S.J., Yun, Y.-S., 2009. Phyto-crystallization of palladium through reduction process using *Cinnamom zeylanicum* bark extract. *Journal of Hazardous Materials* 171, 400–404. https://doi.org/10.1016/j.jhazmat.2009.06.014

Selvi, A.M., Palanisamy, S., Jeyanthi, S., Vinosha, M., Mohandoss, S., Tabarsa, M., You, S., Kannapiran, E., Prabhu, N.M., 2020. Synthesis of Tragia involucrata mediated platinum nanoparticles for comprehensive therapeutic applications: antioxidant, antibacterial and mitochondria-associated apoptosis in HeLa cells. *Process Biochemistry* 98, 21–33. https://doi.org/10.1016/j.procbio.2020.07.008

Shaik, M., Ali, Z., Khan, Mujeeb, Kuniyil, M., Assal, M., Alkhathlan, H., Al-Warthan, A., Siddiqui, M., Khan, Merajuddin, Adil, S., 2017. Green synthesis and characterization of palladium nanoparticles using Origanum vulgare L. extract and their catalytic activity. *Molecules* 22, 165. https://doi.org/10.3390/molecules22010165

Sharmila, G., Farzana Fathima, M., Haries, S., Geetha, S., Manoj Kumar, N., Muthukumaran, C., 2017. Green synthesis, characterization and antibacterial efficacy of palladium nanoparticles synthesized using *Filicium decipiens* leaf extract. *Journal of Molecular Structure* 1138, 35–40. https://doi.org/10.1016/j.molstruc.2017.02.097

Sheny, D.S., Philip, D., Mathew, J., 2013. Synthesis of platinum nanoparticles using dried Anacardium occidentale leaf and its catalytic and thermal applications. *Spectrochimica Acta Part A: Molecular and Biomolecular Spectroscopy* 114, 267–271. https://doi.org/10.1016/j.saa.2013.05.028

Shirahata, S., Hamasaki, T., Haramaki, K., Nakamura, T., Abe, M., Yan, H., Kinjo, T., Nakamichi, N., Kabayama, S., Teruya, K., 2011. Anti-diabetes effect of water containing hydrogen molecule and Pt nanoparticles. *BMC Proceedings* 5, P18, 1753-6561-5-S8-P18. https://doi.org/10.1186/1753-6561-5-S8-P18

Siddiqi, K.S., Husen, A., 2016. Green synthesis, characterization and uses of palladium/platinum nanoparticles. *Nanoscale Research Letters* 11, 482. https://doi.org/10.1186/s11671-016-1695-z

Singh, H., Bamrah, A., Bhardwaj, S.K., Deep, A., Khatri, M., Brown, R.J.C., Bhardwaj, N., Kim, K.-H., 2021. Recent advances in the application of noble metal nanoparticles in colorimetric sensors for lead ions. *Environmental Science: Nano* 8, 863–889. https://doi.org/10.1039/D0EN00963F

Song, J.Y., Kwon, E.-Y., Kim, B.S., 2010. Biological synthesis of platinum nanoparticles using Diopyros kaki leaf extract. *Bioprocess and Biosystems Engineering* 33, 159–164. https://doi.org/10.1007/s00449-009-0373-2

Soundarrajan, C., Sankari, A., Dhandapani, P., Maruthamuthu, S., Ravichandran, S., Sozhan, G., Palaniswamy, N., 2012. Rapid biological synthesis of platinum nanoparticles using Ocimum sanctum for water electrolysis applications. *Bioprocess and Biosystems Engineering* 35, 827–833. https://doi.org/10.1007/s00449-011-0666-0

Surendra, T.V., Roopan, S.M., Arasu, M.V., Al-Dhabi, N.A., Rayalu, G.M., 2016. RSM optimized *Moringa oleifera* peel extract for green synthesis of *M. oleifera* capped palladium nanoparticles with antibacterial and hemolytic property. *Journal of Photochemistry and Photobiology B: Biology* 162, 550–557. https://doi.org/10.1016/j.jphotobiol.2016.07.032

Surya, C., Arul John, N.A., Pandiyan, V., Ravikumar, S., Amutha, P., Sobral, A.J.F.N., Krishnakumar, B., 2019. Costus speciosus leaf extract assisted CS-Pt-TiO2 composites: Synthesis, characterization and their bio and photocatalytic applications. *Journal of Molecular Structure* 1195, 787–795. https://doi.org/10.1016/j.molstruc.2019.06.030

Tahir, K., Nazir, S., Ahmad, A., Li, B., Ali Shah, S.A., Khan, A.U., Khan, G.M., Khan, Q.U., Haq Khan, Z.U., Khan, F.U., 2016a. Biodirected synthesis of palladium nanoparticles using Phoenix dactylifera leaves extract and their size dependent biomedical and catalytic applications. *RSC Advances* 6, 85903–85916. https://doi.org/10.1039/C6RA11409A

Tahir, K., Nazir, S., Ahmad, A., Li, B., Khan, A.U., Khan, Z.U.H., Khan, F.U., Khan, Q.U., Khan, A., Rahman, A.U., 2017. Facile and green synthesis of phytochemicals capped platinum nanoparticles and in vitro their superior antibacterial activity. *Journal of Photochemistry and Photobiology B: Biology* 166, 246–251. https://doi.org/10.1016/j.jphotobiol.2016.12.016

Tahir, K., Nazir, S., Li, B., Ahmad, A., Nasir, T., Khan, A.U., Shah, S.A.A., Khan, Z.U.H., Yasin, G., Hameed, M.U., 2016b. *Sapium sebiferum* leaf extract mediated synthesis of palladium nanoparticles and in vitro investigation of their bacterial and photocatalytic activities. *Journal of Photochemistry and Photobiology B: Biology* 164, 164–173. https://doi.org/10.1016/j.jphotobiol.2016.09.030

Tailor, G., Chaudhary, J., Jandu, S., Chetna, Mehta, C., Yadav, M., Verma, D., 2023. A review on green route synthesized nickel nanoparticles: biological and photo-catalytic applications. *Results in Chemistry* 6, 101195. https://doi.org/10.1016/j.rechem.2023.101195

Thirumurugan, A., Aswitha, P., Kiruthika, C., Nagarajan, S., Christy, A.N., 2016. Green synthesis of platinum nanoparticles using Azadirachta indica—an eco-friendly approach. *Materials Letters* 170, 175–178. https://doi.org/10.1016/j.matlet.2016.02.026

Ullah, S., Ahmad, A., Wang, A., Raza, M., Jan, A.U., Tahir, K., Rahman, A.U., Qipeng, Y., 2017. Bio-fabrication of catalytic platinum nanoparticles and their in vitro efficacy against lungs cancer cells line (A549). *Journal of Photochemistry and Photobiology B: Biology* 173, 368–375. https://doi.org/10.1016/j.jphotobiol.2017.06.018

Vaghela, H., Shah, R., Pathan, A., 2019. Palladium nanoparticles mediated through Bauhinia variegata: potent in vitro anticancer activity against MCF-7 cell lines and antimicrobial assay. *CNM* 3, 168–177. https://doi.org/10.2174/2405461504666190131142303

Veisi, H., Rashtiani, A., Barjasteh, V., 2016. Biosynthesis of palladium nanoparticles using *Rosa canina* fruit extract and their use as a heterogeneous and recyclable catalyst for Suzuki–Miyaura coupling reactions in water. *Applied Organometallic Chemistry* 30, 231–235. https://doi.org/10.1002/aoc.3421

Veisi, H., Rostami, A., Shirinbayan, M., 2017. Greener approach for synthesis of monodispersed palladium nanoparticles using aqueous extract of green tea and their catalytic activity for the Suzuki–Miyaura coupling reaction and the reduction of nitroarenes. *Applied Organometallic Chemistry* 31, e3609. https://doi.org/10.1002/aoc.3609

Velmurugan, P., Shim, J., Kim, K., Oh, B.-T., 2016. *Prunus* × *yedoensis* tree gum mediated synthesis of platinum nanoparticles with antifungal activity against phytopathogens. *Materials Letters* 174, 61–65. https://doi.org/10.1016/j.matlet.2016.03.069

Vinod, V.T.P., Saravanan, P., Sreedhar, B., Devi, D.K., Sashidhar, R.B., 2011. A facile synthesis and characterization of Ag, Au and Pt nanoparticles using a natural hydrocolloid gum kondagogu (*Cochlospermum gossypium*). *Colloids and Surfaces B: Biointerfaces* 83, 291–298. https://doi.org/10.1016/j.colsurfb.2010.11.035

Yang, C., Wang, M., Zhou, J., Chi, Q., 2017. Bio-synthesis of peppermint leaf extract polyphenols capped nano-platinum and their in-vitro cytotoxicity towards colon cancer cell lines (HCT 116). *Materials Science and Engineering: C* 77, 1012–1016. https://doi.org/10.1016/j.msec.2017.04.020

Zhang, C., Xu, C., Gao, X., Yao, Q., 2022. Platinum-based drugs for cancer therapy and anti-tumor strategies. *Theranostics* 12, 2115–2132. https://doi.org/10.7150/thno.69424

10 Plant-Assisted Biogenic Synthesis of Medicinally Significant Zinc Oxide Nanoparticles

Hrishikesh Khodade, Deepak Shelke, Hiralal Sonawane, Mahadev Chambhare, Rahul Kamble, and Pankaj Kumar Chaurasia

10.1 INTRODUCTION

Nanotechnology is a fast growing area that deals with working on very tiny materials, ranging from 1 nm to 100 nm (Ke, 2011). When materials are this small they start showing special behavior that is quite different from how they act in larger form (Zhang et al., 2020). Nanomaterial shows distinct properties mainly because of their substantial surface to volume ration and pronounced quantum mechanical phenomena at the nanoscale (Ni et al., 2018). The advancement in nanotechnology has significantly transformed various sectors, especially medicine, facilitating the creation of advanced diagnostic methods, efficient drug delivery platforms and novel therapeutic strategies. Nanotechnology provides accurate treatments that directly target specific areas in the body. This approach helps the patients recover better and reduce unwanted side effects. Along with this, nanotechnology has led to the creation of new materials and small devices like nanosensors and tiny robots called nanorobots. These innovations have excellent potential for various uses in the future (Liu et al., 2023; Roberto Vega-Baudrit & Villalobos, 2017). Nanoparticles are a basic part of nanotechnology and can be made using different materials. Each material gives certain special properties to the particles. For example, metal nanoparticles like gold, silver, copper and iron are well known because they can help in chemical reactions, fight against microbes and show magnetic effects. Because of these useful properties nanoparticles are commonly used in both medicine and industry (Maťátková et al., 2022). Some nanoparticles are synthesized from ceramic material like silica and titanium dioxide. These are mainly used in medicinal imaging and drug delivery because they are stable and safe for the human body (Moreno-Vega et al., 2012). Nanoparticles made from biodegradable polymers area also useful, especially in delivering medicines and helping in tissue repair. They can hold the medicine inside and release it slowly where needed (Gagliardi et al., 2021). There are also lipid-based nanoparticles, such as liposomes and solid lipid nanoparticles. These are

 DOI: 10.1201/9781003559900-10

biocompatible and can easily pass through cell membranes to carry drugs more effectively (Luiz et al., 2023). Each different type of nanoparticle has its own benefits which is the reason for their use in many areas like science, healthcare and industries.

Among many types of nanoparticles, zinc oxide (ZnO) nanoparticles are special because they have many useful properties and can be used in different ways. They show unique optical, magnetic and catalytic features that are not seen in their bulk form (Parihar et al., 2018). These special features make them helpful in the medical field. ZnO nanoparticles are often used in skin treatment because they reduce inflammation and fight against microbes. This makes them useful for skin problems like acne, dermatitis and psoriasis (Raszewska-Famielec & Flieger, 2022). They are also added to sunscreen because they can block harmful UV rays, which helps protect the skin from damage and lowers the risk of skin cancer (Smijs & Pavel, 2011).

Apart from their use in skincare and sun protection, ZnO nanoparticles (NPs) are also being widely studied for improvement in how medicine is delivered in the body. These nanoparticles are produced in a way that allows them to carry drugs or genes or proteins directly to the specific cells where they are required. This makes the nanoparticles-based treatment work better and helps to reduce side effects. This targeted method is very useful in cancer treatment, where ZnO nanoparticles are able to deliver anticancer drugs directly to the tumor cell while keeping healthy cells unaffected (Chandrakala et al., 2022; Mitchell et al., 2020). Researchers are also testing ZnO NPs in vaccine production, in which they help to stimulate the immune system and make the vaccine work effectively (Kheirollahpour et al., 2019). Because they are safe for the body and not very toxic, ZnO NP are also used in medical implants and devices, where they prevent infections and help with tissue healing (Franco et al., 2022a). ZnO NPs' multiple uses and safe nature make them important tools in the growing field of nanomedicine, offering new ways to solve many health problems.

10.2 ZINC OXIDE NANOPARTICLES SYNTHESIS APPROACH

In late 1900s and early 2000s, ZnO NPs synthesis interest had grown among researchers because of their capacity to be used in agriculture, medical and environmental science purposes. This happens due to the large surface area compared with its volume which is the main reason for their reactivity (Jia et al., 2020).

When it comes to the synthesis approach to nanoparticles, several methods are available. Initially, chemical and physical methods were commonly used. In the chemical method harmful chemicals are generally used as reducing agents, but they also show side effects by producing toxic byproducts. In some physical methods such as ball milling and laser ablation zinc nanoparticles can be produced, but these techniques are very expensive (Basnet et al., 2018; Iravani & Zolfaghari, 2013). To overcome these problems, a third method which is known as the biological method is of interest because it is very cheap and safe. In this method extracts of microorganisms and plants are used as reducing agents to produce ZnO NPs (Ingale et al., 2023; Mahendiran et al., 2017; Thakkar et al., 2010).

To produce ZnO NPs using plant extract the process requires a zinc precursor. Zinc acetate dihydrate ($Zn(CH_3COO)_2{\cdot}2H_2O$) is one of the precursors used, where 0.01 M aqueous solution of Zinc acetate dihydrate is mixed with the filtered plant extract in proportion of 1:1. The plant extract then acts as not only a reducing agent but also as a stabilizing agent for synthesized ZnO NPs (Asia Thabassoom et al., 2018; Ghaffari et al., 2024). This reaction is sensitive to pH and that's why it should be carefully performed in a controlled environment of pH and temperature (around 60° C–70° C) for complete reduction of Zn^{2+} in ZnO NPs (Smijs & Pavel, 2011). Zinc oxide is another precursor generally used for the production of ZnO NPs in which constant steering is needed after the addition of a precursor in plant extract in 1:1 proportion. Temperature should be maintained around 70° C for evenly spread ZnO NPs formation because they tend to clump together. To prevent such a condition polyvinylpyrrolidone (PVP) is generally added in to the solution which gets attached with produced nanoparticles and prevents clumping (Mahendiran et al., 2017; Szczyglewska et al., 2023).

10.3 PLANT-ASSISTED BIOGENIC SYNTHESIS

The biological method is also known as the green synthesis method because of its clean, safe and easy to use protocol. This has several advantages, most importantly with respect to environmental concerns. A variety of microbes and plant extracts contain essential reducing agents that are naturally available and useful for nanoparticles synthesis (Basnet et al., 2018; Chandrakala et al., 2022).

With our top interest, the use of plants in biological methods, many reports are available which are presented in Table 10.1. Phytochemicals consist of many proteins, ions and secondary metabolites such as alkaloids, phenols and flavonoids. These compounds act as natural reducing and stabilizing agents (Dilip Lokare et al., n.d.; Elumalai & Velmurugan, 2015; Mahendiran et al., 2017; Mohamad et al., 2014).

10.4 MEDICINAL SIGNIFICANCE OF ZNO NANOPARTICLES

Acne, dermatitis, eczema and psoriasis are conditions which are related to skin and can be treated with ZnO NPs because of their anti-inflammatory and anti-oxidant potential (Raszewska-Famielec & Flieger, 2022). They are commonly used in sunscreen lotions to block harmful radiations such as UV A and UV B. This helps in avoiding the conditions of skin burn, irritations and in some cases cancer (Smijs & Pavel, 2011). ZnO NPs are also noted for boosting tissue repair and the immune system, and therefore they are useful for promoting the healing process especially in wound healing cases (Nascimento et al., 2023; Singh et al., 2021).

In medical science especially in drug delivery, these ZnO NPs can carry drugs along them to the specific target areas. This targeted drug delivery mechanism is very useful to treat the affected area and keep other non-affected areas safe, which helps in avoiding side effects (Chandrakala et al., 2022). In hospitals various instruments can be coated with the help of ZnO NPs to lower the infection risk, which helps to get rid of bacterial growth over surgical instruments because of ZnO NPs' antibacterial potential (Franco et al., 2022a). In the treatment of brain related conditions like Alzheimer's and Parkinson's, ZnO NPs can be useful because they have shown their potential to travel across the blood brain barrier (Patel et al., 2023).

TABLE 10.1
Plants Used for ZnO Nanoparticles Synthesis

Sr.	Plant Name	Shape	Size (nm)	Biological Applications	Reference
1	*Annona squamosa*	Spherical	20–50	Antimicrobial, anti-inflammatory	(Asia Thabassoom et al., 2018)
2	*Bauhinia variegata*	Spherical	10–40	Wound healing, antimicrobial	(Abbas et al., 2024)
3	*Carica papaya*	Spherical	5–30	Antioxidant, antimicrobial	(Dulta et al., 2022)
4	*Cassia fistula*	Spherical	15–40	Antimicrobial, antioxidant	(Suresh et al., 2015)
5	*Catharanthus roseus*	Spherical	10–30	Antimicrobial, anticancer	(Gupta et al., 2018)
6	*Centella asiatica*	Spherical	20–50	Antimicrobial, antioxidant	(Ramu Mahesh et al., 2023)
7	*Garcinia mangostana*	Spherical	10–50	Antimicrobial, anti-inflammatory	(Chandrakala et al., 2022)
8	*Hibiscus rosa-sinensis*	Spherical	20–40	Antimicrobial, anticancer	(Asia Thabassoom et al., 2018)
9	*Acalypha indica*	Spherical	10–40	Antimicrobial, antioxidant	(Sabir et al., 2014)
10	*Aegle marmelos*	Spherical	20–50	Antimicrobial, anti-inflammatory	(Sriramulu et al., 2018)
11	*Aloe vera*	Spherical	5–30	Antioxidant, wound healing	(Mahendiran et al., 2017)
12	*Azadirachta indica*	Spherical	15–40	Antimicrobial, anticancer	(Elumalai & Velmurugan, 2015)
13	*Calotropis procera*	Spherical	10–40	Antimicrobial, antioxidant	(Sabir et al., 2014)
14	*Cassia auriculata*	Spherical	20–50	Antimicrobial, anti-inflammatory	(Hano & Abbasi, 2021)
15	*Curcuma longa*	Spherical	10–30	Antimicrobial, anticancer	(Jacob, 2019)
16	*Datura metel*	Spherical	20–40	Antimicrobial, anti-inflammatory	(Akila et al., 2024)
17	*Emblica officinalis*	Spherical	10–50	Antioxidant, antimicrobial	(Ramalingam & Namasivayam, 2015)
18	*Gymnema sylvestre*	Spherical	10–40	Antimicrobial, anti-inflammatory	(Kavipriya & Ramasubburayan, 2024)
19	*Hibiscus sabdariffa*	Spherical	20–50	Antioxidant, antimicrobial	(Shaba et al., 2021)
20	*Lantana camara*	Spherical	10–30	Antimicrobial, anticancer	(Shaba et al., 2021)

(Continued)

TABLE 10.1 (*Continued*)
Plants Used for ZnO Nanoparticles Synthesis

Sr.	Plant Name	Shape	Size (nm)	Biological Applications	Reference
21	*Leucas aspera*	Spherical	10–40	Antimicrobial, antioxidant	(Sivaramakrishnan et al., 2019)
22	*Moringa oleifera*	Spherical	20–50	Antioxidant, anti-inflammatory	(Arévalo-Híjar et al., 2018)
23	*Ocimum sanctum*	Spherical	10–30	Antimicrobial, antioxidant	(Sabir et al., 2014)
24	*Vitex negundo*	Spherical	20–40	Antimicrobial, antioxidant	(Shaba et al., 2021)
25	*Withania somnifera*	Spherical	10–50	Antimicrobial, anticancer	(Chandrakala et al., 2022)
26	*Zingiber officinale*	Spherical	10–40	Antioxidant, anti-inflammatory	(Gupta et al., 2018)
27	*Ocimum basilicum*	Spherical	20–50	Antimicrobial, antioxidant	(Shelke et al., 2023)
28	*Coriandrum sativum*	Spherical	10–30	Antimicrobial, antioxidant	(Sabir et al., 2014)
29	*Cymbopogon citratus*	Spherical	20–40	Antimicrobial, antioxidant	(Hano & Abbasi, 2021)
30	*Tridax procumbens*	Spherical	10–30	Antimicrobial, anti-inflammatory	(Kannan et al., 2020)

In skin care, wound healing, infection fighting, drug delivery and even in nervous system treatments these ZnO NPs can offer hope in nanomedicine sciences (Hamdan et al., 2017).

10.4.1 Anticancer Potential of ZnO Nanoparticles (ZnO NPs)

Because of their anticancer properties ZnO NPs are getting a lot of attention. They can destroy cancer cells by a natural process of programmed cell death known as apoptosis. This takes place by different pathways; one of them is by production of reactive oxygen species (ROS), another one by causing damage to the mitochondria, and in some cases by the activation of tumor-suppressing proteins like p53. According to Sharma et al. (2012) ZnO NPs can kill cancer cells without spoiling normal cells, making these nanoparticles useful for targeted cancer therapy. ZnO NPs are also being studied in photodynamic therapy (PDT), a treatment that uses light to activate compounds that produce toxic oxygen molecules to destroy cancer cells. Zhang et al. (2014) revealed that ZnO NPs can work as photo-sensitizers in PDT. Exposure to light results in the production of singlet oxygen that helps kill cancer cells, especially in skin cancer, showing the potential to improve PDT outcome.

Different types of cancer cells such as breast (MCF-7), lung (A549), prostate (PC-3) and colon (HCT-116) are tested. Nanoparticles kill the cancer cells by generating

ROS, damaging the mitochondria, and affecting the DNA (Ansari et al., 2018; Ghaffari et al., 2024; Zhang et al., 2014). ZnO NPs are also useful as drug carriers. For example, Ma et al. (2015) use of folic acid coated ZnO NPs to specifically target cancer cells, which consisting a folate receptors. This method offers proper drug delivery and increased cancer cell death compared to regular ZnO NPs.

Nanoparticles' safety is a reason which makes ZnO NPs more suitable for cancer treatment. They are generally biologically compatible and have very low toxicity, which is important for medical applications. The cancer cells show more metabolic activities and weak antioxidant defense as compared to the normal cells. However, targeted delivery of drugs with the help of ZnO NPs specifically at tumor areas can be responsible for the generation of ROS. This results in to the death of cancer cell while leaving normal cells unharmed (Sharma et al., 2012; Sonawane et al., 2021; Verma et al., 2021; Zhang et al., 2014). In research it has been shown that cancer cells are removed by the apoptosis pathway; where the ROS generally disturbs the mitochondria membrane this results in the escape of molecules like Cytochrome c into the cytoplasm from the mitochondria. The chain reaction takes place which involves cytoplasmic proteins such as Apaf-1 and enzymes like caspase-9 and caspase-3. These apoptotic effects are visible in many cancer cell lines and are directly proportional to the dosage of ZnO NPs to these cancer cells (Majeed et al., 2019; Moghaddam et al., 2017; Patel et al., 2023; Patrón-Romero et al., 2022; Tanino et al., 2020; Virych et al., 2022). This dose dependent behavior allows researchers to fine-tune treatment using ZnO NPs. Anjum et al. (2021) emphasized that by combining ROS generation and mitochondrial damage, ZnO NPs offer a strong and targeted way to fight cancer.

It has been noted that upon the entry of ZnO NPs into the cancer cell they damage DNA and activate p53 protein. p53 protein is well known for regulation of cell cycle and initiation of apoptosis (Ghaffari et al., 2024). Activated p53 can either stop the cell from dividing to allow time for repair or push it toward apoptosis by turning on pro apoptotic genes like Bax and NOXA (Sharma et al., 2012). These genes cause changes in the mitochondria, which help in the release of cytochrome c and trigger cell death. Since normal cells have stronger antioxidant systems and more stable p53 activity, they are less affected by this process (Lee et al., 2008; Michalak et al., 2008; Roufayel et al., 2022). A ZnO nanoparticle induces the p53 which results in apoptosis mainly in cancer cells (while leaving normal cells mostly unharmed) and shows their therapeutic potential. Normal cells possess well-functioning p53 and excellent antioxidant protection; therefore, they are less likely to be damaged by the oxidative stress and DNA damage caused by ZnO NPs. This selective action helps to reduce the side effects and makes ZnO NP-based treatments safer for use (Michalak et al., 2008; Roufayel et al., 2022).

10.4.2 Antimicrobial Potential of ZnO Nanoparticles (ZnO NPs)

10.4.2.1 Antibacterial Potential of Zinc Oxide Nanoparticles

ZnO NPs are widely known for their strong antimicrobial effects. Their activity is mainly due to three key mechanisms: the production of ROS, the disruption of microbial cell membrane and release of Zn^{2+} ions (Elumalai & Velmurugan, 2015).

Like many nanoparticles, ZnO NPs produce ROS which causes oxidative stress and damage to microbial cells. These ROS damage important components such as DNA, protein and lipids, ultimately leading to cell death. Studies have shown that ZnO NPs are effective against bacteria like *S. aureus* and *E. coli* by generating ROS that destroy cell components (Mohammed Sadiq et al., 2010). ZnO NPs also work by damaging microbial cell membranes. Their high surface area and reactivity make it easy for them to attach to the bacterial surface which causes an increase in membrane permeability and results in leakage of vital cell contents like ions and proteins. The damage makes the barrier get weaker and eventually leads to cell death (Siddiqi et al., 2018). Maťátková et al. (2022) reported the same in fungi as ZnO NPs show strong antifungal activity against *Aspergillus niger* due to the disruption of the fungal membrane which is responsible for content leaking out.

In addition to these effects, ZnO NPs release Zn^{2+} ions into the surrounding environment, and these ions interfere with the essential bacterial function such as enzyme activity and protein synthesis, which is additional to their antimicrobial action (Vitasovic et al., 2024). Due to these properties, ZnO NPs have been used in many practical applications. They are used in antimicrobial coating, food packaging and hygiene products (Ahmadi et al., 2020; Ramasamy & Lee, 2016).

10.4.2.2 Antimycotic Potential of Zinc Oxide Nanoparticles

ZnO NPs have shown strong antifungal properties, making them promising agents for treatment of fungal infection. Their antifungal mechanism involves multiple actions such as breaking fungal cell membrane and generating ROS which results in disturbance of fungal metabolism (Slavin et al., 2022). ZnO NPs interact with lipids in the fungal membrane which causes an increase in its permeability which results in a leak of essential protein and ions, which leads to cell death (Li et al., 2020).

Along with this, ZnO NPs produce ROS such as hydrogen peroxide and hydroxyl radicals which cause oxidative stress inside fungal cells. These key parts of the cell including DNA, proteins and lipids are damaged by the ROS (Sabuncuoglu, 2020; Slavin et al., 2022). Fungal cells are fragile to ROS because they have a weaker antioxidant system as compared to human cells (Gomaa, 2022). ZnO NPs are effective against many pathogenic fungi that threaten human health, agriculture and food safety. These include *Candida albicans*, *Aspergillus niger*, *Trichoderma rubrum* and *Fosarium oxysporum* which are the pathogens responsible for infections like candidiasis, aspergillosis, athlete's foot and wilt diseases in crops (Daniel et al., 2023; Mahamuni-Badiger et al., 2023; Parveen et al., 2023).

In the case of *Candida albicans*, which causes common oral and genital infections, ZnO NPs increase the permeability of its cell membrane which causes leakage and cell damage. Additionally, ROS generated by ZnO NPs add to this stress which results in cell death by apoptosis. This combined action makes ZnO NPs very effective against *Candida* infections (Dananjaya et al., 2018; Parveen et al., 2023).

Aspergillus niger, responsible for lung infections and food spoilage, is also sensitive to ZnO NPs. These NPs stop the fungi from growing by damaging their cell wall and spores. The high surface activity of ZnO NPs allows them to get attached to fungal spores and stops them from germinating, which is useful in both clinical and agricultural points of view (Wani et al., 2012).

ZnO NPs show interaction with the hyphae of *Trichophyton rubrum* and restrict their biofilm formation too. This fungus is known for athlete's foot infection, thus ZnO NPs inhibit or prevent skin infection that can be further caused by these fungi (Patiño-Herrera et al., 2018).

Wilt disease is common in the agricultural sector and is caused by the *Fusarium oxysporum* fungus. ZnO NPs application shows the spore's inhibition by interfering with the energy system of these fungi. This leads to a safe alternative over chemical fungicides (González-Merino et al., 2021).

10.4.3 Antioxidant Potential of ZnO Nanoparticles

Reactive oxygen species (ROS) are responsible for oxidative damage of important proteins, lipids and nuclear and mitochondrial DNA. ZnO NPs show good antioxidant activities by scavenging ROS. This property of these nanoparticles helps to reduce the oxidative stress (Asia Thabassoom et al., 2018; Kumar et al., 2020). Damage to cellular molecules can lead to adverse conditions of tissues or organs; in some cases brain cell damage may result in Parkinson's or Alzheimer conditions, and heart cell damage may be responsible for malfunctioning of the heart (Kumar et al., 2020; Patel et al., 2023). ZnO NPs inhibit the formation of superoxide anions, hydroxyl radicals and hydrogen peroxide by inducing and assisting the natural antioxidant enzymes (Mihailovic et al., 2021). Every cell includes a cell membrane; in oxidative stress condition when lipids peroxidation takes place by free radical attack such as ROS this results in the formation of reactive aldehydes which is called malondialdehyde (MDA). MDA indicates how much lipid peroxidation has undertaken. So a lower MDA level means that ZnO NPs are protecting the cell wall (Bayat et al., 2023; Sundaram Sanjay et al., 2014). Another important way ZnO NPs help is by increasing the activity of antioxidant enzymes in the body. These enzymes include glutathione peroxidase (GPx), catalase (CAT) and superoxide dismutase (SOD), and they remove ROS from the cell and keep it healthy. Research shows that ZnO NPs increase both the expression levels and enzymatic activities of these antioxidant enzymes. This gives the body better protection from oxidative stress. In some studies related to brain protection, ZnO NPs were found to improve SOD and CAT activity in nerve cells which helped them to reduce the damage to increase cell survival. In the case of heart problems, this kind of activity is helpful in treating diseases where oxidative stress plays a major role (Hao & Chen, 2012).

10.4.4 Applications of ZnO Nanoparticles in Skincare and Wound Healing

10.4.4.1 Skincare

ZnO NPs have gained popularity in skincare products because of their anti-inflammatory, antioxidant and UV-blocking qualities. Human skin frequently suffers from issues like dark spots, early aging and even skin cancer like conditions, since it is continuously exposed to sunlight, pollution and other damaging environmental factors. ZnO NPs are added to a variety of skincare products including moisturizers, sunscreen lotions and anti-aging lotions, to protect the skin against these issues (Irede et al., 2024).

One of the most common applications for ZnO NPs is in sunscreen. These nanoparticles are highly operational at blocking both UVA and UVB rays. Not like the chemical sunscreens which absorb UV and may cause skin sensitivities, ZnO NPs serve as a physical barrier, reflecting and scattering UV radiation. This makes them a perfect choice for individuals with sensitive skin. ZnO NPs are considered safe because they don't go deep into the skin. They offer a strong sun protection (high SPF) and help to prevent sunburn as well as premature aging and in some cases even skin cancers like melanoma (Fonseca et al., 2023).

Besides offering sun protection, ZnO nanoparticles are also helpful in treating skin problems like acne, rosacea and dermatitis due to their anti-inflammatory and antibacterial effects (Prajapati et al., 2024). For instance, in the case of acne which is caused by excess oil and bacterial infection ZnO NPs help by reducing oil production and stopping the growth of bacteria like *Propionibacterium acnes* (Al-Momani et al., 2024; Prajapati et al., 2024).

10.4.4.2 Wound Healing

In treatment of persistent injuries which are difficult to heal and prone to infections such as diabetic foot ulcers, ZnO NPs have also shown their potential in wound healing, mainly due to their antibacterial, anti-inflammatory and tissue regeneration properties. ZnO NPs speed up the healing by fighting against the infection, reducing the inflammation and promoting the formation of new tissues (Balalakshitha et al., 2021). In serious wound conditions, infection caused by the bacteria is the main reason for a longer duration of the healing process. ZnO NPs show good antibacterial properties against *Staphylococcus aureus* and *Escherichia coli*, commonly found bacteria around the wound, thus accelerating the wound healing process (González-Merino et al., 2021; Vitasovic et al., 2024). ZnO NPs coated bandages and wound dressings can effectively reduce the growth of such harmful bacteria around the wound region. Because of this the additional use of antibiotics can be reduced (Khatami et al., 2018).

Wound formation takes place when a sharp instrument causes skin to cut open and breaks the collagen layer. ZnO NPs induce the fibroblast cells to create more collagen. Collagen is the main protein required for the formation of new tissues. In addition to this, ZnO NPs help in the initiation of re-epithelialization which means new skin formation takes place (Han et al., 2020; Vedhanayagam et al., 2018).

Wound formation and inflammation are two sides of one coin when a wound occurs that results in serious ache in and around that region of the body. Inflammation is aroused due to the ROS generated in those areas, and these ROS are inhibited by the action of ZnO NPs which reduces the inflammation. This helps in faster wound healing and tissue regeneration (Han et al., 2020; Khatami et al., 2018).

10.5 APPLICATIONS OF ZNO NPS IN VARIOUS FIELDS

ZnO NPs have been proven to be the best in many areas such as:

i. **Medical Devices**: ZnO NPs are used in coating of medical and surgical equipment's, this helps to prevent the infection which causes by these instrument while operating the patients (Ramasamy & Lee, 2016)

ii. **Water Disinfection**: due to their antimicrobial activity on various bacteria's and fungi, ZnO NPs are commonly used in water storage containers and filters to keep water clean and fresh by killing germs in it (Ahmadi et al., 2020; Ramasamy & Lee, 2016; Zazouli et al., 2021)
iii. **Food Packaging**: Antimicrobial potential of ZnO NPs are useful in increasing the shelf life of packaged food by inhibiting the growth of fungi and bacteria over longer period of time to maintain food in fresh condition (Ahmadi et al., 2020).
iv. **Personal Care Products**: In the cosmetics industry, ZnO NPs are widely used in various ointments, lotions and creams. They inhibit the growth of skin infection causing bacteria, protect from harmful radiation and provide a soothing effect on skin (Raszewska-Famielec & Flieger, 2022).
v. **Catalysis in Organic Synthesis:** ZnO NPs can act as catalysts to speed up the process of production of pharmaceuticals, fertilizers and other organic compounds. Their use shows more efficient production while fewer side effects are observed (Sabir et al., 2014).
vi. **Environmental Cleanup**: Water pollution is mainly caused by industrial waste water being released directly into fresh water, and polluted water attracting the growth of various microbes and plants in that area. ZnO NPs act in various ways to minimize the growth of such microbes in water and also degrade harmful chemicals present in water and help in restoration (Vasantharaj et al., 2021).
vii. **Enzymatic Mimicry**: Nanozymes are known to mimic natural enzymes; ZnO NPs can act as nanozymes. In one example, ZnO NPs showed colorimetric activity similar to that used in blood sugar testing instruments. These instruments work on the principle of colorimetric assays, where the oxidation of specific compounds occurs in the presence of hydrogen peroxide (H_2O_2). Because of the peroxidase-like property, ZnO NPs function as effective nanozymes and are useful in various biochemical applications (Wu et al., 2019).
viii. **Photocatalysis**: Industrial dyes are major pollutants, responsible for water and soil pollutions. ZnO NPs after absorbing UV light are found to be effective on degradation of these dyes such as methylene blue. This has applications in waste water treatment (Modi et al., 2022; Ong et al., 2018).
ix. **Biocatalysis**: In biosensors, ZnO NPs can act as peroxidase and glucose oxidase, where they can catalyze the conversion of glucose to gluconic acid and hydrogen peroxide. This has an applications in diagnostic kits (Wu et al., 2019).
x. **Catalysis in Chemical Reactions**: In the pharmaceutical industries the production of heterocyclic compounds can be done by the use of ZnO NPs because of their high surface area to low volume and reactivity (Ramalingam & Namasivayam, 2015).
xi. **Gas Sensing**: Gases like hydrogen, methane and carbon monoxide even at very low levels can be detected by the ZnO NPs just because of their sensitivity to gas changes. This makes them useful for safety and monitoring in both industrial and household environments (Franco et al., 2022b).

10.6 FUTURE PROSPECTS AND CHALLENGES

Even though ZnO NPs show great promise, there are still some challenges that are need to be worked on. For practical use it is important to improve their stability and make sure they can be reused. We also need better understanding of how they affect human health and the environment in the long duration. More detailed studies on their toxicity and environmental effects are required to ensure their safe use. Another issue is the cost for making ZnO NPs through affordable and large-scale methods is still a challenge. To move forward we also need proper guidelines for testing their catalytic ability, biocompatibility and safety. This will help in getting approvals and bringing these materials into wider use.

To overcome these challenges, continued research and new ideas are very important. With the proper efforts, researchers can make better use of ZnO NPs in different fields. ZnO NPs used in areas like healthcare, environment and industry can be real beneficial in improving treatments, reducing pollutions and making processes more efficient. For a sustainable future these ZnO NPs are very useful in their proper use.

10.7 CONCLUSION

ZnO NPs are becoming more and more popular because of their safe results. Among various modes of synthesis, green synthesis is cheaper, safer and the best alternative to other hazardous and expensive methodologies. Plant extracts which consist of various phytochemicals are useful for the reduction of zinc in to the nanoparticle form. These ZnO nanoparticles are mostly in spherical shape and average size of 30 nm. Due to high surface area, they are biocompatible to be used in various fields including medicine, agriculture and environmental restoration. To treat adverse medical conditions such as cancer and brain related diseases, these nanoparticles hold a promising solution.

When it comes to cleaning water and improving the environment, these nanoparticles are quite helpful. They can break down harmful chemicals and also kill germs. That's why they are being used in water filters and even in coating water tanks. In agriculture too, they seem useful. They help in making the soil better and can work as a micronutrient, which supports plant growth and increases crop yield. In food-related uses, ZnO NPs help stop bacterial growth in food packaging. This way, food stays fresh for a longer time. This is good for reducing waste. These particles are also used in industries where chemical reactions are needed. Because of their unique surface activity, they help speed up certain reactions. Their enzyme-like nature also makes them useful in medical sensors and test kits.

To characterize these nanoparticles in detail, researchers use different lab techniques. Some of the commonly used are UV-Visible spectroscopy, electron microscopes like SEM and TEM, FTIR and XRD. There are also other methods like DLS and zeta potential testing. These tools help in checking the size, shape, surface charge and overall behavior of the nanoparticles. But of course, there are some issues. Sometimes they are not stable, or they stick together. We also don't fully know what long-term effects they might have on health or nature. And making them in bulk, at low cost, is still a big task. That's why more work needs to be done. Researchers have to keep testing and improving them. Also, some proper safety rules and testing systems should be made. If all this is done step-by-step, these nanoparticles can truly be used in many areas helping people in real life, protecting the environment and also making everyday technology more efficient.

ACKNOWLEDGEMENT

The Mendeley Reference Manager tool is used for the references formatting.

REFERENCES

Abbas, Z., Irshad, M., Ali, S., Summer, M., Rasheed, A., & Jawad, M. (2024). Radical scavenging potential of spectrophotometric, spectroscopic, microscopic, and EDX observed zinc oxide nanoparticles from leaves, buds, and flowers extract of Bauhinia Variegata Linn: A thorough comparative insight. *Microscopy Research and Technique*. https://doi.org/10.1002/JEMT.24587

Ahmadi, A., Ahmadi, P., & Ehsani, A. (2020). Development of an active packaging system containing zinc oxide nanoparticles for the extension of chicken fillet shelf life. *Food Science & Nutrition*, *8*(10), 5461–5473. https://doi.org/10.1002/FSN3.1812

Akila, K., Thambidurai, S., Suresh, N., & Prabu, K. M. (2024). Photocatalytic and bactericidal activity of zinc oxide nanoparticles biosynthesized with Datura metel leaf extract. *Ionics*, *30*(6), 3637–3649. https://doi.org/10.1007/S11581-024-05457-W/METRICS

Al-Momani, H., Massadeh, M. I., Almasri, M., Al Balawi, D., Aolymat, I., Hamed, S., Albiss, B. A., Ibrahim, L., Balawi, H. A., & Al Haj Mahmoud, S. (2024). Anti-bacterial activity of green synthesised silver and zinc oxide nanoparticles against propionibacterium acnes. *Pharmaceuticals*, *17*(2), 255. https://doi.org/10.3390/PH17020255/S1

Anjum, S., Hashim, M., Malik, S. A., Khan, M., Lorenzo, J. M., Abbasi, B. H., & Hano, C. (2021). Recent advances in zinc oxide nanoparticles (ZnO NPs) for cancer diagnosis, target drug delivery, and treatment. *Cancers*, *13*(18), 4570. https://doi.org/10.3390/CANCERS13184570

Ansari, M. T., Sami, F., Khairudiin, F. A., Atan, M. Z., bin Tengku Mohamad, T. A. S., Majeed, S., & Ali, S. (2018). Applications of zinc nanoparticles in medical and healthcare fields. *Current Nanomedicine*, *8*(3), 225–233. https://doi.org/10.2174/2405461503666180709100110

Arévalo-Híjar, L., Aguilar-Luis, M. Á., Caballero-García, S., Gonzáles-Soto, N., & Del Valle-Mendoza, J. (2018). Antibacterial and cytotoxic effects of Moringa oleifera (Moringa) and Azadirachta indica (Neem) methanolic extracts against strains of Enterococcus faecalis. *International Journal of Dentistry*, *2018*(1), 1071676. https://doi.org/10.1155/2018/1071676

Asia Thabassoom, H., AUxILIA RUBY, S., & Felicita Florence, J. (2018). Biosynthesis of Beta vulgaris mediated ZnO nanoparticles and their anti-bacterial, anti-fungal, anti-diabetic and anti-cancer activities. *Oriental Journal of Chemistry, 37*(4), 911–921. https://doi.org/10.13005/ojc/370420

Balalakshitha, M., Kolanjinathan, K., Aishwarya, T., & Vinothini, B. (2021). A review on diabetic wound infection and its treatment using zinc oxide nanoparticle synthesized by green chemistry techniques. *Bulletin of Environment, Pharmacology and Life Sciences*, *10*(10), 1–22.

Basnet, P., Inakhunbi Chanu, T., Samanta, D., & Chatterjee, S. (2018). A review on biosynthesized zinc oxide nanoparticles using plant extracts as reductants and stabilizing agents. *Journal of Photochemistry and Photobiology B: Biology*, *183*, 201–221. https://doi.org/10.1016/J.JPHOTOBIOL.2018.04.036

Bayat, M., Daei, S., Ziamajidi, N., Abbasalipourkabir, R., & Nourian, A. (2023). The protective effects of vitamins A, C, and E on zinc oxide nanoparticles (ZnO NPs)-induced liver oxidative stress in male Wistar rats. *Drug and Chemical Toxicology*, *46*(2), 209–218. https://doi.org/10.1080/01480545.2021.2016809

Chandrakala, V., Aruna, V., & Angajala, G. (2022). Review on metal nanoparticles as nanocarriers: Current challenges and perspectives in drug delivery systems. *Emergent Materials*, *5*(6), 1593–1615. https://doi.org/10.1007/S42247-021-00335-X

Dananjaya, S. H. S., Kumar, R. S., Yang, M., Nikapitiya, C., Lee, J., & De Zoysa, M. (2018). Synthesis, characterization of ZnO-chitosan nanocomposites and evaluation of its antifungal activity against pathogenic Candida albicans. *International Journal of Biological Macromolecules*, *108*, 1281–1288. https://doi.org/10.1016/J.IJBIOMAC.2017.11.046

Daniel, A. I., Keyster, M., & Klein, A. (2023). Biogenic zinc oxide nanoparticles: A viable agricultural tool to control plant pathogenic fungi and its potential effects on soil and plants. *Science of The Total Environment*, *897*, 165483. https://doi.org/10.1016/J.SCITOTENV.2023.165483

Dilip Lokare, P., Kumar, A., Bhaskar, K., Kant Patra, K., Agrawal, S., Jadhav, S., Narhar Patil, D., & Haribhau Khodade, H. (n.d.). *A cognitive behavioral defense mechanism of plants against herbivory*. https://doi.org/10.33451/florafauna.v29i1pp49-55

Dulta, K., Koşarsoy Ağçeli, G., Chauhan, P., Jasrotia, R., & Chauhan, P. K. (2022). Ecofriendly synthesis of zinc oxide nanoparticles by Carica papaya leaf extract and their applications. *Journal of Cluster Science*, *33*(2), 603–617. https://doi.org/10.1007/S10876-020-01962-W/METRICS

Elumalai, K., & Velmurugan, S. (2015). Green synthesis, characterization and antimicrobial activities of zinc oxide nanoparticles from the leaf extract of Azadirachta indica (L.). *Applied Surface Science*, *345*, 329–336. https://doi.org/10.1016/J.APSUSC.2015.03.176

Fonseca, M., Rehman, M., Soares, R., & Fonte, P. (2023). The impact of flavonoid-loaded nanoparticles in the UV protection and safety profile of topical sunscreens. *Biomolecules*, *13*(3), 493. https://doi.org/10.3390/BIOM13030493

Franco, D., Calabrese, G., Guglielmino, S. P. P., & Conoci, S. (2022a). Metal-based nanoparticles: Antibacterial mechanisms and biomedical application. *Microorganisms*, *10*(9), 1778. https://doi.org/10.3390/MICROORGANISMS10091778

Franco, M. A., Conti, P. P., Andre, R. S., & Correa, D. S. (2022b). A review on chemiresistive ZnO gas sensors. *Sensors and Actuators Reports*, *4*, 100100. https://doi.org/10.1016/J.SNR.2022.100100

Gagliardi, A., Giuliano, E., Venkateswararao, E., Fresta, M., Bulotta, S., Awasthi, V., & Cosco, D. (2021). Biodegradable polymeric nanoparticles for drug delivery to solid tumors. *Frontiers in Pharmacology*, *12*, 601626. https://doi.org/10.3389/FPHAR.2021.601626/BIBTEX

Ghaffari, S. B., Sarrafzadeh, M. H., Salami, M., & Alvandi, A. (2024). A comparative study of the action mechanisms and development strategies of different ZnO-based nanostructures in antibacterial and anticancer applications. *Journal of Drug Delivery Science and Technology*, *91*, 105221. https://doi.org/10.1016/J.JDDST.2023.105221

Gomaa, E. Z. (2022). Microbial mediated synthesis of zinc oxide nanoparticles, characterization and multifaceted applications. *Journal of Inorganic and Organometallic Polymers and Materials*, *32*(11), 4114–4132. https://doi.org/10.1007/S10904-022-02406-W

González-Merino, A. M., Hernández-Juárez, A., Betancourt-Galindo, R., Ochoa-Fuentes, Y. M., Valdez-Aguilar, L. A., & Limón-Corona, M. L. (2021). Antifungal activity of zinc oxide nanoparticles in Fusarium oxysporum-Solanum lycopersicum pathosystem under controlled conditions. *Journal of Phytopathology*, *169*(9), 533–544. https://doi.org/10.1111/JPH.13023

Gupta, M., Tomar, R. S., Kaushik, S., Mishra, R. K., & Sharma, D. (2018). Effective antimicrobial activity of green ZnO nano particles of Catharanthus roseus. *Frontiers in Microbiology*, *9*(SEP), 392294. https://doi.org/10.3389/FMICB.2018.02030/BIBTEX

Hamdan, S., Pastar, I., Drakulich, S., Dikici, E., Tomic-Canic, M., Deo, S., & Daunert, S. (2017). Nanotechnology-driven therapeutic interventions in wound healing: Potential uses and applications. *ACS Central Science*, *3*(3), 163–175. https://doi.org/10.1021/ACSCENTSCI.6B00371/ASSET/IMAGES/LARGE/OC-2016-003714_0006.JPEG

Han, B., Fang, W. H., Zhao, S., Yang, Z., & Hoang, B. X. (2020). Zinc sulfide nanoparticles improve skin regeneration. *Nanomedicine: Nanotechnology, Biology and Medicine*, *29*, 102263. https://doi.org/10.1016/J.NANO.2020.102263

Hano, C., & Abbasi, B. H. (2021). Plant-based green synthesis of nanoparticles: Production, characterization and applications. *Biomolecules*, *12*(1), 31. https://doi.org/10.3390/BIOM12010031

Hao, L., & Chen, L. (2012). Oxidative stress responses in different organs of carp (Cyprinus carpio) with exposure to ZnO nanoparticles. *Ecotoxicology and Environmental Safety*, *80*, 103–110. https://doi.org/10.1016/J.ECOENV.2012.02.017

Ingale, S., Gaykar, B., Pagar, K., & Shelke, D. (2023). Green synthesis of silver nanoparticles from Cyathocline purpurea (Don) O. Kuntze and its efficacy against MCF-7 cancer cell line. *BioNanoScience*, *13*(3), 1197–1212. https://doi.org/10.1007/s12668-023-01138-6

Iravani, S., & Zolfaghari, B. (2013). Green synthesis of silver nanoparticles using Pinus eldarica bark extract. *BioMed Research International*, *2013*. https://doi.org/10.1155/2013/639725

Irede, E. L., Awoyemi, R. F., Owolabi, B., Aworinde, O. R., Kajola, R. O., Hazeez, A., Raji, A. A., Ganiyu, L. O., Onukwuli, C. O., Onivefu, A. P., & Ifijen, I. H. (2024). Cutting-edge developments in zinc oxide nanoparticles: Synthesis and applications for enhanced antimicrobial and UV protection in healthcare solutions. *RSC Advances*, *14*(29), 20992–21034. https://doi.org/10.1039/D4RA02452D

Jacob, V. (2019). In vitro analysis: The antimicrobial and antioxidant activity of zinc oxide nanoparticles from curcuma longa. *Asian Journal of Pharmaceutical and Clinical Research*, *12*. https://doi.org/10.22159/ajpcr.2019.v12i1.28808

Jia, X., Liu, C., Neale, Z. G., Yang, J., & Cao, G. (2020). Active materials for aqueous zinc ion batteries: Synthesis, crystal structure, morphology, and electrochemistry. *Chemical Reviews*, *120*(15), 7795–7866. https://doi.org/10.1021/ACS.CHEMREV.9B00628/ASSET/IMAGES/MEDIUM/CR9B00628_0040.GIF

Kannan, S., Subiramaniyam, N. P., & Sathishkumar, M. (2020). A novel green synthesis approach for improved photocatalytic activity and antibacterial properties of zinc sulfide nanoparticles using plant extract of Acalypha indica and Tridax procumbens. *Journal of Materials Science: Materials in Electronics*, *31*(12), 9846–9859. https://doi.org/10.1007/S10854-020-03529-X/METRICS

Kavipriya, R., & Ramasubburayan, R. (2024). Phytofabrication of biocompatible zinc oxide nanoparticle using Gymnema sylvestre and its potent in vitro antibacterial, antibiofilm, and cytotoxicity against human breast cancer cells (MDA-MB-231). *Bioprocess and Biosystems Engineering*, 1–15. https://doi.org/10.1007/S00449-024-03035-Y/METRICS

Ke, C. (2011). Recent advances in nanotechnology. *Recent Advances in Nanotechnology*, 1–281. https://doi.org/10.17352/2455-3492.000053

Khatami, M., Varma, R. S., Zafarnia, N., Yaghoobi, H., Sarani, M., & Kumar, V. G. (2018). Applications of green synthesized Ag, ZnO and Ag/ZnO nanoparticles for making clinical antimicrobial wound-healing bandages. *Sustainable Chemistry and Pharmacy*, *10*, 9–15. https://doi.org/10.1016/J.SCP.2018.08.001

Kheirollahpour, M., Mehrabi, M., Dounighi, N. M., Mohammadi, M., & Masoudi, A. (2019). Nanoparticles and vaccine development. *Pharmaceutical Nanotechnology*, *8*(1), 6–21. https://doi.org/10.2174/2211738507666191024162042

Kumar, H., Bhardwaj, K., Nepovimova, E., KuČa, K., Dhanjal, D. S., Bhardwaj, S., Bhatia, S. K., Verma, R., & Kumar, D. (2020). Antioxidant functionalized nanoparticles: A combat against oxidative stress. *Nanomaterials*, *10*(7), 1334. https://doi.org/10.3390/NANO10071334

Lee, D. H., Kim, C., Zhang, L., & Lee, Y. J. (2008). Role of p53, PUMA, and Bax in wogonin-induced apoptosis in human cancer cells. *Biochemical Pharmacology*, *75*(10), 2020–2033. https://doi.org/10.1016/J.BCP.2008.02.023

Li, Y., Liao, C., & Tjong, S. C. (2020). Recent advances in zinc oxide nanostructures with antimicrobial activities. *International Journal of Molecular Sciences*, *21*(22), 8836. https://doi.org/10.3390/IJMS21228836

Liu, R., Luo, C., Pang, Z., Zhang, J., Ruan, S., Wu, M., Wang, L., Sun, T., Li, N., Han, L., Shi, J., Huang, Y., Guo, W., Peng, S., Zhou, W., & Gao, H. (2023). Advances of nanoparticles as drug delivery systems for disease diagnosis and treatment. *Chinese Chemical Letters*, *34*(2), 107518. https://doi.org/10.1016/J.CCLET.2022.05.032

Luiz, H., Oliveira Pinho, J., & Gaspar, M. M. (2023). Advancing medicine with lipid-based nanosystems—The successful case of liposomes. *Biomedicines*, *11*(2), 435. https://doi.org/10.3390/BIOMEDICINES11020435

Ma, Y. Y., Ding, H., & Xiong, H. M. (2015). Folic acid functionalized ZnO quantum dots for targeted cancer cell imaging. *Nanotechnology*, *26*(30), 305702. https://doi.org/10.1088/0957-4484/26/30/305702

Mahamuni-Badiger, P., Ghare, V., Nikam, C., & Patil, N. (2023). The fungal infections and their inhibition by Zinc oxide nanoparticles: an alternative approach to encounter drug resistance. *The Nucleus*, *67*(2), 291–309. https://doi.org/10.1007/S13237-023-00439-1

Mahendiran, D., Subash, G., Arumai Selvan, D., Rehana, D., Senthil Kumar, R., & Kalilur Rahiman, A. (2017). Biosynthesis of zinc oxide nanoparticles using plant extracts of Aloe vera and Hibiscus sabdariffa: Phytochemical, antibacterial, antioxidant and anti-proliferative Studies. *BioNanoScience*, *7*(3), 530–545. https://doi.org/10.1007/S12668-017-0418-Y/METRICS

Majeed, S., Danish, M., Ismail, M. H. Bin, Ansari, M. T., & Ibrahim, M. N. M. (2019). Anticancer and apoptotic activity of biologically synthesized zinc oxide nanoparticles against human colon cancer HCT-116 cell line – in vitro study. *Sustainable Chemistry and Pharmacy*, *14*, 100179. https://doi.org/10.1016/J.SCP.2019.100179

Maťátková, O., Michailidu, J., Miškovská, A., Kolouchová, I., Masák, J., & Čejková, A. (2022). Antimicrobial properties and applications of metal nanoparticles biosynthesized by green methods. *Biotechnology Advances*, *58*, 107905. https://doi.org/10.1016/J.BIOTECHADV.2022.107905

Michalak, E. M., Villunger, A., Adams, J. M., & Strasser, A. (2008). In several cell types tumour suppressor p53 induces apoptosis largely via Puma but Noxa can contribute. *Cell Death & Differentiation*, *15*(6), 1019–1029. https://doi.org/10.1038/cdd.2008.16

Mihailovic, V., Katanic Stankovic, J. S., Selakovic, D., & Rosic, G. (2021). An overview of the beneficial role of antioxidants in the treatment of nanoparticle-induced toxicities. *Oxidative Medicine and Cellular Longevity*, *2021*(1), 7244677. https://doi.org/10.1155/2021/7244677

Mitchell, M. J., Billingsley, M. M., Haley, R. M., Wechsler, M. E., Peppas, N. A., & Langer, R. (2020). Engineering precision nanoparticles for drug delivery. *Nature Reviews Drug Discovery*, *20*(2), 101–124. https://doi.org/10.1038/s41573-020-0090-8

Modi, S., Yadav, V. K., Gacem, A., Ali, I. H., Dave, D., Khan, S. H., Yadav, K. K., Rather, S. U., Ahn, Y., Son, C. T., & Jeon, B. H. (2022). Recent and emerging trends in remediation of methylene blue dye from wastewater by using zinc oxide nanoparticles. *Water*, *14*(11), 1749. https://doi.org/10.3390/W14111749

Moghaddam, A. B., Moniri, M., Azizi, S., Rahim, R. A., Ariff, A. Bin, Navaderi, M., & Mohamad, R. (2017). Eco-friendly formulated zinc oxide nanoparticles: Induction of cell cycle arrest and apoptosis in the MCF-7 cancer cell line. *Genes*, *8*(10), 281. https://doi.org/10.3390/GENES8100281

Mohamad, N. A. N., Arham, N. A., Jai, J., & Hadi, A. (2014). Plant extract as reducing agent in synthesis of metallic nanoparticles: A review. *Advanced Materials Research*, *832*, 350–355. https://doi.org/10.4028/WWW.SCIENTIFIC.NET/AMR.832.350

Mohammed Sadiq, I., Chandrasekaran, N., & Mukherjee, A. (2010). Studies on effect of TiO2 nanoparticles on growth and membrane permeability of Escherichia coli, Pseudomonas aeruginosa, and Bacillus subtilis. *Current Nanoscience*, *6*(4), 381–387. https://doi.org/10.2174/157341310791658973

Moreno-Vega, A. I., Gómez-Quintero, T., Nuñez-Anita, R. E., Acosta-Torres, L. S., & Castaño, V. (2012). Polymeric and ceramic nanoparticles in biomedical applications. *Journal of Nanotechnology*, *2012*(1), 936041. https://doi.org/10.1155/2012/936041

Nascimento, L. B. dos S., Casanova, L. M., & Costa, S. S. (2023). Bioactive compounds from kalanchoe genus potentially useful for the development of new drugs. *Life*, *13*(3), 646. https://doi.org/10.3390/LIFE13030646

Ni, B., Shi, Y., & Wang, X. (2018). The sub-nanometer scale as a new focus in nanoscience. *Advanced Materials*, *30*(43), 1802031. https://doi.org/10.1002/ADMA.201802031

Ong, C. B., Ng, L. Y., & Mohammad, A. W. (2018). A review of ZnO nanoparticles as solar photocatalysts: Synthesis, mechanisms and applications. *Renewable and Sustainable Energy Reviews*, *81*, 536–551. https://doi.org/10.1016/J.RSER.2017.08.020

Parihar, V., Raja, M., & Paulose, R. (2018). A brief review of structural, electrical and electrochemical properties of zinc oxide nanoparticles. *Reviews on Advanced Materials Science*, *53*(2), 119–130. https://doi.org/10.1515/RAMS-2018-0009/MACHINEREADABLECITATION/RIS

Parveen, J., Sultana, T., Kazmi, A., Malik, K., Ullah, A., Ali, A., Qayyum, B., Raja, N. I., Mashwani, Z. U. R., & Rehman, S. U. (2023). Phytosynthesized nanoparticles as novel antifungal agent for sustainable agriculture: A mechanistic approach, current advances, and future directions. *Journal of Nanotechnology*, *2023*(1), 8011189. https://doi.org/10.1155/2023/8011189

Patel, T. A., Kevadiya, B. D., Bajwa, N., Singh, P. A., Zheng, H., Kirabo, A., Li, Y.-L., Patel, K. P., Patel, T. A., Kevadiya, B. D., Bajwa, N., Singh, A., Zheng, H., Kirabo, A., Li, Y.-L., & Patel, K. P. (2023). Role of nanoparticle-conjugates and nanotheranostics in abrogating oxidative stress and ameliorating neuroinflammation. *Antioxidants*, *12*(10), 1877. https://doi.org/10.3390/ANTIOX12101877

Patiño-Herrera, R., Catarino-Centeno, R., Robles-Martínez, M., Zarate, M. G. M., Flores-Arriaga, J. C., & Pérez, E. (2018). Antimycotic activity of zinc oxide decorated with silver nanoparticles against Trichophyton mentagrophytes. *Powder Technology*, *327*, 381–391. https://doi.org/10.1016/J.POWTEC.2018.01.004

Patrón-Romero, L., Luque-Morales, P. A., Loera-Castañeda, V., Lares-Asseff, I., Leal-Ávila, M. Á., Alvelais-Palacios, J. A., Plasencia-López, I., & Almanza-Reyes, H. (2022). Mitochondrial dysfunction induced by zinc oxide nanoparticles. *Crystals*, *12*(8), 1089. https://doi.org/10.3390/CRYST12081089

Prajapati, B. G., Sharma, D., Elossaily, G. M., Sharma, N., Bilandi, A., & Kapoor, D. U. (2024). Unlocking potential of zinc oxide nanoparticles in enhancing topical drug delivery. *Nano-Structures & Nano-Objects*, *39*, 101302. https://doi.org/10.1016/J.NANOSO.2024.101302

Ramalingam, S., & Namasivayam, E. (2015). Antimicrobial activity of green synthesized zinc oxide nanoparticles from Emblica officinalis. *Article in International Journal of Pharmaceutical Sciences Review and Research*, *43*(2), 110–115. https://www.researchgate.net/publication/282754422

Ramasamy, M., & Lee, J. (2016). Recent nanotechnology approaches for prevention and treatment of biofilm-associated infections on medical devices. *BioMed Research International*, *2016*(1), 1851242. https://doi.org/10.1155/2016/1851242

Ramu Mahesh, A., Leno Jenita Joseph, J., Sood, R., Rathore, S. S., & Leno Jenita, J. (2023). Synthesis of zinc oxide nanoparticles using Centella asiatica. *Advances in Pharmacology and Pharmacy*, *11*(4), 270–278. https://doi.org/10.13189/app.2023.110404

Raszewska-Famielec, M., & Flieger, J. (2022). Nanoparticles for topical application in the treatment of skin dysfunctions—An overview of dermo-cosmetic and dermatological products. *International Journal of Molecular Sciences, 23*(24), 15980. https://doi.org/10.3390/IJMS232415980

Roberto Vega-Baudrit, J., & Villalobos, C. (2017). *Nanobots: Development and Future*. https://doi.org/10.15406/ijbsbe.2017.02.00037

Roufayel, R., Younes, K., Al-Sabi, A., & Murshid, N. (2022). BH3-only proteins noxa and puma are key regulators of induced apoptosis. *Life, 12*(2), 256. https://doi.org/10.3390/LIFE12020256

Sabir, S., Arshad, M., & Chaudhari, S. K. (2014). Zinc oxide nanoparticles for revolutionizing agriculture: Synthesis and applications. *The Scientific World Journal, 2014*(1), 925494. https://doi.org/10.1155/2014/925494

Sabuncuoglu, S. (2020). *Mycotoxins and food safety*. IntechOpen. https://doi.org/10.5772/intechopen.77743

Shaba, E. Y., Jacob, J. O., Tijani, J. O., & Suleiman, M. A. T. (2021). A critical review of synthesis parameters affecting the properties of zinc oxide nanoparticle and its application in wastewater treatment. *Applied Water Science, 11*(2), 1–41. https://doi.org/10.1007/S13201-021-01370-Z

Sharma, V., Anderson, D., & Dhawan, A. (2012). Zinc oxide nanoparticles induce oxidative DNA damage and ROS-triggered mitochondria mediated apoptosis in human liver cells (HepG2). *Apoptosis, 17*(8), 852–870. https://doi.org/10.1007/S10495-012-0705-6/METRICS

Shelke, D. B., Islam, N. F., Chambhare, M. R., Sonawane, H. B., Patowary, R., Prasad, R., & Sarma, H. (2023). Enhancing secondary metabolites and alleviating environmental stress in crops with mycogenic nanoparticles: A comprehensive review. *Biocatalysis and Agricultural Biotechnology, 52*(August), 102805. https://doi.org/10.1016/j.bcab.2023.102805

Siddiqi, K. S., ur Rahman, A., Tajuddin, & Husen, A. (2018). Properties of zinc oxide nanoparticles and their activity against microbes. *Nanoscale Research Letters, 13*(1), 1–13. https://doi.org/10.1186/S11671-018-2532-3

Singh, T. A., Sharma, A., Tejwan, N., Ghosh, N., Das, J., & Sil, P. C. (2021). A state of the art review on the synthesis, antibacterial, antioxidant, antidiabetic and tissue regeneration activities of zinc oxide nanoparticles. *Advances in Colloid and Interface Science, 295*, 102495. https://doi.org/10.1016/J.CIS.2021.102495

Sivaramakrishnan, M., Jagadeesan Sharavanan, V., Karaiyagowder Govindarajan, D., Meganathan, Y., Devaraj, B. S., Natesan, S., Kothandan, R., & Kandaswamy, K. (2019). Green synthesized silver nanoparticles using aqueous leaf extracts of Leucas aspera exhibits antimicrobial and catalytic dye degradation properties. *SN Applied Sciences, 1*(3), 1–8. https://doi.org/10.1007/S42452-019-0221-1/FIGURES/9

Slavin, Y. N., Bach, H., Pallavicini, P., Slavin, Y. N., & Bach, H. (2022). Mechanisms of antifungal properties of metal nanoparticles. *Nanomaterials, 12*(24), 4470. https://doi.org/10.3390/NANO12244470

Smijs, T. G., & Pavel, S. (2011). Titanium dioxide and zinc oxide nanoparticles in sunscreens: Focus on their safety and effectiveness. *Nanotechnology, Science and Applications, 4*(1), 95–112. https://doi.org/10.2147/NSA.S19419

Sonawane, H., Arya, S., Math, S., & Shelke, D. (2021). Myco-synthesized silver and titanium oxide nanoparticles as seed priming agents to promote seed germination and seedling growth of Solanum lycopersicum: A comparative study. *International Nano Letters, 11*(4), 371–379. https://doi.org/10.1007/s40089-021-00346-w

Sriramulu, M., Shukla, D., & Sumathi, S. (2018). Aegle marmelos leaves extract mediated synthesis of zinc ferrite: Antibacterial activity and drug delivery. *Materials Research Express, 5*(11), 115404. https://doi.org/10.1088/2053-1591/AADD88

Sundaram Sanjay, S., Pandey, A. C., Kumar, S., & Pandey, A. K. (2014). Cell membrane protective efficacy of ZnO nanoparticles. *SOP Transactions on Nano-Technology, 1*(1).

Suresh, D., Nethravathi, P. C., Udayabhanu, Rajanaika, H., Nagabhushana, H., & Sharma, S. C. (2015). Green synthesis of multifunctional zinc oxide (ZnO) nanoparticles using Cassia fistula plant extract and their photodegradative, antioxidant and antibacterial activities. *Materials Science in Semiconductor Processing*, *31*, 446–454. https://doi.org/10.1016/J.MSSP.2014.12.023

Szczyglewska, P., Feliczak-Guzik, A., & Nowak, I. (2023). Nanotechnology–general aspects: A chemical reduction approach to the synthesis of nanoparticles. *Molecules*, *28*(13), 4932. https://doi.org/10.3390/MOLECULES28134932

Tanino, R., Amano, Y., Tong, X., Sun, R., Tsubata, Y., Harada, M., Fujita, Y., & Isobe, T. (2020). Anticancer activity of ZnO nanoparticles against human small-cell lung cancer in an orthotopic mouse model. *Molecular Cancer Therapeutics*, *19*(2), 502–512. https://doi.org/10.1158/1535-7163.MCT-19-0018/88336/AM/ANTICANCER-ACTIVITY-OF-ZNO-NANOPARTICLES-AGAINST

Thakkar, K. N., Mhatre, S. S., & Parikh, R. Y. (2010). Biological synthesis of metallic nanoparticles. *Nanomedicine: Nanotechnology, Biology and Medicine*, *6*(2), 257–262. https://doi.org/10.1016/J.NANO.2009.07.002

Vasantharaj, S., Sathiyavimal, S., Senthilkumar, P., Kalpana, V. N., Rajalakshmi, G., Alsehli, M., Elfasakhany, A., & Pugazhendhi, A. (2021). Enhanced photocatalytic degradation of water pollutants using bio-green synthesis of zinc oxide nanoparticles (ZnO NPs). *Journal of Environmental Chemical Engineering*, *9*(4), 105772. https://doi.org/10.1016/J.JECE.2021.105772

Vedhanayagam, M., Unni Nair, B., & Sreeram, K. J. (2018). Collagen-ZnO scaffolds for wound healing applications: Role of dendrimer functionalization and nanoparticle morphology. *ACS Applied Bio Materials*, *1*(6), 1942–1958. https://doi.org/10.1021/ACSABM.8B00491/SUPPL_FILE/MT8B00491_SI_001.PDF

Verma, R., Pathak, S., Srivastava, A. K., Prawer, S., & Tomljenovic-Hanic, S. (2021). ZnO nanomaterials: Green synthesis, toxicity evaluation and new insights in biomedical applications. *Journal of Alloys and Compounds*, *876*, 160175. https://doi.org/10.1016/J.JALLCOM.2021.160175

Virych, P. A., Zadvornyi, T. V., Borikun, T. V., Lykhova, O. O., Chumachenko, V. A., Virych, P. A., Pavlenko, V. A., Kutsevol, N. V., & Lukianova, N. Y. (2022). Effects of dextran-graft-polyacrylamide/ZnO nanoparticles on prostate cancer cell lines in vitro. *Experimental Oncology*, *44*(3), 217–221. https://doi.org/10.32471/EXP-ONCOLOGY.2312-8852.VOL-44-NO-3.18452

Vitasovic, T., Caniglia, G., Eghtesadi, N., Ceccato, M., Bojesen, E. D., Gosewinkel, U., Neusser, G., Rupp, U., Walther, P., Kranz, C., & Ferapontova, E. E. (2024). Antibacterial action of Zn2+ ions driven by the in vivo formed ZnO nanoparticles. *ACS Applied Materials and Interfaces*. https://doi.org/10.1021/ACSAMI.4C04682/SUPPL_FILE/AM4C04682_SI_001.PDF

Wani, A. H., Amin, M., Shahnaz, M., & Shah, M. A. (2012). Antimycotic activity of nanoparticles of MgO, FeO and ZnO on some pathogenic fungi. *Nanotechnology: Concepts, Methodologies, Tools, and Applications*, *2*(4), 59–70. https://doi.org/10.4018/IJMMME.2012100105

Wu, J., Wang, X., Wang, Q., Lou, Z., Li, S., Zhu, Y., Qin, L., & Wei, H. (2019). Nanomaterials with enzyme-like characteristics (nanozymes): Next-generation artificial enzymes (II). *Chemical Society Reviews*, *48*(4), 1004–1076. https://doi.org/10.1039/C8CS00457A

Zazouli, M. A., Eslamifar, M., & Javan, F. (2021). Water disinfection using silver and zinc oxide nanoparticles. *Journal of Nano Research*, *69*, 105–121. https://doi.org/10.4028/WWW.SCIENTIFIC.NET/JNANOR.69.105

Zhang, H., Shan, Y., & Dong, L. (2014). A comparison of TiO2 and ZnO nanoparticles as photosensitizers in photodynamic therapy for cancer. *Journal of Biomedical Nanotechnology*, *10*(8), 1450–1457. https://doi.org/10.1166/JBN.2014.1961

Zhang, X., Zhao, N., & He, C. (2020). The superior mechanical and physical properties of nanocarbon reinforced bulk composites achieved by architecture design—A review. *Progress in Materials Science*, *113*, 100672. https://doi.org/10.1016/J.PMATSCI.2020.100672

11 Plant-Derived Nanocellulose

Unveiling Extraction Techniques and Therapeutic Applications

Jitendra Patel, Surendra Agrawal, Pankaj Kumar Chaurasia, Harish Rajak, and Vijay Kumar Patel

11.1 INTRODUCTION

Cellulose, the most abundant organic polymer on the planet, is a key component of plant cell walls, which provide strength and rigidity. Besides, it is essential to the carbon cycle and also is an ingredient of biomass and ecological sustainability. In nature, cellulose sustains soil health, provides a vital energy source to herbivores via microbial digestion, and is industrially applied in paper, textiles, and biofuels. Its biodegradability makes it environmentally important for sustainable development (Aziz et al., 2022).

Nanocellulose refers to nanoscale cellulose materials derived from plant fibers, bacteria, or tunicates with outstanding mechanical resistance, biodegradability, and high surface area. Its adoption has revolutionized many industries such as biomedical, packaging, and nanocomposites due to its light weight, renewable, and ecofriendly nature. Nanocellulose improves medicine delivery, wound healing, water filtration, and sustainable alternatives to plastics. Due to its importance in the promotion of green technologies, improvement of material performance, and support of a circular bioeconomy, nanocellulose serves as a key sustainable material driving innovation across multiple sectors (Phanthong et al., 2018).

Plant-based nanocellulose is mainly obtained from lignocellulosic biomass including wood, agricultural residues, or non-woody plants like cotton, flax, and hemp. There is a considerable supply of cellulose for nanocellulose extraction from perennial lignocellulosic materials, such as banana peels, coconut husks, wheat straw, and sugarcane bagasse. And algae and bacterial cellulose also enable more sustainable manufacture. Thus, renewable precursor sources yield eco-friendly and biodegradable nanomaterials that find extensive applications in biomedicine, packaging, and nanocomposites, representing key aspects of green technology while reducing dependence on synthetic materials (Curvello et al., 2021).

DOI: 10.1201/9781003559900-11

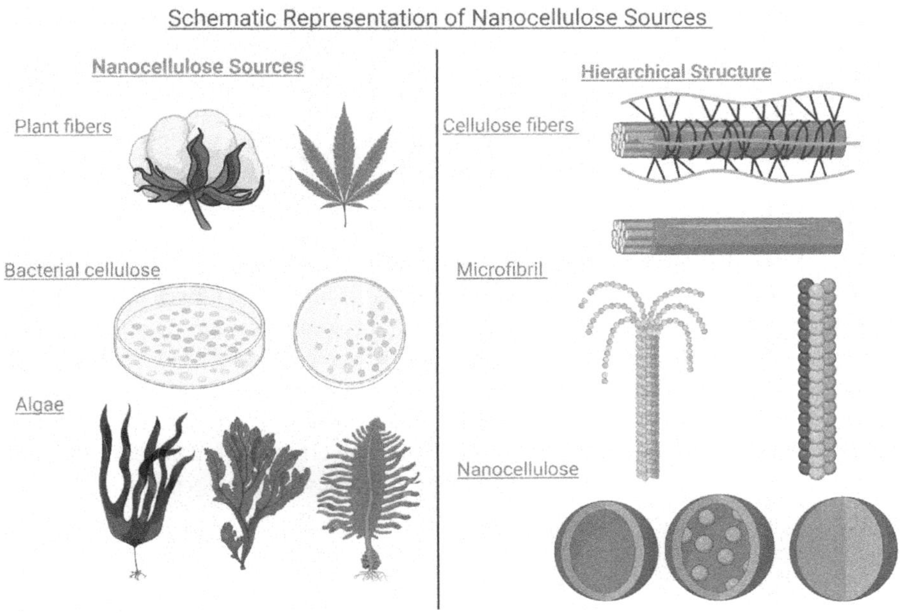

FIGURE 11.1 Different nanocellulose sources.

Nanocellulose is gaining attention for use in medical applications owing to its having biocompatibility, biodegradability, and non-toxicity. With such elevated surface area and mechanical integrity, it is best suited for pharmaceutical delivery, wound repair, and tissue engineering. The nanocellulose-derived hydrogels promote controlled drug release, and their antibacterial properties play a role in infection control. Its role in biosensors and regenerative medicine increases its significance. Nanocellulose opens up new sustainable and efficient biomedical solutions, in turn decreasing reliance on synthetic materials (Das et al., 2022) through nanocellulose research. The various nanocellulose sources are given in Figure 11.1.

11.2 METHODS FOR EXTRACTING NANOCELLULOSE

The major source of nanocellulose is plant fibers, which keep good properties, mechanical strength, and biocompatibilities. The extraction techniques such as enzymatic, chemical, mechanical, and sustainable processes are used to gain nanocellulose from lignocellulosic matters (Shamsuri et al., 2022).

11.2.1 Pre-Treatment of Plant Fibers

Separating noncellulosic materials such as hemicellulose, lignin, and pectin is crucial for better noncellulosic extraction. Mechanical operations like grinding and milling help cellulose fibers lose their size. Chemical treatments including alkali and bleaching are carried out to improve quality and fibrillation activity (Long et al., 2021). Using alkaline sodium hydroxide, cellulose effectiveness and bond breaking of lignin-carbohydrates are investigated. An oxidative mechanism such as hydrogen peroxide and peracetic acid treatment is done to help the fibrillation process. These procedures improve extraction's efficacy (Sabeeh et al., 2020).

11.2.2 Mechanical Techniques

Using a limited aperture in measured pressure fragmentation of fibers into small scale size, a great pressure halogenation is applied. This process gives nanocellulose in a homogenous form with better dispersion quality (Patrignana & Lanciotti, 2016). Shear force is used through microfluidization to convert cellulose fibers into nanocellulose. Colloidal stability and lattice fibrillation through microchannel are achieved (Kavinila et al., 2023). Sound waves with high frequency through ultrasonication make cavitations that rupture fibers and enhance the efficacy of nanocellulose. This enzymatic and chemical treatment enhances efficiency and integrity (Kutlu et. al., 2022).

11.2.3 Chemical Techniques

Acid hydrolysis employing predominantly sulfuric acid (H2SO4) selectively hydrolyzes amorphous cellulose to produce nanocrystalline cellulose NCC. This technique yields a high-purity NCC with regulated particle size although it necessitates at times meticulous handling due to the corrosive properties of acids and possible environmental implications (Cresswell et al., 2019). 2,2,6,6-Tetramethylpiperidine-1-oxyl (TEMPO) mediated oxidation selectively modifies cellulose at the position C6 resulting in the introduction of carboxyl groups. This technique improves water dispersibility and enables nanofibrillation, yielding highly charged cellulose nanofibers (CNF) with enhanced functional characteristics (Signori-Iamin et al., 2022).

11.2.4 Enzymatic Methods

Enzymatic hydrolysis uses cellulases and other hydrolytic enzymes to selectively decompose cellulose yielding nanocellulose while retaining its crystalline structure. This biocatalytic method guarantees environmentally sustainable processing enhanced regulation of fiber dimensions and reduced energy consumption relative to mechanical and chemical techniques (Vasić et al., 2021).

11.2.5 Eco-Friendly and Sustainable Strategies

Innovative eco-friendly plans prioritize reducing ecological damage. Ionic liquids and deep eutectic solvents dissolve cellulose without harsh reagents enabling the extraction of nanocellulose. Biological approaches using microbial fermentation or enzymatic bioprocessing provide sustainable alternatives for the manufacture of nanocellulose (Mathew et al. 2020). These extractions collectively increase the effective nanocellulose and each method has unique advantages and limitations. Current

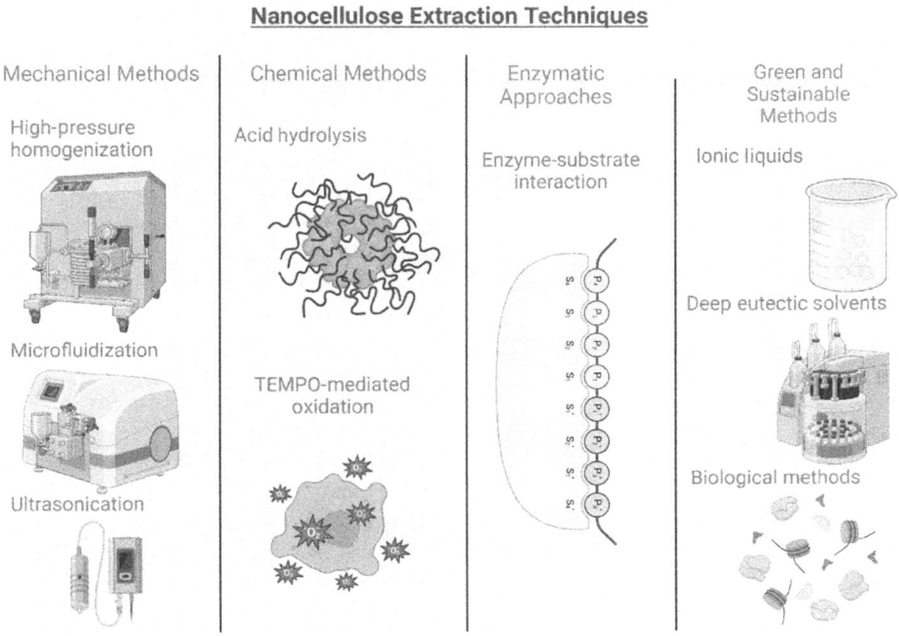

FIGURE 11.2 Nanocellulose extraction technique.

research seeks to increase technologies for scalable, economical, and environmentally sustainable nanocellulose manufacturing.

11.2.5.1 Characterization of Nanocellulose

The challenging characterization of nanocellulose for assessing its structural, physicochemical, and functional efficacy, so as to verify its suitability for various applications. The significant techniques of characterization and nanocellulose application to nanofibers at invention are discussed in this part. The various nanocellulose extraction techniques are given in Figure 11.2.

11.2.6 Structural Characterization

X-ray diffraction provides understanding of the crystallinity index and phase composition of nanocellulose as well as explores its crystal structure. This distinguishes areas with amorphous and crystalline forms influencing mechanical and thermal characteristics (Huang et al., 2022). Accurate images of nanocellulose morphology clarifying surface structure fibrillation and aggregation come via scanning electron microscopy. Understanding particle size and fiber dispersion calls for this method (Datye & DeLaRiva, 2023). Microscopic study of nanocellulose made possible by transmission electron microscopy provides thorough understanding of fiber diameter, shape, and distribution. It works very well to distinguish nanocrystalline cellulose from CNF (Pu et al., 2020).

TABLE 11.1
Summary of Characterization Techniques and Their Purpose

Characterization Technique	Purpose
X-ray Diffraction (XRD)	Determines crystalline structure and degree of crystallinity of nanocellulose
Scanning Electron Microscopy (SEM)	Provides surface morphology and structural details at the micro and nanoscale
Transmission Electron Microscopy (TEM)	Analyzes internal nanostructure, particle size, and dispersion of nanocellulose
Fourier Transform Infrared Spectroscopy (FTIR)	Identifies functional groups and chemical modifications in nanocellulose
Zeta Potential Measurement	Assesses surface charge and colloidal stability of nanocellulose suspensions
Thermogravimetric Analysis (TGA)	Evaluates thermal stability and decomposition temperature of nanocellulose.
Differential Scanning Calorimetry (DSC)	Measures phase transitions, heat capacity, and thermal properties.
Atomic Force Microscopy (AFM)	Provides topographical imaging and mechanical properties like surface roughness and stiffness.
Dynamic Light Scattering (DLS)	Determines particle size distribution and polydispersity index in nanocellulose dispersions.
X-ray Photoelectron Spectroscopy (XPS)	Analyzes surface chemical composition and elemental states.

11.2.7 Physicochemical Characteristics

Zeta potential testing addresses colloidal stability and surface charge of nanocellulose dispersions. Self-assemble behavior, electrostatic interactions, and suspension stability are among the parameters this one influences. Differential scanning calorimetry (DSC) and thermogravimetric analysis (TGA) probe thermal stability. While DSC evaluates changes in heat flow connected to phase transitions, TGA quantifies weight loss relative to temperature and reveals patterns of degradation (Zeng et al., 2022). A summary of characterization techniques and their purpose is given in Table 11.1.

11.2.8 Analysis of Functional Groups

Fourier transform functional groups found in nanocellulose are identified via infrared spectroscopy. It provides understanding of chemical changes, hydroxyl group interactions, and oxidation states, therefore enabling material characterization and modification (Yudianti et al., 2011). Understanding the characteristics of nanocellulose will help one to maximize it for many commercial and biological uses.

11.3 NANOCELLULOSE IN BIOMEDICAL AND PHARMACEUTICAL APPLICATIONS

11.3.1 Nanocellulose as a Pharmaceutical Delivery System

With their biocompatibility tunable porosity and large surface area, nanocellulose hydrogels and nanogels provide a multifarious platform for drug administration. These approaches improve treatment efficacy by means of regulated pharmaceutical distribution, hence lowering adverse effects. Functionalized nanocellulose makes tailored distribution better possible by means of functionalizing nanocellulose, so making it suitable for several therapeutic uses (Liu et al., 2017)

11.3.2 Applications in Wound Healing and Tissue Engineering

Nanocellulose is important in wound healing and tissue engineering offering biocompatible scaffolds that increase cell adhesion and proliferation. Its antibacterial and antioxidant efficacy contribute to infection prevention and the enhancement of tissue regeneration. Nanocellulose-based dressings improve moisture retention and prepare an ideal environment for wound healing.

11.3.3 Antimicrobial Characteristics of Nanocellulose

Nanocellulose has inherent antibacterial properties by compromising microbial membranes, thus impeding bacterial proliferation. Surface changes with antimicrobial agents augment bioactivity, making nanocellulose a viable material for biomedical coatings and packaging. Its biocompatibility and its affinity to combine with other antibacterial agents expand its potential in healthcare applications.

11.3.4 Nanocellulose in Biosensors

Biosensors based on nanocellulose provide enhanced sensitivity and specificity for the detection of biomolecule in medical diagnostics. Its superior electrochemical and optical characteristics augment sensor efficacy. These biosensors are used in the detection of illness, environmental surveillance, and food safety applications. Functionalized nanocellulose enhances signal transduction positioning it as a novel instrument in point-of-care diagnostics (Neubauerova et al., 2020).

11.4 OBSTACLES AND PROSPECTIVE OUTLOOKS

Notwithstanding its potential applications, large-scale nanocellulose production encounters challenges pertaining to energy consumption, expensive extraction techniques, and variability in raw materials. It is imperative to develop efficient, environmentally sustainable, and economically viable processing methods to meet industrial requirements while maintaining consistent quality and yield for extensive biomedical and pharmaceutical applications (Zhang et al., 2023).

11.4.1 Concerns Regarding Stability and Safety in Medical Devices

The long-term biocompatibility, biodegradability, and possible cytotoxicity of nanocellulose are significant concerns. It is essential to ensure safety that the interaction between chemical substances interacting with biological systems is clear. Stability concerns such as aggregation and degradation in physiological environments must be resolved through advanced modification methods to guarantee effective and safe application in drug administration and tissue engineering (Zaini et al., 2023).

11.4.2 Regulatory Considerations and Standardization

The absence of established practices and regulatory requirements impedes the clinical translation of nanocellulose. Establishing explicit frameworks for safety, efficacy, and toxicity evaluation is crucial for regulatory approval. Collaboration is essential to establish rules that promote the commercialization of nanocellulose-based biomedical products (Ji & Wang, 2023).

11.4.3 Future Research Directions in Nanocellulose-Based Biomedicine

Future research directions in nanocellulose biomedicine will be directed at nanocelluloses-based biomedicine and nanotechnology. Future research should concentrate on improving functionalization techniques, augmenting bioavailability, and advancing tailored drug delivery systems. Nanocellulose has been shown in personalized medicine. Regenerative therapy and biosensing uses will reveal new applications. Progression in nanotechnology and integrated biomaterials will propel the emerging generation of nanocellulose-based medicinal advancement (Khalid et al., 2023).

11.5 CONCLUSION

Nanocellulose has emerged as a versatile biomaterial with immense potential in biomedical and pharmaceutical applications. It is derived from plant-based sources through mechanical, chemical, and enzymatic methods. Advanced extraction techniques, such as high-pressure homogenization, acid hydrolysis, and enzymatic hydrolysis, ensure controlled production of nanocellulose with desirable physicochemical properties and enhanced applicability in different fields.

Nanocellulose plays a promising role in biomedical delivery systems, wound healing, tissue engineering, and biosensing. Its biocompatibility, high surface area, and tunable functionalization enable controlled drug release by antibiotics, improved antimicrobial properties, and enhanced diagnostic capabilities. Despite these advantages, challenges such as large-scale production, stability, safety, regulatory concerns, and operational concerns must be addressed to facilitate clinical translation. Future research should focus on optimizing functional modifications, improving sustainability, and exploring novel biomedical applications. Nanocellulose is gaining substantial promise in revolutionizing modern medicine by paving the way for innovative and eco-friendly therapeutic solutions.

REFERENCES

Aziz, T., Farid, A., Haq, F., Kiran, M., Ullah, A., Zhang, K., Li, C., Ghazanfar, S., Sun, H., Ullah, R., Ali, A., Muzammal, M., Shah, M., Akhtar, N., Selim, S., Hagagy, N., Samy, M., & Al Jaouni, S. K. (2022). A review on the modification of cellulose and its applications. *Polymers, 14*(15). https://doi.org/10.3390/polym14153206

Cresswell, S. L., Loughlin, W. A., Coster, M. J., & Green, D. M. (2019). Development and production of interactive videos for teaching chemical techniques during laboratory sessions. *Journal of Chemical Education, 96*(5). https://doi.org/10.1021/acs.jchemed.8b00647

Curvello, R., Kerr, G., Micati, D. J., Chan, W. H., Raghuwanshi, V. S., Rosenbluh, J., Abud, H. E., & Garnier, G. (2021). Engineered plant-based nanocellulose hydrogel for small intestinal organoid growth. *Advanced Science, 8*(1). https://doi.org/10.1002/advs.202002135

Das, S., Ghosh, B., & Sarkar, K. (2022). Nanocellulose as sustainable biomaterials for drug delivery. *Sensors International, 3*. https://doi.org/10.1016/j.sintl.2021.100135

Datye, A., & DeLaRiva, A. (2023). Scanning electron microscopy (SEM). In *Springer handbooks*. Springer, Cham. https://doi.org/10.1007/978-3-031-07125-6_18

Huang, X., Ai, C., Yao, H., Zhao, C., Xiang, C., Hong, T., & Xiao, J. (2022). Guideline for the extraction, isolation, purification, and structural characterization of polysaccharides from natural resources. *EFood, 3*(6). https://doi.org/10.1002/efd2.37

Ji, C., & Wang, Y. (2023). Nanocellulose-stabilized Pickering emulsions: Fabrication, stabilization, and food applications. *Advances in Colloid and Interface Science, 318*. https://doi.org/10.1016/j.cis.2023.102970

Kavinila, S., Nimbkar, S., Moses, J. A., & Anandharamakrishnan, C. (2023). Emerging applications of microfluidization in the food industry. *Journal of Agriculture and Food Research, 12*. https://doi.org/10.1016/j.jafr.2023.100537

Khalid, M. Y., Arif, Z. U., Noroozi, R., Hossain, M., Ramakrishna, S., & Umer, R. (2023). 3D/4D printing of cellulose nanocrystals-based biomaterials: Additives for sustainable applications. *International Journal of Biological Macromolecules, 251*. https://doi.org/10.1016/j.ijbiomac.2023.126287

Kutlu, N., Pandiselvam, R., Kamiloglu, A., Saka, I., Sruthi, N. U., Kothakota, A., Socol, C. T., & Maerescu, C. M. (2022). Impact of ultrasonication applications on color profile of foods. *Ultrasonics Sonochemistry, 89*. https://doi.org/10.1016/j.ultsonch.2022.106109

Liu, J., Willför, S., & Mihranyan, A. (2017). On importance of impurities, potential leachables and extractables in algal nanocellulose for biomedical use. *Carbohydrate Polymers, 172*. https://doi.org/10.1016/j.carbpol.2017.05.002

Long, Y., Zhang, Z., Fu, K., & Li, Y. (2021). Efficient plant fibre yarn pre-treatment for 3D printed continuous flax fibre/poly(lactic) acid composites. *Composites Part B: Engineering, 227*. https://doi.org/10.1016/j.compositesb.2021.109389

Mathew, G. M., Mathew, D. C., Sukumaran, R. K., Sindhu, R., Huang, C. C., Binod, P., Sirohi, R., Kim, S. H., & Pandey, A. (2020). Sustainable and eco-friendly strategies for shrimp shell valorization. *Environmental Pollution, 267*. https://doi.org/10.1016/j.envpol.2020.115656

Neubauerova, K., Carneiro, M. C. C. G., Rodrigues, L. R., Moreira, F. T. C., & Sales, M. G. F. (2020). Nanocellulose-based biosensor for colorimetric detection of glucose. *Sensing and Bio-Sensing Research, 29*. https://doi.org/10.1016/j.sbsr.2020.100368

Patrignani, F., & Lanciotti, R. (2016). Applications of high and ultra high pressure homogenization for food safety. *Frontiers in Microbiology, 7*. https://doi.org/10.3389/fmicb.2016.01132

Phanthong, P., Reubroycharoen, P., Hao, X., Xu, G., Abudula, A., & Guan, G. (2018). Nanocellulose: Extraction and application. *Carbon Resources Conversion, 1*(1). https://doi.org/10.1016/j.crcon.2018.05.004

Pu, S., Gong, C., & Robertson, A. W. (2020). Liquid cell transmission electron microscopy and its applications. *Royal Society Open Science, 7*(1). https://doi.org/10.1098/rsos.191204

Sabeeh, M., Zeshan, Liaquat, R., & Maryam, A. (2020). Effect of alkaline and alkaline-photocatalytic pretreatment on characteristics and biogas production of rice straw. *Bioresource Technology, 309*. https://doi.org/10.1016/j.biortech.2020.123449

Shamsuri, A. A., Siti Nurul, S. N. A., & Abdan, K. (2022). Nanocellulose extraction using ionic liquids: Syntheses, processes, and properties. *Frontiers in Materials, 9*. https://doi.org/10.3389/fmats.2022.919918

Signori-Iamin, G., Santos, A. F., Corazza, M. L., Aguado, R., Tarrés, Q., & Delgado-Aguilar, M. (2022). Prediction of cellulose micro/nanofiber aspect ratio and yield of nanofibrillation using machine learning techniques. *Cellulose, 29*(17). https://doi.org/10.1007/s10570-022-04847-5

Singh, H., Kumar Verma, A., Kumar Trivedi, A., & Gupta, M. K. (2023). Characterization of nanocellulose isolated from bamboo fibers. *Materials Today: Proceedings*. https://doi.org/10.1016/j.matpr.2023.02.300

Vasić, K., Knez, Ž., & Leitgeb, M. (2021). Bioethanol production by enzymatic hydrolysis from different lignocellulosic sources. *Molecules, 26*(3). https://doi.org/10.3390/molecules26030753

Yudianti, R., Onggo, H., Sudirman, Saito, Y., Iwata, T., & Azuma, J. I. (2011). Analysis of functional group sited on multi-wall carbon Nanotube surface. *Open Materials Science Journal, 5*. https://doi.org/10.2174/1874088X01105010242

Zaini, H. M., Saallah, S., Roslan, J., Sulaiman, N. S., Munsu, E., Wahab, N. A., & Pindi, W. (2023). Banana biomass waste: A prospective nanocellulose source and its potential application in food industry—A review. *Heliyon, 9*(8). https://doi.org/10.1016/j.heliyon.2023.e18734

Zeng, Y., Chen, E., Zhang, X., Li, D., Wang, Q., & Sun, Y. (2022). Nutritional value and physicochemical characteristics of alternative protein for meat and dairy—A review. *Foods, 11*(21). https://doi.org/10.3390/foods11213326

Zhang, F., Shen, R., Li, N., Yang, X., & Lin, D. (2023). Nanocellulose: An amazing nanomaterial with diverse applications in food science. *Carbohydrate Polymers, 304*. https://doi.org/10.1016/j.carbpol.2022.120497

12 Regulatory and Ethical Considerations in Plant-Based Nanomedicine

Subham K. Vishwakarma, Vaibhav Singh, Ekta Shirbhate, Rakesh K Kore, Vijay Kumar Patel, and Harish Rajak

12.1 INTRODUCTION

Plant-based nanomedicine is an exciting new area that has gained new life as a Field of Treatment by being physicochemically safe and highly efficacious. These very aspects lend the greatest power to the design and development of novel treatment modalities: while preclinical data could be considered promising assets, the slow pace of clinical translation has begun to gnaw at the heart of its multifaceted regulatory and ethical challenges. To sum up, sustainability, environmental impact, and bio-exploration would be the primary ethical issues concerning plant-based nanomedicine. Plant collecting and their components ought to be done sustainably and without upsetting the natural surroundings. Thirdly, it is also important to protect indigenous knowledge and resources against exploitation without fair compensation for the same to uphold ethical standards. Lastly, plant-based nanomedicines have long-term implications on human health, which need to be studied, and informed consent has to be obtained from participants in the clinical trials.

Pharmaceutical studies often analyze the quality, safety, and efficacy of medications, which has a direct impact on public health and welfare; in fact, product quality assurance is a philosophy that serves as the backbone of pharmaceutical sciences (Kulkarni and Shaw 2016). Initially, the development of a medication commences with the natural, unrefined (crude) drug. The eradication of inferior products and the quality of crude medications used in medical treatment form the foundation of pharmacognosy and pharmaceutical sciences, which have evolved and expanded in tandem with advances in medical science and medicine (Goda 2022). The processing of plants or plant parts for therapeutic reasons is known as phytomedicine, or herbal medicine. The strict requirements of the medicinal product regulation or other relevant legislative frameworks, like the medical devices regulation, must be met by nanomedicines since they constitute a new product class in the healthcare industry.

DOI: 10.1201/9781003559900-12

Once a drug is developed, strict quality control measures are implemented throughout the manufacturing process. This process includes different types of testing processes that ensure the finished product meets the defined standards of excellence. The tests may include raw material and chemical analysis, microbiological tests, and stability parameter evaluations (Neeta Ratanghayra 2022). Pharmaceutical industries have strict standards and rules imposed by government authorities such as the Food and Drug Administration (FDA) in the United States or equivalent bodies worldwide. Regulations cover everything from manufacturing methods and labelling to clinical trials and post-market regulation of products. Since standardization of the active pharmaceutical ingredient and formulation of the phytomedicines from the medicinal plants may not be easy to standardize, the regulatory agencies' concern for standardization of herbal medicines makes it relatively easier to assess important considerations such as efficacy and safety (Barbosa et al. 2012).

A quite recent product category in the healthcare sector is nanomedicines. Laws connected to medicinal products or equivalent legal systems, such as those controlling medical equipment, place rigorous criteria on them. The great increase of worldwide trade in high-value botanical goods has brought the center of the debate questions related to such supply connected with ethnopharmacology and medicinal plant research. Identification and authentication of raw plant material and finished herbal products rely on phytochemical and sensory screening methods, respectively, to reveal species-specific characteristics and components (Parvin et al. 2023).

12.2 GLOBAL REGULATORY FRAMEWORK FOR NANOMEDICINES

The rapidly growing field of nanomedicine presents significant opportunities as well as challenges in the regulation of the activities of the sector at the international level. Regulation agencies around the world face critical challenges of having to develop overall and harmonized frameworks for regulatory oversight of emerging development and adoption of nanomaterials within healthcare (Gupta and Xie 2018). The United States, United Kingdom, European Union, and Australia have greatly contributed to forming the regulatory principles for nanomedicines. These principles aim to consider specific physicochemical properties and associated risks of nanoscale materials, which might be different compared to their corresponding bulk counterparts. Innovative Medicines Initiative in Europe and the National Center for Advancing Translational Sciences in the United States are the emerging powerful platforms that will help to strengthen regulatory sciences for nanomedicines (Dave, Sur, and Gupta 2021).

Development and clinical translation of nanomaterials have faced several challenges that include extensive property diversity of the new materials as well as critical points to identify and control within the manufacturing process. The implementation of quality-by-design concepts like process analytical technologies has been suggested to ensure approaches to online quality assessment. Despite the lack of well-defined, universally accepted protocols for the preclinical development and characterization of nanomedicines, regulatory entities have taken steps forward in proposing comprehensive and harmonized regulatory approaches to bridge this gap (Dave, Sur, and Gupta 2021; Csóka et al. 2021).

12.2.1 Food and Drug Administration (FDA)

The FDA is closely supervising applications of nanotechnology since it is the most powerful organization in charge of approving and regulating pharmaceuticals and medical devices in the USA. The FDA released a guidance paper titled "Nanotechnology Applications to Food and Drug Administration Regulated Products" (Office of the Commissioner 2019) in 2014 which outlined its nanotechnology policy. This policy was important since it was the first action done towards including nanotechnology in product control. Through nanomedicine, changes in USA's regulatory rules towards herbal medical items are emerging progressively through nanotechnology. The FDA's activities are critical in protecting the public from insurgent products as well as making sure that the products work. With respect to the former, significant challenges, however, include the nature of nanomaterials, such as high dimensionality and complexity due to interconnections, which tend to be problematic. Still, the FDA strives to meet these obstacles in this new area of nanomedicine and its way of thinking (Paradise 2019).

The FDA is the leading authority with the general overall primary regulatory responsibility for the approval and monitoring of drugs and medical devices in the United States. In 2014, the FDA issued a guidance document titled "Nanotechnology Applications to Food and Drug Administration Regulated Products," in which the agency clearly stated its policy on nanotechnology (Office of the Commissioner 2019). In that policy, the approach consisted of:

12.2.1.1 Product Characterization

The FDA has stated that the characterization of the properties of nanomaterials is one area that needs to be closely examined in order to make predictions in biological contexts.

12.2.1.2 Safety Assessment

Based on their possible hazard, mode of distribution within the body, and long-term consequences, they support the application of an FDA safety assessment strategy. The FDA says that sponsors are urged to conduct some of their research to address these safety concerns.

12.2.1.3 Clinical Trial Design

Clinical studies on nanomedicines should take particular behaviour at the nanoscale into account, the FDA advises. This entails deciding on suitable dosages and closely observing for any side effects of particular uniqueness. Herbal nanomedicine is progressively being implemented in the USA under the direction of nanotechnology in regulatory policy. In this regard, the efforts of the Food and Drug Administration (FDA) are absolutely vital in controlling the safety and efficacy of such items. Still, the great dimensionality and connections of nanomaterials provide difficulties.

12.2.1.4 Guidance

The FDA has clarified particular aspects of nanotechnology and issued further advice on the need for precision in classifying nanomaterials using often difficult phonemes and other word categories.

12.2.2 Challenges and Future Directions

12.2.2.1 Lack of Consensus

More accurate attribution of a single meaning or regulation to engineered mist-coated particles is needed, which leads to problems with product safety regulation and marketing.

12.2.2.2 Emerging Research

The maturing of herbal nanomedicines in the growing scientific literature demonstrates continuous development that calls for proper regulatory control over the layers and profundo that follow it.

12.2.3 European Medicines Agency (EMA)

12.2.3.1 Defining Nanomedicines in the EU

Nanomedicine, by combining chemical, metabolic, pharmacological, and immunological properties, can provide benefits to patients with new and innovative treatments that traditional medicines may not be able to. The European Union has recognized nanotechnology as a Key Enabling Technology that can help patients by offering novel and inventive medical options for diagnostics and therapies. Additionally, nanotechnology is an area of investment and technology that will help European industries stay competitive and open up new markets.

The European Medicines Agency is responsible for the scientific assessment, monitoring, and safety of medicines in the EU. In 2014, the EMA published a reflection paper titled "Nanomedicines," which provides guidance on the development and evaluation of nanomedicines. The evaluation of the quality, safety, and effectiveness of nanomedicines was one of the key aspects focusing on the need to perform extensive characterization of nanomaterials and their potential impact. The key points are:

12.2.3.1.1 Definition and Scope

The word "nanomedicine" refers to some medicinal formulation employing nanotechnology, though an all-encompassing definition is still lacking.

12.2.3.1.2 Regulatory Challenges

Aspects of nanomedicine make it difficult to incorporate them into the traditional regulatory framework; thus, a new regulation is necessary to mitigate the risks presented.

12.2.3.1.3 Quality Assurance

The EMA suggests that there are advanced quality control and quality assurance procedures which need to be implemented in the course of development.

12.2.3.1.4 Collaboration

Proper regulation will improve the safety of patients while enhancing innovation which is essential for the industry by harnessing industrial, academic, and regulatory collaborations.

12.2.4 Regulatory Framework of India

The regulatory and ethical framework for nanomedicine-based phytomedicine in India is evolving, reflecting the complexities of integrating nanotechnology with traditional herbal practices. Nanomedicine, mainly herbal nanomedicines (HNMs), presents both opportunities and challenges in regulation and governance. While HNMs offer enhanced bioavailability and reduced toxicity compared to conventional drugs (Teja et al. 2022), the lack of a clear regulatory framework in India poses governance gaps (Bhatia and Chugh 2017). A multi-level national governance system has been proposed to address these challenges involving regulatory space, policy regime, and stakeholder inclusion.

The Central Drugs Standard Control Organization (CDSCO) is crucial in overseeing the safety, efficacy, and quality of herbal nanomedicines in India. India, which is the second-largest accustomed drug producer globally, is currently carrying out the certainly the most generous nanotech and herbal practices collaboration. Complementarity between nanoparticles and traditional herbal medicine contributes to the therapeutic efficacy and the bioavailability of the approved drugs, meanwhile, it limits the toxicity. The regulatory set-up of the drug controller is key in herbal medicine, particularly in the field of nano. This functional relationship raises particular regulatory and ethical difficulties that will only be solved by a complete understanding of the current guidelines and practices. The legal framework of nanotechnology in traditional medicine is the main reason for that (DGHS 2024; Ashok kumar et al. 2023).

CDSCO monitors the approval of new drugs, particularly those with an herbal nature, which must pass efficacy and safety tests and elaborate clinical trials. On the contrary, there is evidence of the earmarking of drugs that did not prove their effectiveness or safety because of the insufficiency of the clinical data. The bioavailability of herbal nanomedicines develops a serious dilemma for those employing nanotechnology without proper authorization. To ensure the efficacy and safety of nanomedicines, a strict regulatory framework is essential. This will also make it easier for these products to go from preclinical research to clinical trials (McGettigan et al. 2015). There are some ethical problems with nanomedicine that need to be resolved. These moral dilemmas include informed consent, the safe use of nanomaterials, and the potential for adverse effects from their usage in medicinal applications. With the help of ethical frameworks that guide studies and help protect the rights and welfare of secondary stakeholders, the fields of nanomedicine can be led towards the right path. Furthermore, handling specific nanomedicine-related issues including the health and safety risks of nanomaterials requires integrating ethical impact assessment into the regulatory process (Hall, Sun, and Ferrari 2012; Allon et al. 2017).

Herbal medicines in India are not only regulated by CDSCO but also by the Ministry of AYUSH which is controlled by traditional systems of medicine (Singh and Shukla 2021). These regulatory bodies need to work in collaboration for bringing out new products taking full advantage of the plant kingdom production technologies through nanotechnology which comes under the broad category of phytopharmaceuticals. Herbal medicines have recently been embraced more and more because they are thought to be safer and work better. Thus, there is a necessity for strict quality control for the products not to be outdated (Sahane and Basarkar 2023; Nirmal et al. 2022).

12.2.4.1 The Regulatory and Ethical Considerations for Nanomedicine in India Involve Several Key Aspects

The Drugs and Cosmetics Act of 1940 is the main regulatory framework in India that governs nanomedicines. Nanotechnology-based medications are subject to approval and control by the Central Drugs Standard Control Organization (CDSCO) (DGHS 2024; Nirmal et al. 2022).

In India, the government and the Department of Science and Technology formed a working committee to regulate and formulate guidelines for nanotechnology-based products. The Indian government announced the first guidelines for nanomedicine regulation in 2019, incorporating the development of novel therapeutics and comparing them to current drugs (Đorđević et al. 2022).

12.2.4.2 Critical Points Regarding the Regulatory Framework for Nanotechnology in India

Certain important challenging aspects influence the regulatory framework for nanotechnological development in India (Malik and Patil 2020; Bhatia and Chugh 2017).

12.2.4.2.1 Lack of Consensus on Definitions

Lack of an internationally recognized definition for nanomaterials and nanomedicines creates difficulties for Indian regulatory initiatives. This lack of clarity makes it hard to label and control products, which leads to the same safety tests and rules that might not work well for modern technology. The inconsistency in definitions hinders the development of clear and effective regulatory guidelines, impacting the evaluation and approval processes for nanotechnology-based products.

12.2.4.2.2 Existing Regulatory Framework Limitations

In India current regulatory frameworks are insufficient to address the specific need of nanomedicine without requisite changes or supportive initiatives.

12.2.4.2.3 Multidisciplinary Nature

The multidisciplinary nature of nanomedicine presents a challenge in the development of a regulated framework that can effectively manage the field's numerous facets.

12.2.4.2.4 Cooperation of Stakeholders

In order to ensure that safe and efficient nanomedicines are developed in the future, industry and regulatory bodies must work together. Improved collaboration between the government, academia, and industry is essential for nanomedicine innovation and effective governance.

12.2.4.2.5 Finance and Research Integration

The integration of finance and research to accelerate research funding and increase research undertakings is yet another inhibiting factor in building a proper regulatory framework.

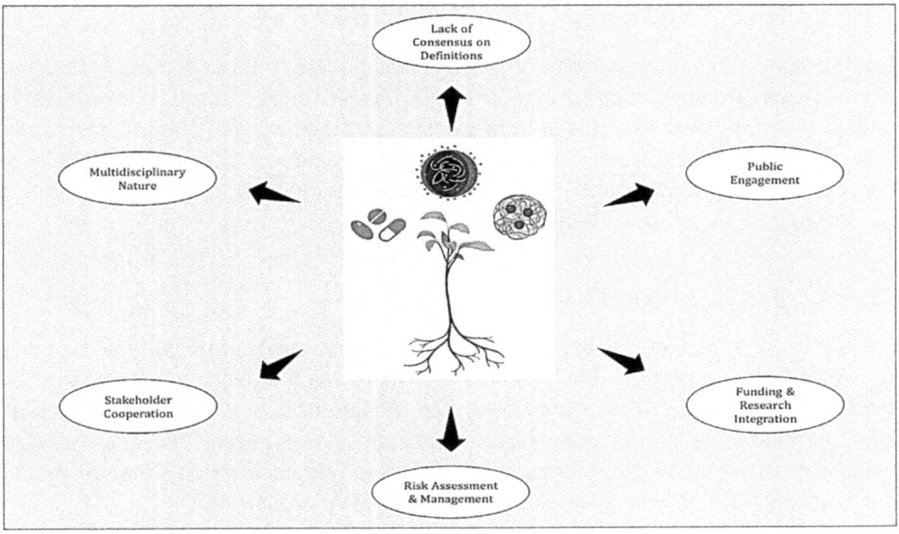

FIGURE 12.1 Key challenges in regulating plant-based nanomedicine in India.

12.2.4.2.6 Public Engagement

The risk versus advantage debate over nanotechnology may shape public opinion and also affect government decisions or regulatory action. Figure 12.1 demonstrates the main difficulties in regulating plant-based nanomedicine in India.

12.3 SAFETY CONCERNS AND RISK ASSESSMENT

12.3.1 Toxicological Evaluations

Material scientists, medics, and toxicologists will collaborate to evaluate nanomedicines technically. Comprehensive in vitro and in vivo investigations are required for this assessment of potential risk to human health. A thorough toxicokinetics evaluation can reveal the patterns of nanomaterial absorption, distribution, metabolism, and excretion in biological systems (Oberdörster 2010; Chen, Riviere, and Lin 2022).

12.3.2 Environmental Concerns

Understanding the ecological effect of nanomedicines depends on knowledge of environmental safety assessment. Early stage of development use of "Safety-by-Design" ideas helps reduce environmental hazards (Zielińska et al. 2020). This entails analyzing possible routes of environmental exposure and doing extensive risk studies to guard ecosystems against any risks connected to nanotechnology (Ma et al. 2024).

12.3.3 Long-Term Effects

Furthermore, the comprehension of long-term exposure to nanomedicine is dependent on careful safety monitoring. Research has to assess potential accumulative effects over long time periods and dose-response associations (Chen, Riviere, and Lin 2022). Therefore, effective monitoring strategies and continuous assessments of both foreseen and unintended effects of applications of nanomedicine are imperative (Accomasso, Cristallini, and Giachino 2018).

12.3.4 Batch-to-Batch Variability Assessment

For safety and effectiveness in nanomedicine manufacturing, uniformity and quality control play an important role. According to Scimeca and Verron (2022), specific safety studies for nano-engineered biomaterials are necessary to ensure robustness and reproducibility of risk assessment procedures. Maintaining the quality takes place over different manufacturing batches and should ensure that standard procedures are used for characterizing the physicochemical properties.

12.3.5 Drug-Nanocarrier Interactions

Safe and efficacious development of nanomedicine is become a reality only through the proper understanding of drug-nanocarriers interaction. The rapidly-expanding uses of nanomaterials need careful assessment of these interactions to avoid any negative consequences (Chen, Riviere, and Lin 2022). This includes analyzing possible changes in therapeutic efficacy in conjunction with nanocarriers, along with their release kinetics and stability of drug-carrier complexes (Patel et al. 2017).

A complete system assessing safety must cover short-term and long-term impacts on human health and environmental safety (Ma et al. 2024). Standardized methods and continuous monitoring systems provide the safe evolution and implementation of plant-based nanomedicines (Scimeca and Verron 2022).

12.4 COMMERCIALIZATION CHALLENGES

12.4.1 Manufacturing Standards

Manufacturing nanomedicines involves both strict attention to quality requirements and scalable manufacturing methods. Small laboratory batch size increased to large industrial volumes makes maintaining product consistency more challenging (Đorđević et al. 2022). Manufacturing methods must provide repeatability even if regulatory criteria for purity and safety of the final nanoparticle-based compositions are satisfied. Using Quality by Design (QbD) ideas will help to establish robust production techniques that provide constantly high-quality products (Paliwal, Jayachandra Babu, and Palakurthi 2014; Csóka et al. 2021).

12.4.2 Quality Assurance

In the manufacture of nanomedicine, quality assurance calls for thorough control systems and uniform testing procedures. Product quality evaluation with an eye towards physicochemical properties, quality control, and manufacturing process validation takes front stage among the most important factors. Enough quality control policies must be followed by manufacturers to guarantee batch-to-batch uniformity and product stability all throughout the shelf life. From the early phases of R&D, a risk-based approach should be used to guarantee effectiveness and safety (Đorđević et al. 2022; Csóka et al. 2021).

12.4.3 Market Authorization Requirements

Authorization of markets for nanomedicines requires careful regulatory paths and extensive documentation standards. Carefully balanced are various demands from patients (clinical and therapeutic use), industry (production), and regulatory bodies (authorization process). With particular attention to provide consistent methods for evaluating critical quality criteria, the authorization process demands complete information on quality, safety, and efficacy (Csóka et al. 2021).

12.4.4 Patent Protection

Although commercialization depends on the protection of intellectual property, in the field of nanomedicine it presents particular challenges. Under the promise of commercial sustainability, innovators may deliberately apply nanomedicine technology using a "commercialization by design approach." Patent strategies should take into account both the expected lengths of market exclusivity and technological improvements (Hemmrich et al. 2024).

12.4.5 Cost Considerations

The effectiveness of cost management can guarantee the twin aims of economic sustainability and accessibility. Balancing research and manufacturing costs with constraints in the healthcare system and commercial possibilities equals successful work (Hemmrich et al. 2024). Scale-up needs to be relatively inexpensive while maintaining product quality and legal compliance. To effectively permit translocation within nanomedicines, this puts variables of research cost against final product price to ascertain the economic feasibility within current healthcare systems (Zhang et al. 2020).

Not all variables yet have been taken into consideration within the process of commercialization, especially with the due diligence on the forepast as in stating, making sure that the final products will stay somehow affordable and commercially viable under present-day healthcare systems. Success would entail thorough scrutiny of processes in production, quality assurance, regulatory compliance, and market entry dynamics (Hemmrich et al. 2024; Wu, Wang, and Li 2020).

12.5 FUTURE PERSPECTIVES

12.5.1 Emerging Regulatory Trends

Different problems and prospects arise leading to an ever-changing regulatory environment for nanomedicines. Thus, the new regulation must be tailored to the specific features of therapeutic agents based on nanotechnology. The focus has now turned towards instituting particular sets of rules for plant-based nanomedicines taking into account their peculiar characteristics and manufacturing methods. Some current advancements point towards a move away from enforced safety measures into a more uniform system of regulatory evaluation (Ali, Neha, and Parveen 2023; Halamoda-Kenzaoui et al. 2019).

12.5.2 Harmony of Global Standards

The process of global standardization is ever more important to facilitate international product development and market access. The emerging nanomedicine sector stands to benefit from cross-sector collaborations and an examination of standards made for a variety of industrial applications (Halamoda-Kenzaoui et al. 2019). More emphasis is now focused on establishing common global standards to facilitate an accelerated approval process in many locales, with stringent requirements on safety and efficacy (Ali, Neha, and Parveen 2023).

12.5.3 Technological Advancements

New technologies in nanomedicine are changing the landscape and creating avenues for new therapies. Accordingly, nanotechnology-based therapeutic agents are providing trailblazing solutions to current healthcare challenges (Ali, Neha, and Parveen 2023). Enhanced delivery has been targeted toward improved drug efficacy and bioavailability, while new manufacturing options will increase production capability. Plant-based nanotechnology appears to be an environmentally friendly approach lowering ecological footprints while raising therapeutic potential (Karnwal et al. 2024).

12.5.4 Ethical Challenges in Future Applications

The future application of nanomedicine would demand a careful consideration of several ethical issues. Rapid developments in nanotechnology often outstrip the establishment of ethical standards and regulatory frameworks (Ali, Neha, and Parveen 2023). Key concerns include:

- Patient safety and informed consent
- Environmental impact
- Access to treatment
- Data privacy and security (Singh et al. 2024).

12.5.5 Research and Development Directions

Development in plant-based nanomedicine seems to be evolving toward more sophisticated and targeted approaches. The development of nanomedicine has enabled more precise disease identification and combined diagnosis and treatment capabilities (Singh et al. 2024). Future research directions focus on:

- Enhanced drug delivery systems
- Improved biocompatibility
- Sustainable production methods
- Integration of artificial intelligence and machine learning (Karnwal et al. 2024).

Nanomedicine in the development can much more accurately provide disease identification and diagnosis compared to any other medical technology. This technology has truly melded diagnosis and therapy into a single unit. Ultimately, the future of plant-based nanomedicine will require a tempered approach towards innovation duly backed by regulation to ensure that the technology reaches the market in the safest and soundest way. Here, researchers, industries, and regulatory bodies must continue to collaborate and address issues that arise while upholding stringent safety and efficacy requirements (Ali, Neha, and Parveen 2023; Halamoda-Kenzaoui et al. 2019). As Table 12.1 shows, an elaborate insight into the ethical and regulatory issues in the evolution of plant-based nanomedicine is developed with challenges and strategies in mind.

12.6 RECOMMENDATIONS AND GUIDELINES

On another note, if desired therapeutic outcomes are to be guaranteed, the development of plant-based nanomedicines will, however, have to depend on total implementation of best practices and regulatory compliance procedures. Quality by Design (QbD) concepts should stress consistent characterizing techniques and validated analytical procedures from the earliest research through to commercialization. Regulation should address assessment of safety and product quality, particularly in the domains of physicochemical characterization and manufacturing process optimization, from a multidisciplinary angle (Csóka et al. 2021; Bremer-Hoffmann, Halamoda-Kenzaoui, and Borgos 2018). Ethical models should address traditional medical ethics and contemporary nanoethics so that they can promote environmental and societal benefits, patient safety, and advancement. In the meantime, they should also learn about quality, safety, and efficacy parameters to develop uniform methods to test important quality parameters (Halamoda-Kenzaoui et al. 2019; Ali, Neha, and Parveen 2023). Supported by constant monitoring and evaluation systems, risk mitigating strategies demand early identification of possible hazards, thorough assessment of risk degree, and application of suitable control measures (Ali, Neha, and Parveen 2023). Supported by consistent methods and risk-based approaches, the efficient use of these guidelines calls for a combined approach between creativity and compliance. Keeping in mind the need for continuous evaluation and updating of the relevant practices with emerging scientific knowledge and regulatory requirements,

TABLE 12.1
Regulatory and Ethical Considerations in Development of Plant-Based Nanomedicine Challenges and Strategies

Aspect	Description	Challenges	Mitigations	References
Regulatory Framework	Governmental guidelines and protocols ensuring safety and efficacy	Varies by region, often ambiguous	Develop international standards	Mohi-ud-din et al. 2020
Approval Process	Steps to market approval for plant-based nanomedicines	Lengthy and costly process	Streamlining and harmonizing protocols	Mohi-ud-din et al. 2020
Ethical Concerns	Patient's rights and transparency	Ensuring informed consent and privacy	Strengthening ethical guidelines	Wasti et al. 2023
Research and Development	Steps involved in bringing plant-based nanomedicine to market	High R&D costs	Increased funding and collaboration	Wasti et al. 2023
Environmental Impact	Effects of nanomedicine on the environment	Potential toxicity	Conduct thorough ecological assessments	Wasti et al. 2023
Cost and Accessibility	Affordability of treatment and widespread availability	High costs and limited access	Subsidies and global distribution	Wasti et al. 2023
Quality Control	Ensuring consistency in production	Varying concentration and quality	Strict regulations and monitoring	Mohi-ud-din et al. 2020
Ethical Manufacturing	Responsible sourcing of plant materials	Potential exploitation of resources	Implement sustainability practices	Mohi-ud-din et al. 2020)
Patient Safety	Minimizing adverse effects and risks	Unknown long-term impact	Comprehensive clinical trials	(Mohi-ud-din et al. 2020)
Legal Issues	Intellectual property and patent laws	Complex and evolving legal landscape	Standardize global IP laws	(Mohi-ud-din et al. 2020)

stakeholder involvement and the building of public awareness is, indeed, of great help towards promoting transparency and informed decision-making. This is how such an umbrella framework ensures ethically sound research and regulatory compliance with the very strong promotion of plant nanomedicines safely and effectively (Bremer-Hoffmann, Halamoda-Kenzaoui, and Borgos 2018).

12.7 CONCLUSION

Apparently, with challenges arising in the area of ethics and law combined, plant nanomedicine gracefully drifts on the path between contemporary nanotechnology and traditional herbal medicine. The international regulatory scene, including institutions like the FDA, EMA, and CDSCO, underlines the significant need for harmonized international standards concerning the development and commercialization, showcasing the dynamic nature of this field of regulation. The mainstay remains safety; such risk analysis systems closely draw attention to both longer-term and acute impacts on human health and environmental health. Particularly, the Indian regulatory scenario puts into relevant context the important need for standardized toxicological evaluations, environmental impact assessments, and strong quality control measures to ensure safe development and applications of plant-based nanomedicine. It also shows the inherent challenges that arise in juxtaposing traditional medicine practice with modern nanotechnology within the walls of a defined regulatory system.

The commercialization challenges including manufacturing standards, quality assurance, and market authorization requirements compound the complexity of putting plant-based nanomedicines on the market while the balance between solid intellectual property protection and ensuring accessibility and affordability is necessary. For the on-coming plant-based nanomedicine to thrive, several factors will be critical: sustainable development of comprehensive and harmonious sets of regulations capable of adaptation to technological advances; implementation of strong safety assessment protocols; development of environmentally friendly, sustainable, and scalable manufacturing processes; and a set of ethical guidelines, which, while able to protect patients' rights, will actually promote equitable access through which plant-based nanomedicine thrives. While every development in technologies and methodologies aiding the movement of plant-based nanomedicine has the potential to change the therapeutic approaches while ensuring safety, efficacy, and ethical consideration in health care service provision, an effective way forward will require periodic collaboration among regulatory bodies, researchers, stakeholders, and health service providers in this specific area.

12.8 DECLARATION OF COMPETING INTEREST

The authors declare that they have no known competing financial interests or personal relationships that could have appeared to influence the work reported in this manuscript.

ACKNOWLEDGEMENTS

One of the authors, Ekta Shirbhate is thankful to the Department of Science and Technology (DST), New Delhi for providing research fellowship under DST-Inspire Fellowship Programme (Fellowship No: IF210097). The corresponding author, Harish Rajak is indebted to ICMR, New Delhi for providing financial support in the form of Extra-Mural research project (Grant number: 58/33/2020/PHA/BMS).

REFERENCES

Accomasso, Lisa, Caterina Cristallini, and Claudia Giachino. 2018. "Risk Assessment and Risk Minimization in Nanomedicine: A Need for Predictive, Alternative, and 3Rs Strategies." *Frontiers in Pharmacology* 9: 228. https://doi.org/10.3389/fphar.2018.00228.

Ali, Faraat, Kumari Neha, and Sana Parveen. 2023. "Current Regulatory Landscape of Nanomaterials and Nanomedicines: A Global Perspective." *Journal of Drug Delivery Science and Technology* 80: 104118. https://doi.org/10.1016/j.jddst.2022.104118.

Allon, Irit, Ahmi Ben-Yehudah, Raz Dekel, Jan-Helge Solbakk, Klaus-Michael Weltring, and Gil Siegal. 2017. "Ethical Issues in Nanomedicine: Tempest in a Teapot?" *Medicine, Health Care and Philosophy* 20 (1): 3–11. https://doi.org/10.1007/s11019-016-9720-7.

Ashok kumar, P., P. M. Divija, B. R. Reshma, K. R. Bhoomika, and R. Ramya Shree. 2023. "Regulatory Requirements for Approval of Biological Products as per CDSCO in Indian Comparison with South Korea." *World Journal of Advanced Research and Reviews* 20 (1): 758–67. https://doi.org/10.30574/wjarr.2023.20.1.2079.

Barbosa, Ramos, Wagner Luiz, Myrth Soares do Nascimento, Lucianna do Nascimento Pinto, Fernando Luiz Costa Maia, Antonio Jorge Ataide Sousa, Jose Otavio Carrera Silva Junior, Mauricia Melo, and Danilo Ribeiro de Oliveir. 2012. "Selecting Medicinal Plants for Development of Phytomedicine and Use in Primary Health Care." In *Bioactive Compounds in Phytomedicine*. InTech. https://doi.org/10.5772/26078.

Bhatia, Pooja, and Archana Chugh. 2017. "A Multilevel Governance Framework for Regulation of Nanomedicine in India." *Nanotechnology Reviews* 6 (4): 373–82. https://doi.org/10.1515/ntrev-2016-0083.

Bremer-Hoffmann, Susanne, Blanka Halamoda-Kenzaoui, and Sven Even Borgos. 2018. "Identification of Regulatory Needs for Nanomedicines." *Journal of Interdisciplinary Nanomedicine* 3 (1): 4–15. https://doi.org/10.1002/jin2.34.

Chen, Qiran, Jim E. Riviere, and Zhoumeng Lin. 2022. "Toxicokinetics, Dose–Response, and Risk Assessment of Nanomaterials: Methodology, Challenges, and Future Perspectives." *Wiley Interdisciplinary Reviews: Nanomedicine and Nanobiotechnology* 14 (6): 1808. https://doi.org/10.1002/wnan.1808.

Csóka, Ildikó, Ruba Ismail, Orsolya Jójárt-Laczkovich, and Edina Pallagi. 2021. "Regulatory Considerations, Challenges and Risk-Based Approach in Nanomedicine Development." *Current Medicinal Chemistry* 28 (36): 7461–76. https://doi.org/10.2174/0929867328666210406115529.

Dave, Vivek, Srija Sur, and Nikita Gupta. 2021. "Current Framework, Ethical Consideration and Future Challenges of Regulatory Approach for Nano-Based Products." In *Nanopharmaceutical Advanced Delivery Systems*, 447–72. Wiley. https://doi.org/10.1002/9781119711698.ch19.

DGHS. 2024. "Central Drugs Standard Control Organization Function." *Cdsco.Gov.In*. Accessed September 29. https://cdsco.gov.in/opencms/opencms/en/About-us/Functions/.

Đorđević, Snežana, María Medel Gonzalez, Inmaculada Conejos-Sánchez, Barbara Carreira, Sabina Pozzi, Rita C. Acúrcio, Ronit Satchi-Fainaro, Helena F. Florindo, and María J. Vicent. 2022. "Current Hurdles to the Translation of Nanomedicines from Bench to the Clinic." *Drug Delivery and Translational Research*, 1–26. https://doi.org/10.1007/s13346-021-01024-2.

Goda, Yukihiro. 2022. "Regulatory Science of Natural Products." *Journal of Natural Medicines* 76 (4): 732–47. https://doi.org/10.1007/s11418-022-01639-w.

Gupta, Ritu, and Huan Xie. 2018. "Nanoparticles in Daily Life: Applications, Toxicity and Regulations." *Journal of Environmental Pathology, Toxicology and Oncology* 37 (3): 209–30. https://doi.org/10.1615/JEnvironPatholToxicolOncol.2018026009.

Halamoda-Kenzaoui, Blanka, Uwe Holzwarth, Gert Roebben, Alessia Bogni, and Susanne Bremer-Hoffmann. 2019. "Mapping of the Available Standards Against the Regulatory Needs for Nanomedicines." *Wiley Interdisciplinary Reviews: Nanomedicine and Nanobiotechnology* 11 (1): 1531. https://doi.org/10.1002/wnan.1531.

Hall, Rebecca M., Tong Sun, and Mauro Ferrari. 2012. "A Portrait of Nanomedicine and Its Bioethical Implications." *Journal of Law, Medicine & Ethics* 40 (4): 763–79. https://doi.org/10.1111/j.1748-720X.2012.00705.x.

Hemmrich, Eva, and Scott McNeil. 2024. "Strategic Aspects for the Commercialization of Nanomedicines." *Journal of Controlled Release* 369: 617–21. https://doi.org/10.1016/j.jconrel.2024.04.003.

Karnwal, Arun, Amar Yasser Jassim, Ameer Abbas Mohammed, Vikas Sharma, Abdel Rahman Mohammad Said Al-Tawaha, and Iyyakkannu Sivanesan. 2024. "Nanotechnology for Healthcare: Plant-Derived Nanoparticles in Disease Treatment and Regenerative Medicine." *Pharmaceuticals* 17 (12): 1711. https://doi.org/10.3390/ph17121711.

Kulkarni, Vitthal S., and Charles Shaw. 2016. "An Overview of Regulatory Aspects for Pharmaceutical Semisolid Dosages." In *Essential Chemistry for Formulators of Semisolid and Liquid Dosages*, 223–37. Elsevier. https://doi.org/10.1016/B978-0-12-801024-2.00012-1.

Ma, Xiaohan, Yaxin Tian, Ren Yang, Haowei Wang, Latifa W. Allahou, Jinke Chang, Gareth Williams, Jonathan C. Knowles, and Alessandro Poma. 2024. "Nanotechnology in Healthcare, and Its Safety and Environmental Risks." *Journal of Nanobiotechnology* 22 (1): 715. https://doi.org/10.1186/s12951-024-02901-x.

Malik, Rashi, and Sharvil Patil. 2020. "Nanotechnology: Regulatory Outlook on Nanomaterials and Nanomedicines in United States, Europe and India." *Applied Clinical Research, Clinical Trials and Regulatory Affairs* 7 (3): 225–36. https://doi.org/10.2174/2213476X06666191129094236.

McGettigan, Patricia, Peter Roderick, Rushikesh Mahajan, Abhay Kadam, and Allyson M. Pollock. 2015. "Use of Fixed Dose Combination (FDC) Drugs in India: Central Regulatory Approval and Sales of FDCs Containing Non-Steroidal Anti-Inflammatory Drugs (NSAIDs), Metformin, or Psychotropic Drugs." *PLOS Medicine* 12 (5): e1001826. https://doi.org/10.1371/journal.pmed.1001826.

Mohi-ud-din, Roohi, Reyaz Hassan Mir, Faheem Hyder Pottoo, Gifty Sawhney, Mubashir Hussain Masoodi, and Zulfiqar Ali Bhat. 2020. "Nanophytomedicine Ethical Issues, Regulatory Aspects, and Challenges." *Nanophytomedicine: Concept to Clinic*, 173–92. https://doi.org/10.1007/978-981-15-4909-0_10.

Neeta Ratanghayra. 2022. "Quality Control During Drug Development." *Drug Discovery from Technology Networks*, December. https://www.technologynetworks.com/drug-discovery/articles/importance-of-quality-control-during-drug-development–368119.

Nirmal, Puneet, Rashmi Singh, Nitin Kumar, and Shalini Sharma. 2022. "Phytopharmaceutical Regulated New Class: An Industrial Initiative of Ayurvedic Drugs towards the Advancement of India System of Medicine." *World Journal of Advanced Research and Reviews* 15 (3): 407–15. https://doi.org/10.30574/wjarr.2022.15.3.0949.

Oberdörster, Gunter. 2010. "Safety Assessment for Nanotechnology and Nanomedicine: Concepts of Nanotoxicology." *Journal of internal medicine* 267 (1): 89–105. https://doi.org/10.1111/j.1365-2796.2009.02187.x.

Office of the Commissioner. 2019. "Considering Whether an FDA-Regulated Product Involves the Application of Nanotechnology." *U.S. Food and Drug Administration*, April. https://www.fda.gov/regulatory-information/search-fda-guidance-documents/considering-whether-fda-regulated-product-involves-application-nanotechnology.

Paliwal, Rishi, R. Jayachandra Babu, and Srinath Palakurthi. 2014. "Nanomedicine Scale-Up Technologies: Feasibilities and Challenges." *AapsPharmscitech* 15 (6): 1527–34. https://doi.org/10.1208/s12249-014-0177-9.

Paradise, Jordan. 2019. "Regulating Nanomedicine at the Food and Drug Administration." *AMA Journal of Ethics* 21 (4): E347–55. https://doi.org/10.1001/amajethics.2019.347.

Parvin, Shohana, Anika Reza, Sridebi Das, Md. Main Uddin Miah, and Shanjida Karim. 2023. "Potential Role and International Trade of Medicinal and Aromatic Plants in the World." *European Journal of Agriculture and Food Sciences* 5 (5): 89–99. https://doi.org/10.24018/ejfood.2023.5.5.701.

Patel, Parmi, and Jigna Shah. 2017. "Safety and Toxicological Considerations of Nanomedicines: The Future Directions." *Current Clinical Pharmacology* 12 (2): 73–82. https://doi.org/10.2174/1574884712666170509161252.

Sahane, Manish B., and Ganesh D. Basarkar. 2023. "A Review of Herbal Regulations and Approval Process in India and Europe." *International Journal of Drug Regulatory Affairs* 11 (4): 25–33. https://doi.org/10.22270/ijdra.v11i4.629.

Scimeca, J. C., and E. Verron. 2022. "Nano-Engineered Biomaterials: Safety Matters and Toxicity Evaluation." *Materials Today Advances* 15: 100260. https://doi.org/10.1016/j.mtadv.2022.100260.

Singh, Kalpana, Shiwani Singhal, Shilpa Pahwa, Vandana Arora Sethi, Shashank Sharma, Preeti Singh, R. D. Kale, S. Wazed Ali, and Suresh Sagadevan. 2024. "Nanomedicine and Drug Delivery: A Comprehensive Review of Applications and Challenges." *Nano-Structures & Nano-Objects* 40: 101403. https://doi.org/10.1016/j.nanoso.2024.101403.

Singh, Saloni, and Vikesh Kumar Shukla. 2021. "Current Regulations for Herbal Medicines in India." *International Journal of Drug Regulatory Affairs* 9 (2): 30–34. https://doi.org/10.22270/ijdra.v9i2.466.

Soares, Sara, João Sousa, Alberto Pais, and Carla Vitorino. 2018. "Nanomedicine: Principles, Properties, and Regulatory Issues." *Frontiers in Chemistry* 6: 360. https://doi.org/10.3389/fchem.2018.00360.

Teja, Parusu Kavya, Jinal Mithiya, Abhijeet S. Kate, Khemraj Bairwa, and Siddheshwar K. Chauthe. 2022. "Herbal Nanomedicines: Recent Advancements, Challenges, Opportunities and Regulatory Overview." *Phytomedicine* 96 (February): 153890. https://doi.org/10.1016/j.phymed.2021.153890.

Wasti, Sophia, Il Ho Lee, Sumin Kim, Jae-Hyun Lee, and Hannah Kim. 2023. "Ethical and Legal Challenges in Nanomedical Innovations: A Scoping Review." *Frontiers in Genetics* 14: 1163392. https://doi.org/10.3389/fgene.2023.1163392.

Wu, Lin-Ping, Danyang Wang, and Zibiao Li. 2020. "Grand Challenges in Nanomedicine." *Materials Science and Engineering* 106: 110302. https://doi.org/10.1016/j.msec.2019.110302.

Zhang, Chenyang, Liang Yan, Xin Wang, Shuang Zhu, Chunying Chen, Zhanjun Gu, and Yuliang Zhao. 2020. "Progress, Challenges, and Future of Nanomedicine." *Nano Today* 35: 101008. https://doi.org/10.1016/j.nantod.2020.101008.

Zielińska, Aleksandra, Beatriz Costa, Maria V. Ferreira, Diogo Miguéis, Jéssica M. S. Louros, Alessandra Durazzo, Massimo Lucarini et al. 2020. "Nanotoxicology and Nanosafety: Safety-by-Design and Testing at a Glance." *International Journal of Environmental Research and Public Health* 17 (13): 4657. https://doi.org/10.3390/ijerph17134657.

13 Bio-Based Nanoemulsions in Food Preservation

Muyideen Olaitan Bamidele,
Sandoval Jose Cortes, María Liliana Flores-López,
Olga Berenice Álvarez Pérez, Monica Lizeth Chavez,
and Cristóbal Noe Aguilar[1]

13.1 INTRODUCTION

Nanoemulsions have emerged as transformative technologies in various industries due to their characteristic fine oil-in-water or water-in-oil dispersions stabilized by surfactant molecules (Wang and Guo 2025). Their droplet sizes vary between 20 and 200 nm which enhances their properties such as surface area, optical and stability (Pintea et al. 2025). There are promising opportunities offered by nanoemulsion as a food preservative specially when fabricated by a bio-based component (Bose et al. 2025). The growing demand for natural and sustainable food additives by consumers has led to the integration of bio-based ingredients into nanoemulsions. These bio-based nanoemulsions offer not only profound eco-friendly alternatives to synthetic additives but also offer innovative solutions for extending the shelf life of perishable products (Bose et al. 2025). They also offer antioxidant and antimicrobial properties obtained from the components used in the formulation. These include essential oils, plants extracts and biopolymers (Sharma et al. 2024). The nano scale size of these nanoemulsions facilitates enhanced dispersion and bioavailability of the active components, ensuring protection against spoilage and microorganisms within complex food matrices (Sharma et al. 2024). Bio-based nanoemulsion also help to minimize the reliance on synthetic additive as a preservative while showing function ability and sustainability goals (Islam et al. 2024). Other ingredients used in the formulation of bio-based nanoemulsion include whey, soy, chitosan and alginate (Gupta et al. 2024). The application of these components into fabrication of bio-based nanoemulsions significantly reduces the environmental footprint of conventional food preservative methods while offering a greener pathway for enhancing food safety and quality (Karnwal and Malik 2024). Despite the aforementioned advantages, bio-based nanoemulsions face challenges which limit their adoption in the food industries, such as high cost of production, scalability concerns and regulatory problems (Çakmakçı, Polatoğlu and Çakmakçı 2024). The recent advancement in the formulation methods involving green approaches such as low energy emulsification and smart

DOI: 10.1201/9781003559900-13

delivery system have paved the way to overcoming these challenges (Pereira-Rojas et al. 2025). Interdisciplinary approaches involving biotechnological innovations and artificial intelligence promise to hasten the development of cost-effective nanoemulsions (Chaudhary, Vasistha and Rai 2024). The chapter explores the transformative potential offered by bio-based nanoemulsion in revolutionizing food preservation. By exploring fabrication, functionality and future perspectives, this study aims to provide a comprehensive understanding of how these cutting-edge technologies can enhance food safety, shelf life and sustainability in the face of global challenges.

13.2 MECHANISMS OF FOOD PRESERVATION

Food safety, shelf-life extension and other challenges in the food industries have necessitated the use of bio-based nanoemulsions as a powerful tool in food preservation while they also address consumer demands for natural and sustainable alternatives (Sharma et al. 2024). The primary mechanism for their effectiveness includes antimicrobial activity, antioxidant properties and barrier function (Figure 13.1). Each of these contributes uniquely to the preservation process (Karnwal and Malik 2024). Nanoemulsion disrupts the cell membranes leading to the inhibition of microorganisms and their death (Jiang et al. 2025). It also combats oxidation through free radical scavenging, forms protective hurdles to reduce moisture loss and slows the deterioration of food (Bose et al. 2025). These mechanisms do not only protect the safety and quality of foods, but also support environmental and health consciousness of food preservation practices (Maddaloni et al. 2025).

13.2.1 Antimicrobial Activity

The need to prevent microbial spoilage and foodborne pathogens has necessitated the use of bio-based nanoemulsion (Panwar et al. 2024). These nanoemulsions enhance

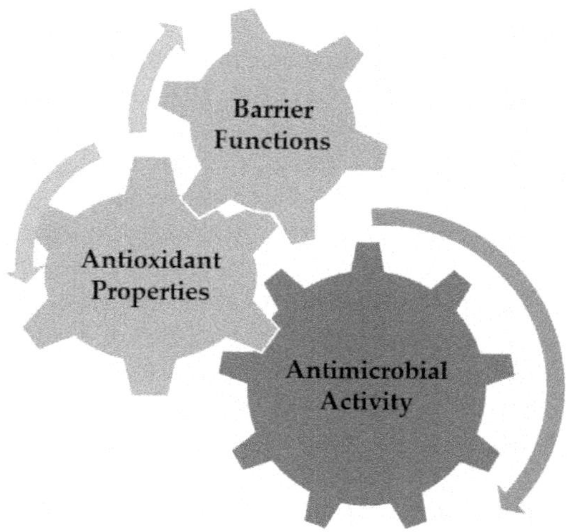

FIGURE 13.1 Mechanism of food preservation.

the stability, efficacy and delivery of the antimicrobial agent through encapsulation. The controlled release of the active ingredient increases its interaction with the cells of the microbes, biofilm penetration and synergistic actions of the nanoemulsion actions (Rathod et al. 2024). They have proven to be an essential tool in modern food preservation because of their ability to fight food spoilage organisms, such as *Listeria monocytogenes* and *Salmonella* spp., at low concentrations (Sharma et al. 2024). The death of the microorganisms is mostly due to the high surface area to volume ratio of the nanoemulsion droplets, which enhances the interaction between the cell and the antimicrobial agent leading to the leakage of intracellular components (Figure 13.2). This interaction has been revealed to effectively combat some microorganisms such as *Listeria monocytogenes* and *Salmonella* spp (Zhang et al. 2024). Bio-based nanoemulsions are able to overcome the protective matrix of the microorganisms by penetrating and destabilizing the cell to exact their effects. Rathod et al. (2024) reported over 70% reductions in biofilm biomass when nanoemulsion formulated with cinnamaldehyde was applied to the surface of fresh produce to improve microbial safety. Hasheminya and Dehghannya (2024) developed active packaging films using *Zhumeria majdae* essential oil nanoemulsion (ZMEO-NE). The ZMEO-NE reduced permeability, tensile strength, light transmission and thickness. The films also showed antioxidant capabilities and antibacterial efficacy against *Staphylococcus aureus* and *Escherichia coli*. Islam et al. (2024) combined clove and thyme essential oils in a nanoemulsion formulation, which enhances the inhibitory effects against multi-drug resistant *E. coli*, demonstrating their potential for tackling emerging food borne pathogens. These components work synergistically by enhancing the

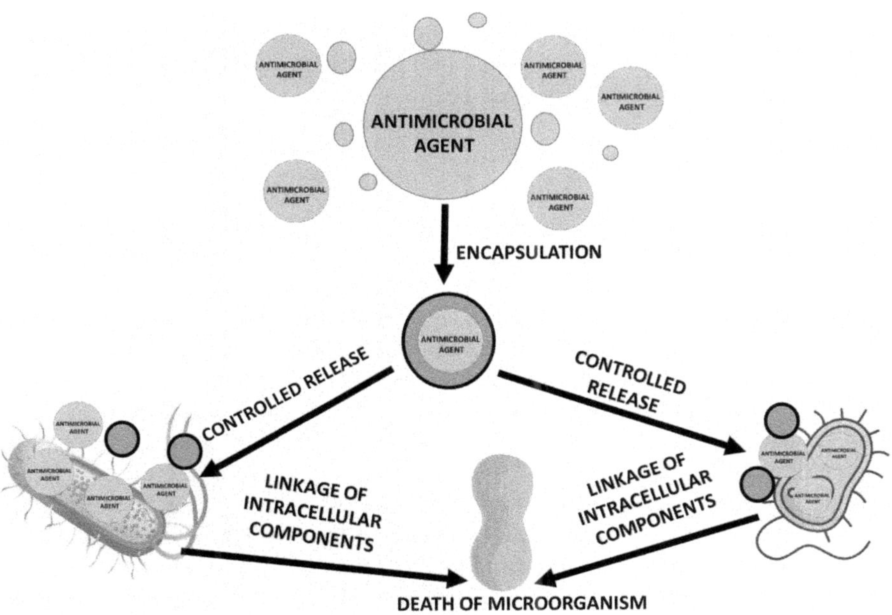

FIGURE 13.2 Mode of antimicrobial activity.

overall efficacy of the formulation as an antimicrobial agent beyond individual formulations with the essential oils.

13.2.2 Antioxidant Properties

Food spoilage has generally been linked to oxidative degradation resulting in discoloration, nutrient loss and off flavor. Antioxidant properties possessed by bio-based nanoemulsion offer a solution to mitigate this spoilage (Bandyopadhyay et al. 2023). These mechanisms include encapsulating natural antioxidants, free scavenger, metal chelators and regeneration of antioxidant system (Pei et al. 2024). Antioxidants such as tocopherols, ascorbic acids, phenolic compounds and carotenoids when encapsulated in form bio-based nanoemulsion shield them from environmental stressors and amplify antioxidant defense within the food matrix (Sharma et al. 2022). This encapsulation improves the stability of these antioxidants and facilitates their controlled release, ensuring sustained activity throughout the shelf life of the product (Panwar et al. 2024). Multilayer nanoemulsions and co-encapsulation strategies which are advanced techniques have been shown to improve antioxidant retention and functionality in complex food matrices (Martins et al. 2024). Bio-based nanoemulsions interrupt the oxidative chain reactions by acting as powerful scavengers of reactive oxygen species and free radicals (Rui et al. 2024). High fat systems such as in meat products have been protected with advanced formulations of bio-based nanoemulsion with radical scavenging capacity (Gohar et al. 2024). Fenton and Haber-Weiss reactions that facilitate lipid peroxidation have been inhibited by incorporating polyphenolic antioxidant compounds in bio-based nanoemulsion formulations (Alfei 2020). Synergistic combinations of different antioxidants within a single nanoemulsion system have been shown to outperform individual components and improve the oxidative stability in fortified oils and spreads (Ponnampalam et al. 2022).

13.2.3 Barrier Functions

Bio-based nanoemulsions as a preservative physically and chemically shield against moisture loss, oxygen infiltration and microbial contamination. They act as dual barriers for food safety and shelf-life extension forming edible coatings or thin films (Yashwanth et al. 2025). Recent advancements in nanoemulsion technology have led to composite coatings that integrate polysaccharides, proteins and lipids, offering superior flexibility and resistance (Islam et al. 2024). Chitosan and alginate when blended in nanoemulsion formulations enhance the performance of fresh cut fruits preservation and shelf-life extension (Panwar et al. 2024). Moisture retention and texture preservation are also improved with a nanoemulsion-based coating, which reduces the trans epidermal water loss from vegetables, fruits and baked goods (Sharma et al. 2022). A hydrophilic-hydrophobic layer which is an advanced formulation further enhances moisture retention capabilities. Lipid rich foods are also protected by high performance oxygen barriers, which prevent oxidative spoilage by minimizing oxygen exposure to food (Plyduang et al. 2021). When antioxidants such as polyphenols are incorporated into the nanoemulsion they also double the preservative effect. Active compounds in nanoemulsions when controlled and target released, ensure prolonged protection against spoilage. Innovative encapsulation technologies which include nano capsules and multilayer emulsion also enhance the release rate precision (Dailin et al. 2024).

13.3 ADVANCEMENTS IN FORMULATION TECHNIQUES

Advancements in bio-based nanoemulsion have been significant owing to the needs for efficient, sustainable and robust systems. This innovation focused on reducing energy consumption and improving stability and usage of environmentally friendly materials. These include advance methods such as high-pressure homogenization, ultrasonication, Phase Inversion Temperature (PIT) and spontaneous emulsification. Green surfactants are also used in hybrid techniques, and advanced characterization tools are central to these developments.

13.3.1 High-Energy Methods

13.3.1.1 High-Pressure Homogenization (HPH)

HPH when used in the formulation of bio-based nanoemulsion involves forcing an oil-water mixture through a narrow gap under high pressure. This generates intense shear forces, turbulence and cavitation resulting in uniformly dispersed nanoscale droplets (Pal et al. 2024). The industrial application of HPH has been optimized to reduce the energy demand while maintaining high kinetic stability (Malik et al. 2023). This method is valuable for the encapsulation of bioactive compounds and enhances their stability, solubility and bioavailability (Zhu et al. 2024). Integration of low energy emulsification and a multistage system are an innovation in the HPH method, which further improves the efficiency and quality of the products. The scalability and reliability of the product make HPH a preferred choice across industries, most importantly in food and pharmaceutical applications.

13.3.1.2 Ultrasonication

Ultrasonication usually generates shock waves and microjets that disrupt oil droplets into nanoscale sizes, and these high frequency sound waves create cavity that collapse the bubbles (Jiao et al. 2024). Nanoemulsions with uniform droplet sizes and preserved functional integrity for the bioactive compounds can be effectively produced using this technique (Nascimento et al. 2024). Programmable ultrasonic devices that allow the precise control of parameters such as amplitude, frequency and time are a recent advancement in the ultrasonication method (Huang et al. 2024). Thermal degradation of sensitive bioactive ingredients such as extracts and essential oils are minimized in the ultrasonication method of nanoemulsion formulation (Sharma et al. 2024). Ultrasonication can be combined with other techniques, such as microfluidization or enzymatic emulsification, to produce complex, multifunctional nanoemulsions with tailored properties and improved stability (Chen et al. 2024).

13.3.2 Low-Energy Methods

13.3.2.1 Phase Inversion Temperature (PIT) Method

Optimization of nanoemulsion in the PIT method involves the shifting of the hydrophilic-lipophilic balance of surfactants, which results in reverse phase inversion between oil in water and water in oil emulsion (Kumar et al. 2024). It's a thermodynamically stable technique for formulating nanoemulsion with reduced energy input. The sustainability of this method is owing to the use of bio-based surfactants such as saponins and phospholipids. This offers improved biodegradability, low toxicity and reduced environmental impact (Pereira-Rojas et al. 2025).

13.3.2.2 Spontaneous Emulsification

This is a low energy method for fabricating nanoemulsion with the use of amphiphilic molecules at the oil-water interface. It is driven by interfacial tension gradients and solvent diffusion, eliminating the need for external energy input such as heat or mechanical shear. Bio-based amphiphiles such as saponins and lecithin are used in the formation, reducing the impact on the environment. It also reduces energy consumption and chemical byproducts while ecofriendly solvents are used (Pereira-Rojas et al. 2025; Sharma et al. 2024).

13.3.3 Hybrid Techniques

This technique for nanoemulsion fabrication involves combining different methodologies taking advantage of individual advantages and mitigating their limitations. It is mostly beneficial for the achievement of superior control droplet size, stability and active ingredient encapsulation. The products are highly suitable for cosmetic, pharmaceutical and food applications (Gul et al. 2024).

13.3.3.1 Combined Ultrasound and HPH Techniques

These involve the combination of HPH and ultrasonication methods in the formulation of nanoemulsion with improved control droplet size, stability and active ingredient encapsulation (Anjum et al. 2024). They are particularly used in pharmaceutical and food grade formulations, due to the ability to load bioactive compounds and prolongation of the stability of the nanoemulsion. Ultrasonication being an emerging technique for nanoemulsion formulation makes use of renewable energy sources that reduce its impact on the environment. The combination of this makes the process control more adaptive with real-time monitoring of the emulsion properties to ensure quality. It is usually tailored to specific properties of the target nanoemulsion such as highly stable and functional products that require reduced energy (Burke et al. 2024).

13.3.3.2 Microwave-Assisted Nano Emulsification

The acceleration produced by the microwave is used in the emulsification process for reducing interfacial tension to enhance oil droplet dispersion. It is particularly more effective for thermally sensitive bioactive compounds, as controlled heating improves their efficiency without denaturing the compounds (Chaudhary 2024).

13.3.4 Computational Design and AI Integration

The design and optimization of bio-based nanoemulsion formulation with machine learning (ML) and artificial intelligence (AI) is transforming the field of food preservation. This emerging technique allows more efficient and precise formulations of the nanoemulsion by modeling complex behavior, predicting stability under storage conditions and optimizing ingredient ratios thereby reducing extensive trial and error (Gorde et al. 2024; Vats and Kumar 2025).

13.3.4.1 Modeling Emulsion Behavior for Food Preservation

An AI and ML algorithms model can be used to control dynamic behavior of nanoemulsion such as droplet formation, coalescence and phase transition under different environmental conditions like temperature, pH and humidity (Patil et al. 2024). This is crucial for understanding the behavior of food formulation throughout their shelf life. It is an essential tool for ensuring effective preservation of nanoemulsion while maintaining food quality. Researchers can fine-tune behaviors of nanoemulsion by simulation to achieve optimal food preservation (Lohita and Srijaya 2024).

13.3.4.2 Predicting Stability and Shelf Life

The stability of nanoemulsion can be predicted using an AI-driven model which analyzes factors such as droplet size, viscosity and surface charge (Meng et al. 2024). Its predictive functions are invaluable in food preservation as it ensures effectiveness in extending shelf life without compromising the texture, flavor or nutritional quality of food by the nanoemulsion (Islam et al. 2024). ML models can be used by manufacturers to forecast potential degradation over time of nanoemulsion and develop formulations that can withstand varying storage conditions from cold storage to ambient temperatures (Shi et al. 2024).

13.3.4.3 Optimizing Ingredient Ratios and Bioactive Delivery

Bio-based nanoemulsion composition ratios which include derived surfactant or essential oils can be optimized using an AI and ML model. This model can predict the ideal combinations of the components in the formulation by inputting data of the bioactive compounds, surfactants and oils (Boudesocque-Delaye et al. 2024). It also predicts the combinations that would enhance the antimicrobial, antioxidant and preservative properties of nanoemulsions. The optimization improves the effectiveness of nanoemulsion's ability to preserve active ingredients and ensure the formulation is sustainable using minimal resources to achieve maximum impact (Mahmud et al. 2024).

13.3.4.4 Sustainability and Customization for Specific Food Matrices

The sustainability of bio-based nanoemulsion can be enhanced using an AI and ML model, which is possible by identifying the most efficient renewable and biodegradable components in the formulation (Nayak et al. 2024). Bio-based nanoemulsion for specific food products such as dairy, bakery and fresh can be customized using this model. Nanoemulsion interaction properties of each food matrix can be predicted with an AI model. This allows for tailored formulations optimization of preservation while meeting consumer demand for natural and eco-friendly products (Gorde et al. 2024).

13.3.4.5 Real-Time Optimization and Scaling

The real-time optimization of nanoemulsion formulation to ensure consistency with quality and performance during scale up can be achieved with AI and ML (Vats and Kumar 2025). They can dynamically adjust component ratios, process parameter and packaging condition to respond to fluctuations in production or environmental factors. This ability is important for large-scale production of bio-based nanoemulsion used for food preservation (Ghoshal 2018).

13.4 COMPONENTS OF BIO-BASED NANOEMULSION

Bio-based nanoemulsion offers transformative solutions for food preservation as a sustainable technology. It is usually fabricated from biodegradable and renewable ingredients which align with sustainable developmental goals. This nanoemulsion addresses growing consumer demands for clean and ecofriendly food products. The functionality of the nanoemulsion depends on three major ingredients (Chaudhary 2024).

13.4.1 Oil Phase

The oil phase is a hydrophobic core that encapsulates lipophilic bioactive compounds to enhance solubility, improve bioavailability and enable controlled release in bio-based nanoemulsion (Majeed and Rather 2025). Nanoemulsion used for food preservation has drawn inspiration from nature to create suitable and efficient products (Wagermaier, Razghandi and Fratzl 2025).

13.4.2 Aqueous Phase

The aqueous phase transcends as a solvent acting as a multifunctional medium that enhances the delivery of bioactive ingredients, system stability and sustainability of the bio-based nanoemulsion. It also emerged as the key player in food preservation technologies as bifunctional agents in advanced formulations and natural hydrosols (Gupta et al. 2024).

13.4.3 Emulsifiers

They are the cornerstone of bio-based nanoemulsions by providing interfacial stability, preventing phase separation and enabling precise control of droplet size. The revolution in food preservation has led to the advancement of emulsifiers by improving the performance, sustainability and functionality of the nanoemulsion (Cheng et al. 2024). Different emulsifiers have been used in the formulations of bio-based nanoemulsions (Table 13.1).

TABLE 13.1
Use of Natural Emulsifiers in Bio-Based Nanoemulsion Formulations

Emulsifier	Source	Chemical Composition	Applications	Advantages	Challenges	Reference
Lecithin	Soybeans, sunflower seeds	Phospholipids, fatty acids	Food (beverages, dressings), cosmetics, pharmaceuticals	Biodegradable, GRAS (Generally Recognized as Safe), versatile	Sensitive to oxidation, limited pH stability	(Ekonomou et al. 2024)
Saponins	Quillaja bark, soapwort	Glycosides, sugar moieties	Food, pharmaceuticals, pesticides	Natural foaming agent, good emulsification properties	Bitterness, potential allergenicity	(Jiang et al. 2024)
Gum Arabic	Acacia trees	Polysaccharides, glycoproteins	Food, cosmetics, pharmaceuticals	High emulsification capacity, excellent stability	High cost, limited solubility at low pH	(Mudgil and Mudgil 2024)
Pectin	Citrus peels, apple pomace	Galacturonic acid, polysaccharides	Food, pharmaceuticals	Low-cost, readily available, forms stable emulsions	Requires calcium ions for gelation in some formulations	(Jia et al. 2024)
Protein Isolates	Whey, soy, pea	Amino acids, proteins	Food, nutraceuticals	Excellent nutritional profile, strong interfacial activity	Denaturation under heat, allergenic potential	(Pandey, Kumar, and Upadhyay 2024)
Chitosan	Shellfish, fungal cell walls	Polyglucosamine	Antimicrobial coatings, drug delivery systems	Antimicrobial properties, film-forming ability	Solubility issues, extraction costs	(Gallo et al. 2016)
Starch Derivatives	Corn, potato, tapioca	Amylose, amylopectin	Food, cosmetics, pharmaceuticals	Biodegradable, high availability, low cost	Retrogradation and syneresis in some formulations	(Chakraborty, Kalita, and Sen 2019)
Xanthan Gum	Fermentation by Xanthomonas sp.	Polysaccharides	Food, oil recovery, cosmetics	High viscosity, excellent emulsion stability	Shear-thinning behavior, microbial contamination risk	(Rashidi et al. 2023)

(Continued)

TABLE 13.1 (*Continued*)
Use of Natural Emulsifiers in Bio-Based Nanoemulsion Formulations

Emulsifier	Source	Chemical Composition	Applications	Advantages	Challenges	Reference
Agar-Agar	Red algae	Polysaccharides	Food, pharmaceuticals	High gel strength, excellent stabilizing properties	Brittleness in some applications	(Thiviya et al. 2023)
Pullulan	Aureobasidium pullulans	Polysaccharides	Edible films, cosmetics, pharmaceuticals	Biocompatible, film-forming, antioxidant properties	High production costs	(Rai et al. 2021)
Alginates	Brown algae	Alginic acid, polysaccharides	Food, pharmaceuticals, wound dressings	Excellent gel-forming properties, biodegradability	Requires calcium ions for stability	(Rahman et al. 2024)
Cellulose Derivatives	Plant cell walls	Cellulose, glucose	Food, cosmetics, pharmaceuticals	Abundant, non-toxic, cost-effective	Limited emulsification efficiency alone	(Bushra et al. 2023)
Casein	Milk	Phosphoproteins	Food, nutraceuticals	Good surface activity, stable emulsion formation	Allergenic potential, limited availability for vegan products	(Nikbakht Nasrabadi, Sedaghat Doost, and Mezzenga 2021)

13.5 APPLICATION IN FOOD PRESERVATION

The unique physiochemical properties of bio-based nanoemulsion led to their application in food preservation. They are formed from the dispersal of two immiscible liquids with nanoscale droplet size. They are typically formed from biodegradable and renewable resources (Cvanić et al. 2023). These emulsions provide a versatile platform for enhancing safety, quality and shelf life of food products. Some of their applications are included in Figure 13.3.

13.5.1 Shelf-Life Extension

The ability to extend the shelf life of perishable food such as yogurt, fresh cut fruits, raw meats and leafy vegetables by bio-based nanoemulsion has been transformative in food preservation. These nanoemulsions encapsulate active ingredients with

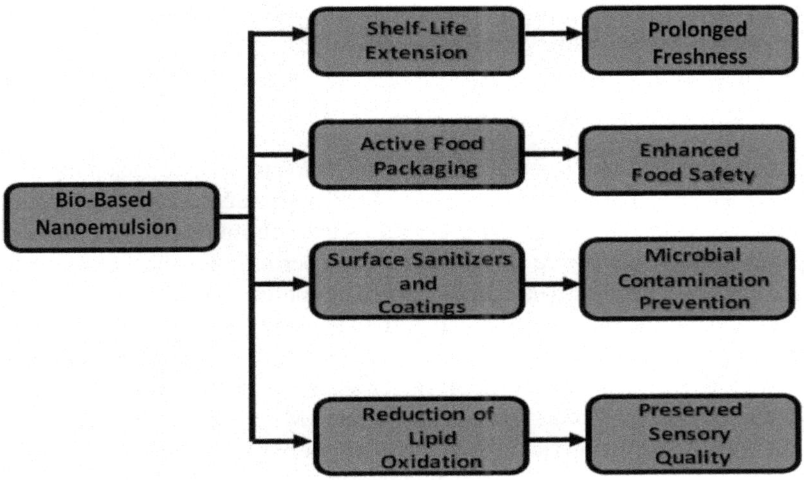

FIGURE 13.3 Applications of bio-based nanoemulsion in food preservation.

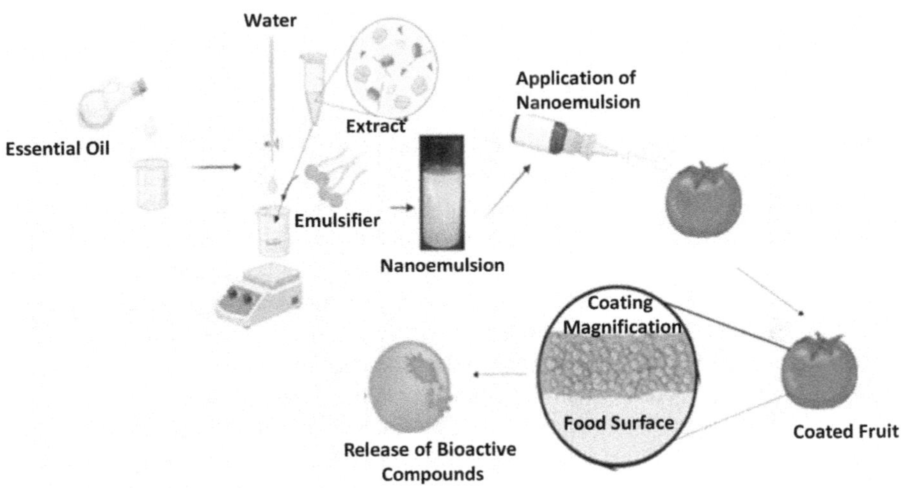

FIGURE 13.4 Extension of shelf life by nanoemulsion.

antioxidant and antimicrobial properties by creating a barrier against the microorganism causing spoilage and oxidation degradation (Figure 13.4). Karnwal and Malik (2024) reported that nanoemulsion formulated with rosemary and oregano essential oil inhibited bacterial growth in meat products, while Shikha Sharma et al. (2024) revealed that encapsulation of ascorbic acid reduces enzymatic browning in fresh fruits. Food waste had been minimized by these nanoemulsion as they delayed spoilage and thereby preserved nutritional quality. They also reduce the reliance on synthetic preservatives, aligning with sustainable food systems. Their other ability of controlled release of active ingredients also enhances the product safety and quality throughout the supply chain (Cvanić et al. 2023).

13.5.2 Active Food Packaging

The enhancement of food safety and quality by nanoemulsion as packaging material represents a cutting-edge approach in active food preservation. These nanoemulsions enriched with essential oils and/or tocopherols have antimicrobial and antioxidants properties respectively. These nanoemulsions can perform dual benefits if contained in the same formulations. They form a physical and chemical barrier against contamination while slowing down oxidative spoilage (Sharma et al. 2024; Cheng et al. 2024). Nanoemulsions when infused in edible films have been revealed to extend the of fruits, vegetables and dairy products by releasing active ingredients in a controlled release manner (Negi, Vanmathi Mugasundari and Moses 2024). The biodegradability nature of the nanomaterial also aligns with the growing consumer demand for sustainable and environmentally friendly materials to reduce the impact of plastic on the environment (Govere et al. 2024). These innovations not only reduce waste but also contribute to the development of smart packaging systems capable of responding to environment conditions such as temperature and humidity changes (Drago et al. 2020).

13.5.3 Surface Sanitizer and Coatings

Surface coating and spraying fruits and vegetables with nanoemulsion are sustainable alternatives to the conventional chemical additive. These formulations made with essential oils (e.g., thyme, oregano) and natural antioxidant (e.g., ascorbic acid) are effective towards target microbial contaminants such as *E. coli*, *Salmonella* and *Listeria monocytogenes* (Rout et al. 2022; Shahidi and Hossain 2022). The nanoscale delivery system ensures uniform coating with prolonged activity which enhances penetration into microstructures on fruits surfaces (Liu, Zhang and Bhandari 2020). This reduces the microbial loads and extends the shelf life of the produce. The sensory attributes of produce which include vibrant color, crisp texture and fresh flavor are preserved by nanoemulsion while ensuring consumer satisfaction and meeting the demands for clean-label solutions (Gali et al. 2024). This technology promotes sustainable postharvest management by reducing reliance on synthetic chemicals and contributes to reduction in food waste. These nanoemulsions can be tailored for specific production types, further optimizing their efficacy and market applicability (Benyam, Soma, and Fraser 2021).

13.5.4 Reduction of Lipid Oxidation

Oxidative rancidity of food is one major challenge of preserving lipid rich food. It usually results in degradation of fats and oils via oxidative reactions. These often cause off flavor, nutrient loss and reduced shelf life of the lipid rich food. The infusion of natural antioxidants such as tocopherols, carotenoids or polyphenols in bio-based nanoemulsion offers cutting edge solutions to lipid oxidation. When these active ingredients are encapsulated in nanoemulsion it enhances their anti-oxidant stability, improves solubility in lipid matrices and enables targeted action by free radical scavenging and interrupting oxidation routes (Sheibani et al. 2024; Padmaja et al. 2024). This technology has effectiveness in extending the shelf life of products like butter, margarine, cheese and snack foods while preserving sensory quality (Maddaloni et al. 2025). It also reduces the dependence on synthetic preservatives usage which aligns with the growing demand for clean and environmental food preservative solutions (Hussain et al. 2024). The scalability and versatility of the nanoemulsion-based system further position it as a transformative tool in the food industry that addresses both sustainability goals and consumer preferences (Kaur et al. 2024).

13.6 CASE STUDIES ON THE APPLICATION OF BIO-BASED NANOEMULSIONS IN PRESERVING DIFFERENT FOOD PRODUCTS

Perishable food such as diary, fresh food and meat can be preserved using innovative approaches such as bio-based nanoemulsions. They give thermal stability enhancement, superior emulsification capacity and multifunctional bioactivities. When essential oils are infused into the formulation, they can inhibit microorganisms that cause spoilage. They can also delay lipid oxidation in dairy products. These nanoemulsions form protective coatings and reduce reliance on synthetic additives. Case studies of nanoemulsion usage, methods and results obtained are highlighted in the following (Table 13.2).

13.7 COMPARATIVE ANALYSIS OF BIO-BASED NANOEMULSION OVER TRADITIONAL SYNTHETIC FOOD PRESERVATIVES

The demand for more sustainable, safer and more efficient food preservatives has led to the development of an innovative alternative to the conventional synthetic preservatives known as bio-based nanoemulsion. The comparative analysis investigates the safety, environmental impact and potential industrial adoption of the products (Table 13.3).

13.8 CHALLENGES OF BIO-BASED NANOEMULSION IN FOOD PRESERVATION

Several challenges have been identified in the use of bio-based nanoemulsion as a preservative, though it offers a promising solution to food sustainability. These problems include cost-effectiveness, formulation stability, regulatory compliance and consumer acceptance. The full potential of nanoemulsion would be unlocked if these obstacles were addressed. This will help revolutionize the use of nanoemulsion in food preservation practices (Cheng et al. 2024).

TABLE 13.2
Case Studies of Nanoemulsion as Preservative for Different Products

Case Study	Objective	Method	Results	Conclusion	Reference
Fruits (Preservation of Fresh Strawberries)	To extend the shelf life of strawberries using nanoemulsions containing essential oils.	A bio-based nanoemulsion was formulated using clove oil as the active antimicrobial agent, emulsified with lecithin. The strawberries were coated with the nanoemulsion and stored at 4°C.	(i) **Microbial Load Reduction**: The nanoemulsion-coated strawberries significantly reduced microbial load, with clove oil's antimicrobial activity inhibiting the growth of common spoilage bacteria and fungi compared to the untreated control group. **Extended Shelf Life**: The shelf life of strawberries was extended by up to 10 days, as the coating prevented rapid microbial growth and maintained the fruit's freshness. **Sensory Quality**: The sensory properties (taste, aroma and texture) of the strawberries were preserved, with no adverse effects reported from the nanoemulsion coating.	The use of clove oil-based nanoemulsions effectively extends the shelf life of strawberries by inhibiting microbial growth and maintaining sensory quality.	(Panwar et al. 2024; Sharma et al. 2022)
Vegetables (Preservation of Fresh-Cut Lettuce)	To enhance the quality and safety of fresh-cut lettuce using bio-based nanoemulsions.	Nanoemulsions were prepared using thyme essential oil and chitosan as the natural emulsifier. Fresh-cut lettuce was dipped in the nanoemulsion and stored at 4°C.	**Antimicrobial Effectiveness**: The nanoemulsion exhibited strong antimicrobial activity against *E. coli* and *Listeria monocytogenes*, reducing the microbial count on the lettuce. **Oxidative Stability**: The antioxidant properties of thyme essential oil prevented browning and maintained the color of the lettuce. **Quality Maintenance**: The treated lettuce retained its crisp texture and fresh appearance for up to 14 days, compared to 7 days for the untreated control.	Thyme essential oil-based nanoemulsions provide an effective means of preserving fresh-cut lettuce by offering antimicrobial protection and preventing oxidative browning.	(Liñán-Atero et al. 2024; Islam et al. 2024)

Dairy Products (Fortification and Preservation of Yogurt)	To fortify yogurt with bioactive compounds and extend its shelf life using bio-based nanoemulsions.	A nanoemulsion containing omega-3 fatty acids and green tea extract was prepared using whey protein as the emulsifier. The nanoemulsion was incorporated into yogurt during the production process.	**Nutrient Fortification**: The incorporation of omega-3 fatty acids and green tea extract enriched the yogurt with additional health benefits, including anti-inflammatory and antioxidant properties. **Microbial Stability**: The green tea extract provided antimicrobial effects, reducing the growth of spoilage microorganisms in the yogurt. **Shelf-Life Extension**: The nanoemulsion-fortified yogurt exhibited a longer shelf life, maintaining quality and safety for up to 30 days, compared to 21 days for the control yogurt.	Bio-based nanoemulsions can be used to fortify yogurt with beneficial bioactive compounds while extending its shelf life through antimicrobial and antioxidant effects.	(Paidari et al. 2024; Ghosh et al. 2024)
Meat Products (Preservation of Chicken Breast Fillets)	To extend the shelf life of chicken breast fillets using bio-based nanoemulsions.	A nanoemulsion was formulated using oregano essential oil and soy protein as the emulsifier. Chicken breast fillets were coated with the nanoemulsion and stored at 4°C.	**Antimicrobial Activity**: The oregano oil-based nanoemulsion significantly reduced the microbial load on the chicken fillets, including common spoilage bacteria and pathogens such as *Salmonella* and *Staphylococcus aureus*. **Oxidative Stability**: The antioxidant properties of oregano oil prevented lipid oxidation, maintaining the freshness and flavor of the chicken. **Shelf-Life Extension**: The coated chicken fillets had an extended shelf life of up to 15 days, compared to 7 days for the untreated fillets.	Oregano essential oil-based nanoemulsions effectively preserve chicken breast fillets by providing antimicrobial protection and preventing oxidative spoilage, thereby extending the product's shelf life.	(KovaČević et al. 2024; Rathod et al. 2024)

TABLE 13.3
Comparative Analysis of Bio-Based Nanoemulsion vs. Synthetic Food Preservatives

Criteria	Bio-Based Nanoemulsions	Synthetic Food Preservatives	Advantages of Bio-Based Nanoemulsions	Reference
Safety	Derived from natural, non-toxic sources (e.g., plant oils, proteins, polysaccharides). Minimal risk of allergies or toxic accumulation	May contain synthetic chemicals like benzoates, nitrates or sorbates, which can pose health risks, including allergies or carcinogenic concerns	Higher consumer acceptance due to natural origin; no harmful residues	(Islam et al. 2024; Chaudhary 2024)
Efficacy	Encapsulation enhances bioactive stability and controlled release of antimicrobial/antioxidant agents. Effective against a broad range of pathogens	Immediate effect but limited long-term stability due to degradation or resistance development	Sustained protection due to controlled release; multifunctional properties (antimicrobial, antioxidant and nutritional)	(Farasati Far et al. 2024; Zhang et al. 2024)
Environmental Impact	Fully biodegradable and derived from renewable resources. Production often follows green chemistry principles	Non-biodegradable residues and potential environmental toxicity. Manufacturing processes often rely on petrochemical derivatives	Lower ecological footprint, aligning with circular economy principles	(Biehl and Zhang 2024; Zhang et al. 2024)
Consumer Perception	Perceived as clean-label and natural; aligns with the growing demand for organic and sustainable products	Often viewed negatively due to artificial origin and potential health risks	Enhanced marketability in health-conscious and eco-conscious markets	(Karnwal and Malik 2024; Ul Islam, Khan, and Ahmad 2024)
Regulatory Compliance	Generally, GRAS (Generally Recognized as Safe) and easier to meet organic certification standards	Subject to strict regulations and restrictions due to health and environmental risks	Facilitates regulatory approval and market entry for organic and natural food products	(Dailin et al. 2024; Ghosh et al. 2024)
Stability	High stability under various environmental conditions due to nano-scale formulation and encapsulation	Prone to degradation under heat, light or pH fluctuations	Greater shelf life and functionality across diverse food matrices	(Rathod et al. 2024; Panwar et al. 2024)
Mode of Action	Multifunctional: provides antioxidant, antimicrobial, and emulsifying properties in one system	Primarily focused on antimicrobial or antioxidant activity with limited multifunctionality	Reduces the need for multiple additives, simplifying formulations	(Nayak et al. 2024; Revutskaya et al. 2024)

13.8.1 Stability Issues

Bio-based nanoemulsion is sensitive to environmental factors such as temperature, pH and ionic strength which can cause phase separation, reduced efficacy and degradation of bioactive compounds encapsulated in the formulation. Formulations with stabilizing agents or improved stability, that is resistance to these factors, would overcome these challenges and ensure consistent performance (Kaur et al. 2024).

13.8.2 Cost and Scalability

Components of bio-based nanoemulsions are usually natural compounds with high purity, and they often require costly processes of extraction and purifications. Their use industrially presents some challenges such as emulsion stability and efficiency during scale up. Large-scale production will require innovative cost-effective raw materials, green extraction and energy efficient methods. This would make them economically viable for large-scale productions (Al-Gethami et al. 2024; Islam et al. 2024).

13.8.3 Regulatory Barriers

The safety regulations and approval of bioactive components in nanoemulsions are time consuming and complex, particularly natural extracts owing to their allergenicity, composition variability and limited data (Sharma et al. 2024). Commercial globalization is complicated by regulatory and regional differences, but these challenges can be addressed by robust standardization protocol, safety assessment technologies and collaboration between stakeholders in the industry (Zhang et al. 2024).

13.8.4 Limited Compatibility with Food Matrices

Food matrices contain a wide range of components which can disrupt nanoemulsions, and this can influence the taste, odor and mouthfeel negatively. Researchers are tailoring formulations to improve the compatibility and stability of the nanoemulsions to various food products using emulsifiers that are resistant to pH or optimizing the oil-to-water ratio especially for foods like acidic beverages and high-fat processed foods (Lakshmayya et al. 2023).

13.8.5 Perception and Acceptance

The doubt of consumers about nano technologies in food stems from a lack of comprehension about safety and health implications. To overcome this problem, effective communication strategies like education, clear labeling and third-party safety certifications are important. Benefits, such as enhanced preservation without synthetic additives, can also be promoted to increase the acceptability (Tawiah, Ofori and George 2024).

13.8.6 Bioactive Compound Stability

The exploration of advanced encapsulation techniques such as nanoparticle coatings and controlled-release systems to protect bioactive components of the nanoemulsion are being researched. These bioactive ingredients are degraded over time, which can reduce their effectiveness due to environmental factors like light, temperature fluctuations and oxygen (Mondéjar-López et al. 2024).

13.8.7 Environmental Impact of Production

Bio-based nanoemulsion is usually marketed as environmentally friendly and faces significant environmental costs due to its intensive energy techniques and non-renewable inputs. To enhance its sustainability, efforts are being made to use renewable energy sources, biodegradable solvents and low-energy emulsification techniques. Integration of circular economy principles could further enhance the environmental footprint of bio-based nanoemulsion (Yashwanth et al. 2025; Mursal et al. 2024).

13.8.8 Potential Interactions with Food Components

Food components such as proteins, lipids and polysaccharides interact with nanoemulsion, and they affect the sensory characteristics of the final product. The texture, mouthfeel, viscosity and appearance of the food would be affected by the interaction between the food and nanoemulsion. Some other interaction may enhance sensory qualities such as flavor release and texture. Optimization of these formulations requires care and testing with no negative interactions and meeting consumer expectations without compromising quality (Goindi et al. 2016).

13.9 FUTURE PERSPECTIVES OF BIO-BASED NANOEMULSION IN FOOD PRESERVATION

Consumer preference for a natural alternative to the synthetic additive used in food preservation has led to the advancement of nanotechnology sustainability initiatives and a driving for the future of bio-based nanoemulsion for food preservation (Figure 13.5). Formulation stability, scalability, renewable materials and environmentally friendly fabrication methods, consumer education, promoting widespread adoption and reducing environmental impact are part of the advancement (Çakmakçı, Polatoğlu and Çakmakçı 2024).

13.9.1 Enhanced Formulations for Targeted Functionality

Natural compounds like algae, fungi and extremophiles are used for their antioxidant, antimicrobial and preservative properties in future formulations of bio-based nanoemulsions. These compounds if incorporated into formulations would enhance their safety. The controlled drug release, extended shelf life and optimized bioavailability are advancement of nanoengineering (Roy et al. 2023).

13.9.2 Smart Packaging Technologies Integration

The incorporation of nanoemulsion in smart packaging brings real-time monitoring and enhanced protection for food preservation. It integrates pH changes and

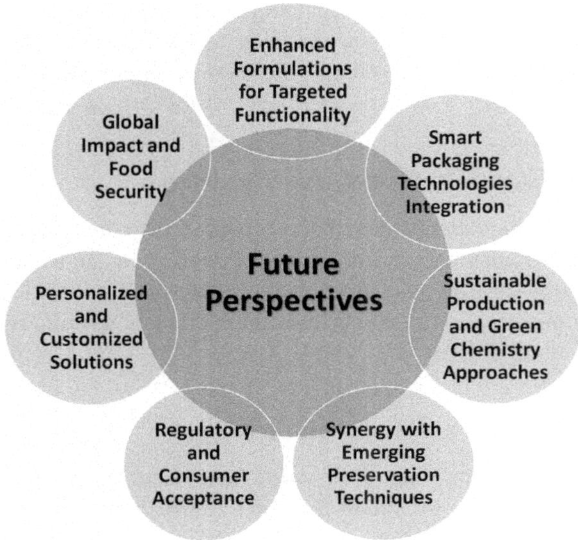

FIGURE 13.5 Future perspectives of bio-based nanoemulsion in food preservation.

microbial growth sensors, provides instant feedback on food quality and enhances active packaging functionality, leading to efficient storage and waste reduction (Bai, Liu and Sun 2023).

13.9.3 Sustainable Production and Green Chemistry Approaches

The circular economy principle will be promoted if bio-based nanoemulsions are formulated with renewable materials and techniques with energy efficient are utilized. The introduction of this method would reduce reliance on synthetic chemicals and lower energy consumption during the production of bio-based nanoemulsions. This would align with the principle of green chemistry for sustainable food preservation (Al-Gethami et al. 2024).

13.9.4 Synergy with Emerging Preservation Techniques

Hybrid preservative techniques can be used to enhance the preservative ability of bio-based nanoemulsion. This method involves the use of cold plasma, pulse light or high-pressure processing. This incorporation can also enhance the antimicrobial activity, reduce spoilage and maintain the integrity of the bioactive compounds. Addressing challenges like oxidative stability, moisture retention and microbial control are necessary for future advancement of bio-based nanoemulsion (Ozkan et al. 2024).

13.9.5 Regulatory and Consumer Acceptance

The commercialization and acceptability of bio-based nanoemulsion technologies require regulatory frameworks such as safety, efficacy and approval. Transparent labeling of products and educational campaigns are beneficial, safe and ecofriendly advantages. Wider adoption of nanoemulsion technologies for food preservation can be achieved by addressing regulatory problems and the education of consumers (Kaur et al. 2024).

13.9.6 Personalized and Customized Solutions

The effectiveness of nanoemulsion as a preservative can be improved by customizing it to suit varying food matrices. Formulations optimization can be achieved by using ML and AI, using real-time data from food production environments. This would help to enhance preservation methods and adaptability of specific industry requirements (Mushtaq et al. 2023).

13.9.7 Global Impact and Food Security

One of the ways of improving food waste and security is the introduction of nanoemulsion which helps to extend the shelf life of perishable foods, especially in regions with limited or unreliable cold chain infrastructure. Production advancement which includes high-throughput emulsification techniques is expected to make nanoemulsion more accessible and affordable to producers in low-resource settings. However, scalability, consumer adoption and regulatory approval are challenges that need to be addressed for nanoemulsion to have a significant impact on food security (Kiran et al. 2024).

13.10 CONCLUSION

The sustainability and effective solution offered by bio-based nanoemulsion in maintaining food quality and extension of shelf life make it an ideal food preservative. These nanoemulsions not only enhance the shelf life of foods but also offer unique properties like antimicrobial and antioxidant activities, improved thermal stability and the ability to encapsulate active ingredients. However, challenges such as formulation refinement, consumer acceptance, regulatory approval and scalability limit the full potential of these bio-based nanoemulsions. Nanotechnology with innovative formulation techniques is an advancement in green chemistry, since synthetic chemicals are hazardous to humans. One of the global goals is sustainability, which is achieved using nanoemulsions to extend shelf life of food, reduce food waste and promote ecofriendly production processes. Consumer demand for clean-label, natural, and health-oriented products can be met through the adoption of bio-based nanoemulsions as food preservatives. Effective education of the end use consumer and transparent communication regarding safety and benefits of bio-based nanoemulsion are important for its widespread acceptance.

ACKNOWLEDGEMENT

Muyideen Olaitan Bamidele appreciates National Council of Humanities, Sciences and Technologies (CONACYT, Mexico) for Ph.D. fellowship support (1245491).

NOTE

1 cristobal.aguilar@uadec.edu.mx

REFERENCES

Al-Gethami, Wafa, Muhammad Azam Qamar, Mohammad Shariq, Abdel-Nasser M. A. Alaghaz, Ahmad Farhan, Ashwaq A. Areshi, and M. Hisham Alnasir. 2024. "Emerging Environmentally Friendly Bio-Based Nanocomposites for the Efficient Removal of Dyes and Micropollutants from Wastewater by Adsorption: A Comprehensive Review." *RSC Advances* 14 (4): 2804–34. https://doi.org/10.1039/D3RA06501D.

Alfei, Silvana. 2020. "Nanotechnology Applications to Improve Solubility of Bioactive Constituents of Foods for Health-Promoting Purposes." In, 189–257. https://doi.org/10.1007/978-3-030-44552-2_8.

Anjum, Varisha, Uday Bagale, Ammar Kadi, Artem Malinin, Irina Potoroko, Amal H. Alharbi, Doaa Sami Khafaga, et al. 2024. "Process Optimization of Tinospora Cordifolia Extract-Loaded Water in Oil Nanoemulsion Developed by Ultrasound-Assisted Homogenization." *Molecules* 29 (8): 1797. https://doi.org/10.3390/molecules29081797.

Bai, Lin, Minghao Liu, and Ying Sun. 2023. "Overview of Food Preservation and Traceability Technology in the Smart Cold Chain System." *Foods* 12 (15): 2881. https://doi.org/10.3390/foods12152881.

Bandyopadhyay, Anupriya, Tuyelee Das, Samapika Nandy, Synudeen Sahib, Subham Preetam, Abilash Valsala Gopalakrishnan, and Abhijit Dey. 2023. "Ligand-Based Active Targeting Strategies for Cancer Theranostics." *Naunyn-Schmiedeberg's Archives of Pharmacology* 396 (12): 3417–41. https://doi.org/10.1007/s00210-023-02612-4.

Benyam, Addisalem (Addis), Tammara Soma, and Evan Fraser. 2021. "Digital Agricultural Technologies for Food Loss and Waste Prevention and Reduction: Global Trends, Adoption Opportunities and Barriers." *Journal of Cleaner Production* 323 (November): 129099. https://doi.org/10.1016/j.jclepro.2021.129099.

Biehl, Philip, and Kai Zhang. 2024. "Introduction to Advances in Bio-Based Polymers: Chemical Structures and Functional Properties at the Interface." In *Green by Design*, 1-1–1-69. IOP Publishing. https://doi.org/10.1088/978-0-7503-6184-2ch1.

Bose, Ipsheta, Mustafa Ahmed, Gaytri Mahajan, Vivek Chauhan, and Swarup Roy. 2025. "Bionanostructured Materials in Food Packaging and Preservation." In *Sustainable Materials for Food Packaging and Preservation*, 127–47. Elsevier. https://doi.org/10.1016/B978-0-443-13567-5.00006-X.

Boudesocque-Delaye, Leslie, Iron Mike Ardeza, Alexis Verger, Roxane Grard, Isabelle Théry-Koné, Xavier Perse, and Emilie Munnier. 2024. "Natural Deep Eutectic Solvents as a Novel Bio-Based Matrix for Ready-to-Use Natural Antioxidants-Enriched Ingredients: Extraction and Formulation Optimization." *Cosmetics* 11 (1): 17. https://doi.org/10.3390/cosmetics11010017.

Braich, Amandeep Kaur, and Gurkirat Kaur. 2024. "Nanostructured Materials for the Development of Bio-Based Plastics for Food Applications." In *Starch Based Nanomaterials for Food Packaging*, 23–50. Elsevier. https://doi.org/10.1016/B978-0-443-18967-8.00005-0.

Burke, Inga, Sven Salzer, Sebastian Stein, Tom Olatomiwa Olakunle Olusanya, Ole Fabian Thiel, and Norbert Kockmann. 2024. "AI-Based Integrated Smart Process Sensor for Emulsion Control in Industrial Application." *Processes* 12 (9): 1821. https://doi.org/10.3390/pr12091821.

Bushra, Rani, Mehraj Ahmad, Farzad Seidi, Qurtulen, Junlong Song, Yongcan Jin, and Huining Xiao. 2023. "Polysaccharide-Based Nanoassemblies: From Synthesis Methodologies and Industrial Applications to Future Prospects." *Advances in Colloid and Interface Science* 318 (August): 102953. https://doi.org/10.1016/j.cis.2023.102953.

Çakmakçı, Songül, Bilgehan Polatoğlu, and Ramazan Çakmakçı. 2024. "Foods of the Future: Challenges, Opportunities, Trends, and Expectations." *Foods* 13 (17): 2663. https://doi.org/10.3390/foods13172663.

Chakraborty, Raja, Pratap Kalita, and Saikat Sen. 2019. "Natural Starch in Biomedical and Food Industry: Perception and Overview." *Current Drug Discovery Technologies* 16 (4): 355–67. https://doi.org/10.2174/1570163815666181003143732.

Chaudhary, Sahil. 2024. "Sustainable Chitosan Nanoemulsion Coatings/Films with Agri-Food Byproducts: Advances, Composition, Production Methods and Applications in Food Preservation." *Journal of Food Measurement and Characterization* 18 (3): 1627–49. https://doi.org/10.1007/s11694-023-02293-w.

Chaudhary, Vaishali, Shrasti Vasistha, and Monika Prakash Rai. 2024. "Nanobiotechnology: An Applicable Approach for Sustainable Future." In *Emerging Sustainable Nanomaterials for Biomedical Applications*, 243–68. Cham: Springer Nature Switzerland. https://doi.org/10.1007/978-3-031-63961-6_10.

Chen, Chao, Xiao Wang, Wenqi Chen, Qin Liu, and Lifeng Wang. 2024. "Encapsulation of Phenolic Acids within Food-Grade Carriers Systems: A Systematic Review." *Critical Reviews in Food Science and Nutrition*, May, 1–20. https://doi.org/10.1080/10408398.2024.2350616.

Cheng, Yiyi, Xuemei Cai, Xin Zhang, Yadong Zhao, Ru Song, Yan Xu, and Haiyan Gao. 2024. "Applications in Pickering Emulsions of Enhancing Preservation Properties: Current Trends and Future Prospects in Active Food Packaging Coatings and Films." *Trends in Food Science & Technology* 151 (September): 104643. https://doi.org/10.1016/j.tifs.2024.104643.

Cvanić, Teodora, Olja Šovljanski, Senka Popović, Tamara Erceg, Jelena Vulić, Jasna Čanadanović-Brunet, Gordana Ćetković, and Vanja TraviČić. 2023. "Progress in Fruit and Vegetable Preservation: Plant-Based Nanoemulsion Coatings and Their Evolving Trends." *Coatings* 13 (11): 1835. https://doi.org/10.3390/coatings13111835.

Dailin, Daniel Joe, Fahim Rithwan, Nur Izyan Wan Azelee, Nurhidayah Zainan, Luo Zaini Mohd Izwan Low, Dayang Norulfairuz Abang Zaidel, and Hesham El Enshasy. 2024. "Trends in Bio-Based Cosmetic Ingredients." In *Biomass-Based Cosmetics*, 27–47. Singapore: Springer Nature Singapore. https://doi.org/10.1007/978-981-97-1908-2_2.

Drago, Emanuela, Roberta Campardelli, Margherita Pettinato, and Patrizia Perego. 2020. "Innovations in Smart Packaging Concepts for Food: An Extensive Review." *Foods* 9 (11): 1628. https://doi.org/10.3390/foods9111628.

Ekonomou, Sotirios I., Anastasia Kyriakoudi, Saliha Saad, Ioannis Mourtzinos, and Alexandros Ch. Stratakos. 2024. "Microfluidics for Food and Nutrition Applications." In, 335–57. https://doi.org/10.1007/978-3-031-60717-2_14.

Farasati Far, Bahareh, Mehdi Jahanbakhshi, Leila Jameie, Faezeh Zolfigol, Parsa Taromi, and Yavuz Nuri Ertas. 2024. "Chitosan-*Graft*-Pomegranate Extract Hydrogel: A Dual-Functional Pad for Antibacterial and Antioxidant Enhancement for Shelf Life Extension in Food Packaging." *ACS Applied Polymer Materials* 6 (16): 9545–58. https://doi.org/10.1021/acsapm.4c01240.

Gali, Swetha Priya, R. G. Reshma, Srinivasan Krishnamoorthy, and R. Meenatchi. 2024. "Coat the Flavor, Preserve the Quality." In *Food Coatings and Preservation Technologies*, 211–58. Wiley. https://doi.org/10.1002/9781394237623.ch7.

Gallo, Monica, Daniele Naviglio, Arturo Armone Caruso, and Lydia Ferrara. 2016. "Applications of Chitosan as a Functional Food." In *Novel Approaches of Nanotechnology in Food*, 425–64. Elsevier. https://doi.org/10.1016/B978-0-12-804308-0.00013-3.

Ghosh, Joyeta, Sudrita Roy Choudhury, Khushboo Singh, Madan Mohan Gupta, and Deepak Sharma. 2024. "Biobased Nanomaterials in Nutraceuticals." In *Biobased Nanomaterials*,

395–448. Singapore: Springer Nature Singapore. https://doi.org/10.1007/978-981-97-0542-9_13.

Ghoshal, Gargi. 2018. "Recent Trends in Active, Smart, and Intelligent Packaging for Food Products." In *Food Packaging and Preservation*, 343–74. Elsevier. https://doi.org/10.1016/B978-0-12-811516-9.00010-5.

Gohar, Osama, Muhammad Zubair Khan, Iram Bibi, Nadia Bashir, Urooj Tariq, Manahil Bakhtiar, Muhammad Ramzan Abdul Karim, Farman Ali, Muhammad Bilal Hanif, and Martin Motola. 2024. "Nanomaterials for Advanced Energy Applications: Recent Advancements and Future Trends." *Materials & Design* 241 (May): 112930. https://doi.org/10.1016/j.matdes.2024.112930.

Goindi, Shishu, Amanpreet Kaur, Randeep Kaur, Atin Kalra, and Pratibha Chauhan. 2016. "Nanoemulsions: An Emerging Technology in the Food Industry." In *Emulsions*, 651–88. Elsevier. https://doi.org/10.1016/B978-0-12-804306-6.00019-2.

Gorde, Pratik Madhukar, Dibya Ranjan Dash, Sushil Kumar Singh, and Poonam Singha. 2024. "Advancements in Sustainable Food Packaging: A Comprehensive Review on Utilization of Nanomaterials, Machine Learning and Deep Learning." *Sustainable Chemistry and Pharmacy* 39 (June): 101619. https://doi.org/10.1016/j.scp.2024.101619.

Govere, Ephraim, David Chikwere, Munyaradzi Chibaro, Mugove Mashingaidze, and Tendai Silvaziso Mandere. 2024. "Enhancing Sustainability in the Modern Transport Through Eco-Friendly Packaging." In, 227–53. https://doi.org/10.4018/979-8-3693-3755-4.ch009.

Gul, Shanzay, Tahseen Fatima Miano, Anees Mujeeb, Muzafaruddin Chachar, Memoona Islam Majeedano, Ghazal Murtaza, Wazeer Ahmed, Yasha Abid Khanzada, and Mariam Ansari. 2024. "Advancements in Nutraceutical Delivery: Integrating Nanotechnology and Microencapsulation for Enhanced Efficacy and Bioavailability." *Matrix Science Pharma* 8 (1): 1–6. https://doi.org/10.4103/mtsp.mtsp_1_24.

Gupta, Divyanshu, Arshiya Lall, Sachin Kumar, Tejaswini Dhanaji Patil, and Kirtiraj K. Gaikwad. 2024. "Plant-Based Edible Films and Coatings for Food-Packaging Applications: Recent Advances, Applications, and Trends." *Sustainable Food Technology* 2 (5): 1428–55. https://doi.org/10.1039/D4FB00110A.

Hasheminya, S.-M., and J. Dehghannya. 2024. Development and Characterization of Kefiran-Gelatin Bio-Nanocomposites Containing Zhumeria majdae Essential Oil Nanoemulsion to Use as Active Food Packaging in Sponge Cakes. *International Journal of Biological Macromolecules* 279: 135120. https://doi.org/10.1016/j.ijbiomac.2024.135120.

Hong, Xin, Qiaoli Zhao, Yuanfa Liu, and Jinwei Li. 2023. "Recent Advances on Food-Grade Water-in-Oil Emulsions: Instability Mechanism, Fabrication, Characterization, Application, and Research Trends." *Critical Reviews in Food Science and Nutrition* 63 (10): 1406–36. https://doi.org/10.1080/10408398.2021.1964063.

Huang, Hui, Yi Zheng, Meiqi Chang, Jun Song, Lili Xia, Chenyao Wu, Wencong Jia, Hongze Ren, Wei Feng, and Yu Chen. 2024. "Ultrasound-Based Micro-/Nanosystems for Biomedical Applications." *Chemical Reviews* 124 (13): 8307–8472. https://doi.org/10.1021/acs.chemrev.4c00009.

Hussain, Shokat, Raheela Akhter, and Shrikant S. Maktedar. 2024. "Advancements in Sustainable Food Packaging: From Eco-Friendly Materials to Innovative Technologies." *Sustainable Food Technology* 2 (5): 1297–1364. https://doi.org/10.1039/D4FB00084F.

Islam, Rayees Ul, Tariq Khan, Pir Mohammad Junaid, and Swaleha Naseem. 2024. "Recent Developments in Sustainable Bio-Based Wax Nanoemulsions as Edible Coatings for Food Commodities: A Comprehensive Review." *Packaging Technology and Science* 37 (12): 1175–94. https://doi.org/10.1002/pts.2854.

Jia, Yangyang, Chuang Wang, Ibrahim Khalifa, Yingheng Zhu, Zhen Wang, Haoyu Chen, Xinhong Liang, Hao Zhang, Lanlan Hu, and Wei Yang. 2024. "Pectin: A Review with Recent Advances in the Emerging Revolution and Multiscale Evaluation Approaches of

Its Emulsifying Characteristics." *Food Hydrocolloids* 157 (December): 110428. https://doi.org/10.1016/j.foodhyd.2024.110428.

Jiang, Liming, Zhiheng Zhang, Chao Qiu, and Jinsheng Wen. 2024. "A Review of Whey Protein-Based Bioactive Delivery Systems: Design, Fabrication, and Application." *Foods* 13 (15): 2453. https://doi.org/10.3390/foods13152453.

Jiang, Yu-Hang, Yu-Ting Zhao, Wei-Gang Xin, Ming Liang, Jia-Jia Song, and Hua-Yi Suo. 2025. "Synergistic Inactivation Effect and Mechanism of Ultrasound Combined with Zanthoxylum Schinifolium Essential Oil Nanoemulsions against Escherichia Coli O157:H7 and Its Application on Fresh-Cut Cucumber." *International Journal of Food Microbiology* 428 (January): 110962. https://doi.org/10.1016/j.ijfoodmicro.2024.110962.

Jiao, Haorong, Qiulian Mao, Noman Razzaq, Rinat Ankri, and Jiabin Cui. 2024. "Ultrasound Technology Assisted Colloidal Nanocrystal Synthesis and Biomedical Applications." *Ultrasonics Sonochemistry* 103 (February): 106798. https://doi.org/10.1016/j.ultsonch.2024.106798.

Karnwal, Arun, and Tabarak Malik. 2024. "Exploring the Untapped Potential of Naturally Occurring Antimicrobial Compounds: Novel Advancements in Food Preservation for Enhanced Safety and Sustainability." *Frontiers in Sustainable Food Systems* 8 (February). https://doi.org/10.3389/fsufs.2024.1307210.

Kaur, Gurveer, Chirasmita Panigrahi, Swati Agarwal, Anjali Khuntia, and Monalisa Sahoo. 2024. "Recent Trends and Advancements in Nanoemulsions: Production Methods, Functional Properties, Applications in Food Sector, Safety and Toxicological Effects." *Food Physics* 1 (September): 100024. https://doi.org/10.1016/j.foodp.2024.100024.

Kiran, Venkatakrishnan, Karthick Harini, Anbazhagan Thirumalai, Koyeli Girigoswami, and Agnishwar Girigoswami. 2024. "Nanotechnology's Role in Ensuring Food Safety and Security." *Biocatalysis and Agricultural Biotechnology* 58 (June): 103220. https://doi.org/10.1016/j.bcab.2024.103220.

KovaČević, Zorana, Ivana Čabarkapa, Ljubiša Šarić, Marko Pajić, Dragana Tomanić, Bojana Kokić, and Dragana D. Božić. 2024. "Natural Solutions to Antimicrobial Resistance: The Role of Essential Oils in Poultry Meat Preservation with Focus on Gram-Negative Bacteria." *Foods* 13 (23): 3905. https://doi.org/10.3390/foods13233905.

Kumar, Mohit, Nitasha Chauhan, Devesh Kumar, Syed Mahmood, Shruti Chopra, and Amit Bhatia. 2024. "Revolutionizing Nanomedicine: Expanding Horizons of Nanoemulsions in Drug Delivery and Beyond." *Journal of Dispersion Science and Technology*, June, 1–26. https://doi.org/10.1080/01932691.2024.2369835.

Lakshmayya, N. S. V., Awdhesh Kumar Mishra, Yugal Kishore Mohanta, Jibanjyoti Panda, Bindu Naik, Bishwambhar Mishra, and Rajender S. Varma. 2023. "Essential Oils-Based Nano-Emulsion System for Food Safety and Preservation: Current Status and Future Prospects." *Biocatalysis and Agricultural Biotechnology* 53 (October): 102897. https://doi.org/10.1016/j.bcab.2023.102897.

Liñán-Atero, Rafael, Fatemeh Aghababaei, Samuel Rodríguez García, Zahra Hasiri, Dimitrios Ziogkas, Andres Moreno, and Milad Hadidi. 2024. "Clove Essential Oil: Chemical Profile, Biological Activities, Encapsulation Strategies, and Food Applications." *Antioxidants* 13 (4): 488. https://doi.org/10.3390/antiox13040488.

Liu, Wenchao, Min Zhang, and Bhesh Bhandari. 2020. "Nanotechnology—A Shelf Life Extension Strategy for Fruits and Vegetables." *Critical Reviews in Food Science and Nutrition* 60 (10): 1706–21. https://doi.org/10.1080/10408398.2019.1589415.

Lohita, Beesetti, and Mupparapu Srijaya. 2024. "Novel Technologies for Shelf-Life Extension of Food Products as a Competitive Advantage: A Review." In, 285–306. https://doi.org/10.1007/978-3-031-51647-4_24.

Maddaloni, Lucia, Laura Gobbi, Giuliana Vinci, and Sabrina Antonia Prencipe. 2025. "Natural Compounds from Food By-Products in Preservation Processes: An Overview." *Processes* 13 (1): 93. https://doi.org/10.3390/pr13010093.

Mahmud, Jumana, Peter Muranyi, Shiv Shankar, Elham Sarmast, Stephane Salmieri, and Monique Lacroix. 2024. "Physiological and Antimicrobial Properties of a Novel Nanoemulsion Formulation Containing Mixed Surfactant and Essential Oils: Optimization Modeling by Response Surface Methodology." *Colloids and Surfaces A: Physicochemical and Engineering Aspects* 686 (April): 133405. https://doi.org/10.1016/j.colsurfa.2024.133405.

Majeed, Massarat, and Mushtaq Ahmad Rather. 2025. "Enhancing Shelf Life and Bioavailability of Vitamin D Through Encapsulation: A Comprehensive Review." *Food Biophysics* 20 (1): 15. https://doi.org/10.1007/s11483-024-09906-x.

Malik, Tanu, Ruchi Sharma, Kashif Ameer, Omar Bashir, Tawheed Amin, Sobiya Manzoor, and Isam A. Mohamed Ahmed. 2023. "Potential of High-Pressure Homogenization (HPH) in the Development of Functional Foods." *International Journal of Food Properties* 26 (1): 2509–31. https://doi.org/10.1080/10942912.2023.2249262.

Martins, Valter F. R., Manuela E. Pintado, Rui M. S. C. Morais, and Alcina M. M. B. Morais. 2024. "Recent Highlights in Sustainable Bio-Based Edible Films and Coatings for Fruit and Vegetable Applications." *Foods* 13 (2): 318. https://doi.org/10.3390/foods13020318.

Meng, Hong, Huanfan He, Yifan He, Xiaojing Pei, Yinmao Dong, Shujing Li, and Ze Zhang. 2024. "Enhanced Stability, Permeability, and Bioavailability of Hydrophobic Compounds in Cosmetics through a Time-Efficient and Convenient Prenanoemulsion Technology." *Journal of Dispersion Science and Technology*, December, 1–9. https://doi.org/10.1080/01932691.2024.2444985.

Mondéjar-López, María, Maria Paz García-Simarro, Pablo Navarro-Simarro, Lourdes Gómez-Gómez, Oussama Ahrazem, and Enrique Niza. 2024. "A Review on the Encapsulation of 'Eco-Friendly' Compounds in Natural Polymer-Based Nanoparticles as Next Generation Nano-Agrochemicals for Sustainable Agriculture and Crop Management." *International Journal of Biological Macromolecules* 280 (November): 136030. https://doi.org/10.1016/j.ijbiomac.2024.136030.

Mudgil, Deepak, and Sheweta Mudgil. 2024. "Acacia Gum: Chemistry, Properties & Food Applications." *Food and Humanity* 2 (May): 100264. https://doi.org/10.1016/j.foohum.2024.100264.

Mursal, Mohd, Arun Kumar, Syed Misbahul Hasan, Sahil Hussain, Kuldeep Singh, Shom Prakash Kushwaha, Muhammad Arif, et al. 2024. "Role of Natural Bioactive Compounds in the Management of Neurodegenerative Disorders." *Intelligent Pharmacy* 2 (1): 102–13. https://doi.org/10.1016/j.ipha.2023.09.006.

Mushtaq, Abeeda, Sajad Mohd Wani, A. R. Malik, Amir Gull, Seema Ramniwas, Gulzar Ahmad Nayik, Sezai Ercisli, Romina Alina Marc, Riaz Ullah, and Ahmed Bari. 2023. "Recent Insights into Nanoemulsions: Their Preparation, Properties and Applications." *Food Chemistry: X* 18 (June): 100684. https://doi.org/10.1016/j.fochx.2023.100684.

Nascimento, Joice Farias do, Flavia Oliveira Monteiro da Silva Abreu, Taysse Holanda, Raquel Oliveira dos Santos Fontenelle, Júlio César Sousa Prado, Emmanuel Silva Marinho, Matheus Nunes da Rocha, et al. 2024. "Evaluation of Emulsification Techniques to Optimize the Properties of Chalcone Nanoemulsions for Antifungal Applications." *Pharmaceuticals* 17 (11): 1442. https://doi.org/10.3390/ph17111442.

Nayak, Sneha, Roopa B. Hegde, Abhishek S. Rao, and H. K. Sachidananda. 2024. "Revitalizing Agriculture with the Potential of Cashew Nutshell Liquid: A Comprehensive Exploration and Synergy with AI." *Discover Applied Sciences* 6 (11): 557. https://doi.org/10.1007/s42452-024-06258-6.

Negi, Aditi, A. Vanmathi Mugasundari, and Jeyan A. Moses. 2024. "Nanoemulsion." In *Food Coatings and Preservation Technologies*, 339–65. Wiley. https://doi.org/10.1002/9781394237623.ch10.

Nikbakht Nasrabadi, Maryam, Ali Sedaghat Doost, and Raffaele Mezzenga. 2021. "Modification Approaches of Plant-Based Proteins to Improve Their Techno-Functionality and Use in Food Products." *Food Hydrocolloids* 118 (September): 106789. https://doi.org/10.1016/j.foodhyd.2021.106789.

Ozkan, Gulay, Sebahattin Serhat Turgut, Busra Gultekin Subasi, Esra Capanoglu, Tuba Esatbeyoglu, and Aberham Hailu Feyissa. 2024. "Food Processing and Preservation in the Food Industry 4.0 Era." In *Food Industry 4.0*, 99–120. Elsevier. https://doi.org/10.1016/B978-0-443-15516-1.00006-2.

Padmaja, P. G., R. Venkateswarlu, S. P. Singh, and Vilas A. Tonapi. 2024. "Enhancing Shelf Life of Pearl Millet Flour." In *Pearl Millet in the 21st Century*, 549–66. Singapore: Springer Nature Singapore. https://doi.org/10.1007/978-981-99-5890-0_21.

Paidari, Saeed, Yasaman Esmaeili, Salam A. Ibrahim, Shima Vahedi, Sawsan Ali Al-Hilifi, and Nafiseh Zamindar. 2024. "Application of Nanoparticles to Enhance the Microbial Quality and Shelf Life of Food Products." In *Microbial Biotechnology in the Food Industry*, 75–102. Cham: Springer International Publishing. https://doi.org/10.1007/978-3-031-51417-3_4.

Pal, Priti, Akhilesh Kumar Singh, Rajesh Kumar Srivastava, Saurabh Singh Rathore, Uttam Kumar Sahoo, Sanjukta Subudhi, Prakash Kumar Sarangi, and Piotr Prus. 2024. "Circular Bioeconomy in Action: Transforming Food Wastes into Renewable Food Resources." *Foods* 13 (18): 3007. https://doi.org/10.3390/foods13183007.

Pandey, Neha, Nishant Kumar, and Ashutosh Upadhyay. 2024. "Soy Protein Modification Strategies, Functional Interplay and Industrial Relevance: A Review." *Journal of Food Measurement and Characterization*, December. https://doi.org/10.1007/s11694-024-02989-7.

Panwar, Anika, Vikas Kumar, Atul Dhiman, Priyanka Thakur, Vishal Sharma, Ajay Sharma, and Satish Kumar. 2024. "Nanoemulsion Based Edible Coatings for Quality Retention of Fruits and Vegetables-Decoding the Basics and Advancements in Last Decade." *Environmental Research* 240 (January): 117450. https://doi.org/10.1016/j.envres.2023.117450.

Patil, Samadhan P., Sammed A. Bhalerao, Yogeshsing N. Rajput, and Amit P. Pratap. 2024. "Production of Rhamnolipids Using WFSO and Its Application in the Development of Antifungal Nanoemulsion Using Hydnocarpus Wightiana, Garcinia Cambogia—Seed Oils." *Industrial Crops and Products* 219 (November): 118885. https://doi.org/10.1016/j.indcrop.2024.118885.

Pei, Jinjin, Chella Perumal Palanisamy, Guru Prasad Srinivasan, Mani Panagal, Sathish Sundar Dhilip Kumar, and Monica Mironescu. 2024. "A Comprehensive Review on Starch-Based Sustainable Edible Films Loaded with Bioactive Components for Food Packaging." *International Journal of Biological Macromolecules* 274 (August): 133332. https://doi.org/10.1016/j.ijbiomac.2024.133332.

Pereira-Rojas, Juan, Víctor Pérez, José G. Parra, Nelson Barrios, Aarón Muñoz, Jimmy Castillo, Lokendra Pal, and Juan Pereira. 2025. "Interfacial Behavior of Neem Oil (Azadirachta Indica): Experimental and Computational Insights." *Colloids and Surfaces A: Physicochemical and Engineering Aspects* 707 (February): 135944. https://doi.org/10.1016/j.colsurfa.2024.135944.

Pintea, Andrada, Andrei Manea, Cezara Pintea, Robert-Alexandru Vlad, Magdalena Bîrsan, Paula Antonoaea, Emöke Margit Rédai, and Adriana Ciurba. 2025. "Peptides: Emerging Candidates for the Prevention and Treatment of Skin Senescence: A Review." *Biomolecules* 15 (1): 88. https://doi.org/10.3390/biom15010088.

Plyduang, Thipapun, Apichart Atipairin, Attawadee Sae Yoon, Namfa Sermkaew, Pajaree Sakdiset, and Somchai Sawatdee. 2021. "Formula Development of Red Palm (Elaeis Guineensis) Fruit Extract Loaded with Solid Lipid Nanoparticles Containing Creams and Its Anti-Aging Efficacy in Healthy Volunteers." *Cosmetics* 9 (1): 3. https://doi.org/10.3390/cosmetics9010003.

Ponnampalam, Eric N., Ali Kiani, Sarusha Santhiravel, Benjamin W. B. Holman, Charlotte Lauridsen, and Frank R. Dunshea. 2022. "The Importance of Dietary Antioxidants on Oxidative Stress, Meat and Milk Production, and Their Preservative Aspects in Farm Animals: Antioxidant Action, Animal Health, and Product Quality—Invited Review." *Animals* 12 (23): 3279. https://doi.org/10.3390/ani12233279.

Rahman, Md. Mostafizur, Md. Abdus Shahid, Md. Tanvir Hossain, Md. Sohan Sheikh, Md. Sunjidur Rahman, Nasir Uddin, Abdur Rahim, Ruhul Amin Khan, and Imam Hossain. 2024. "Sources, Extractions, and Applications of Alginate: A Review." *Discover Applied Sciences* 6 (8): 443. https://doi.org/10.1007/s42452-024-06151-2.

Rai, Mahendra, Magdalena Wypij, Avinash P. Ingle, Joanna Trzcińska-Wencel, and Patrycja Golińska. 2021. "Emerging Trends in Pullulan-Based Antimicrobial Systems for Various Applications." *International Journal of Molecular Sciences* 22 (24): 13596. https://doi.org/10.3390/ijms222413596.

Rashidi, Ahmad Ramli, Nur Izyan Wan Azelee, Dayang Norulfairuz Abang Zaidel, Lai Fatt Chuah, Awais Bokhari, Hesham Ali El Enshasy, and Daniel Joe Dailin. 2023. "Unleashing the Potential of Xanthan: A Comprehensive Exploration of Biosynthesis, Production, and Diverse Applications." *Bioprocess and Biosystems Engineering* 46 (6): 771–87. https://doi.org/10.1007/s00449-023-02870-9.

Rathod, Nikheel Bhojraj, Raciye Meral, Shahida Anusha Siddiqui, Nilesh Nirmal, and Fatih Ozogul. 2024. "Nanoemulsion-Based Approach to Preserve Muscle Food: A Review with Current Knowledge." *Critical Reviews in Food Science and Nutrition* 64 (19): 6812–33. https://doi.org/10.1080/10408398.2023.2175347.

Revutskaya, Natalia, Ekaterina Polishchuk, Ivan Kozyrev, Liliya Fedulova, Valentina Krylova, Viktoriya Pchelkina, Tatyana Gustova, et al. 2024. "Application of Natural Functional Additives for Improving Bioactivity and Structure of Biopolymer-Based Films for Food Packaging: A Review." *Polymers* 16 (14): 1976. https://doi.org/10.3390/polym16141976.

Rout, Srutee, Srushti Tambe, Ram Kumar Deshmukh, Suraj Mali, Jorddy Cruz, Prem Prakash Srivastav, Purnima D. Amin, Kirtiraj K. Gaikwad, Eloisa Helena de Aguiar Andrade, and Mozaniel Santana de Oliveira. 2022. "Recent Trends in the Application of Essential Oils: The Next Generation of Food Preservation and Food Packaging." *Trends in Food Science & Technology* 129 (November): 421–39. https://doi.org/10.1016/j.tifs.2022.10.012.

Roy, Swarup, Ram Kumar Deshmukh, Shefali Tripathi, Kirtiraj K. Gaikwad, Sabya Sachi Das, and Devanshi Sharma. 2023. "Recent Advances in the Carotenoids Added to Food Packaging Films: A Review." *Foods* 12 (21): 4011. https://doi.org/10.3390/foods12214011.

Rui, Litong, Ying Li, Xiaodan Wu, Ying Wang, and Xiufang Xia. 2024. "Effect of Clove Essential Oil Nanoemulsion on Physicochemical and Antioxidant Properties of Chitosan Film." *International Journal of Biological Macromolecules* 263 (April): 130286. https://doi.org/10.1016/j.ijbiomac.2024.130286.

Shahidi, Fereidoon, and Abul Hossain. 2022. "Preservation of Aquatic Food Using Edible Films and Coatings Containing Essential Oils: A Review." *Critical Reviews in Food Science and Nutrition* 62 (1): 66–105. https://doi.org/10.1080/10408398.2020.1812048.

Sharma, Himani, Atin Kumar, Rajat Singh, Anurag Malik, and Rachna Juyal. 2024. "Enhancing Food Security and Sustainability: Harnessing Nano-Omics-Based Technologies for Quality Monitoring." In *Sustainable Agriculture*, 191–212. De Gruyter. https://doi.org/10.1515/9783111369754-009.

Sharma, Kanika, Azadeh Babaei, Khyati Oberoi, Krishna Aayush, Ruchi Sharma, and Somesh Sharma. 2022. "Essential Oil Nanoemulsion Edible Coating in Food Industry: A Review." *Food and Bioprocess Technology* 15 (11): 2375–95. https://doi.org/10.1007/s11947-022-02811-6.

Sharma, Neha, Harpreet Kaur, Gurkirat Kaur, Arashdeep Singh, and Sunny Sharma. 2024. "Appraisal of Cutting-Edge Techniques for Prolonging Fresh Berries Shelf Life:

Innovations in Essential Oil Nanoemulsion-Based Edible Coatings." *Scientia Horticulturae* 337 (November): 113564. https://doi.org/10.1016/j.scienta.2024.113564.

Sharma, Ramesh, Pinku Chandra Nath, Puja Das, Sarvesh Rustagi, Minaxi Sharma, Natarajan Sridhar, Tridip Kumar Hazarika, Priya Rana, Prakash Kumar Nayak, and Kandi Sridhar. 2024. "Essential Oil-Nanoemulsion Based Edible Coating: Innovative Sustainable Preservation Method for Fresh/Fresh-Cut Fruits and Vegetables." *Food Chemistry* 460 (December): 140545. https://doi.org/10.1016/j.foodchem.2024.140545.

Sharma, Shikha, Kohei Nakano, Santosh Kumar, and Vimal Katiyar. 2024. "Edible Packaging to Prolong Postharvest Shelf-Life of Fruits and Vegetables: A Review." *Food Chemistry Advances* 4 (June): 100711. https://doi.org/10.1016/j.focha.2024.100711.

Sharma, Shubham, Lilly Mulrey, Megan Byrne, Amit K. Jaiswal, and Swarna Jaiswal. 2022. "Encapsulation of Essential Oils in Nanocarriers for Active Food Packaging." *Foods* 11 (15): 2337. https://doi.org/10.3390/foods11152337.

Sheibani, Samira, Shima Jafarzadeh, Zeinab Qazanfarzadeh, M. M. Jeevani Osadee Wijekoon, Norazatul Hanim Mohd Rozalli, and Abdorreza Mohammadi Nafchi. 2024. "Sustainable Strategies for Using Natural Extracts in Smart Food Packaging." *International Journal of Biological Macromolecules* 267 (May): 131537. https://doi.org/10.1016/j.ijbiomac.2024.131537.

Shi, Ce, Zhiyao Zhao, Zhixin Jia, Mengyuan Hou, Xinting Yang, Xiaoguo Ying, and Zengtao Ji. 2024. "Artificial Neural Network-Based Shelf Life Prediction Approach in the Food Storage Process: A Review." *Critical Reviews in Food Science and Nutrition* 64 (32): 12009–24. https://doi.org/10.1080/10408398.2023.2245899.

Tawiah, Benjamin, Emmanuel Abankwah Ofori, and Soney C. George. 2024. "Nanotechnology in Societal Development." In, 1–64. https://doi.org/10.1007/978-981-97-6184-5_1.

Thiviya, Punniamoorthy, Ashoka Gamage, Anuradhi Liyanapathiranage, Madhubhashini Makehelwala, R. S. Dassanayake, Asanga Manamperi, Othmane Merah, Sudhargar Mani, Janardhan Reddy Koduru, and Terrence Madhujith. 2023. "Algal Polysaccharides: Structure, Preparation and Applications in Food Packaging." *Food Chemistry* 405 (March): 134903. https://doi.org/10.1016/j.foodchem.2022.134903.

Tripathi, Abhishek Dutt, Ruchi Sharma, Aparna Agarwal, and Rizwana Haleem. 2021. "Nanoemulsions Based Edible Coatings with Potential Food Applications." *International Journal of Biobased Plastics* 3 (1): 112–25. https://doi.org/10.1080/24759651.2021.1875615.

Ul Islam, Rayees, Tariq Khan, and Mohd Danish Ahmad. 2024. "Nano-Emulsions for Edible Coating." In *Organic-Based Nanomaterials in Food Packaging*, 173–208. Cham: Springer Nature Switzerland. https://doi.org/10.1007/978-3-031-63829-9_10.

Vats, Diksha, and Vimal Kumar. 2025. "Enhanced Nanoemulsion Engineering Through Continuous Ultrasonication and Integrated Machine Learning Algorithms." *Industrial & Engineering Chemistry Research* 64 (1): 357–68. https://doi.org/10.1021/acs.iecr.4c02489.

Vignesh, V., Dibyakanta Seth, Chingakham Ngotomba Singh, Madhuresh Dwivedi, Shyam Kumar Singh, and Samlesh Kumari. 2024. "Electrohydrodynamic Encapsulation: A Novel Technique to Enhance the Stability of Bioactive Compounds in Food." *Trends in Food Science & Technology* 151 (September): 104626. https://doi.org/10.1016/j.tifs.2024.104626.

Wagermaier, Wolfgang, Khashayar Razghandi, and Peter Fratzl. 2025. "A Bio-Inspired Perspective on Materials Sustainability." *Advanced Materials*, January. https://doi.org/10.1002/adma.202413096.

Wang, Cuina, and Mingruo Guo. 2025. "Methods and Technologies to Improve Bioavailability of Bioactive Substances." In *Functional Foods*, 287–305. Elsevier. https://doi.org/10.1016/B978-0-443-19100-8.00003-8.

Yashwanth, Arcot, Rundong Huang, Monica Iepure, Minchen Mu, Wentao Zhou, Angela Kunadu, Courtney Carignan, et al. 2025. "Food Packaging Solutions in the Post-per- and Polyfluoroalkyl Substances (PFAS) and Microplastics Era: A Review of Functions, Materials, and Bio-based Alternatives." *Comprehensive Reviews in Food Science and Food Safety* 24 (1). https://doi.org/10.1111/1541-4337.70079.

Zhang, Junyi, Jianxiong Hao, Jingyi Wang, Huiying Li, and Dandan Zhao. 2024. "Strategic Manipulation of Biofilm Dispersion for Controlling *Listeria Monocytogenes* Infections." *Critical Reviews in Food Science and Nutrition*, October, 1–10. https://doi.org/10.1080/10408398.2024.2409340.

Zhang, Liang, Wenze Wang, Wen He, Ting Du, Shaochi Wang, Puyuan Hu, Bing Pan, Jingjing Jin, Lizhi Liu, and Jianlong Wang. 2024. "A Tailored Slow-Release Film with Synergistic Antibacterial and Antioxidant Activities for Ultra-Persistent Preservation of Perishable Products." *Food Chemistry* 430 (January): 136993. https://doi.org/10.1016/j.foodchem.2023.136993.

Zhang, Quanyu, Baohua Kong, Haotian Liu, Xin Du, Fangda Sun, and Xiufang Xia. 2024. "Nanoscale Pickering Emulsion Food Preservative Films/Coatings: Compositions, Preparations, Influencing Factors, and Applications." *Comprehensive Reviews in Food Science and Food Safety* 23 (1). https://doi.org/10.1111/1541-4337.13279.

Zhu, Xianglu, Rahel Suchintita Das, Mysore Lokesh Bhavya, Marco Garcia-Vaquero, and Brijesh K. Tiwari. 2024. "Acoustic Cavitation for Agri-Food Applications: Mechanism of Action, Design of New Systems, Challenges and Strategies for Scale-Up." *Ultrasonics Sonochemistry* 105 (May): 106850. https://doi.org/10.1016/j.ultsonch.2024.106850.

14 Revolutionizing Cancer Treatment

Nanoparticle-Enhanced Noscapine for Management of Cancer

Jyotirmayee, Pratyush Pragyandipta, Pradeep K Naik, and Malkhey Verma

14.1 INTRODUCTION

For a multicellular organism to live, its cells must operate in harmony or follow the laws of cell growth and reproduction. But in some instances, these regular cells become "rebellious" all of a sudden, violate the rules of cell division, multiply uncontrollably, and intrude into the other tissues by usurping resources. The proto-oncogenes are the genes that control normal cell growth, but mutations in proto-oncogenes or oncogenes can result in uncontrollable cell growth, which leads to cancer. Though treatments such as chemotherapy, radiotherapy, and surgery have been improved, cancer is still a significant issue in contemporary oncology, particularly with an impact on survival rates (Shams et al., 2023). Drugs targeting microtubules, such as vinca alkaloids and taxanes derivatives, are commonly utilized in the treatment of cancer, but they are hindered by severe toxicity and off-targeting (Exertier et al., 2024; Liu et al., 2023). Investigating these limitations, a promising solution emerged in the form of noscapine, an opium alkaloid purified from the poppy of Papaver somniferum. Normal arrested cells, meanwhile, can resume mitosis after drug withdrawal owing to metabolic clearance mechanisms (Landen et al., 2002). In addition, noscapine neither polymerizes tubulin nor depolymerizes tubulin, thus inducing no or little toxicity (Ke et al., 2000). Due to some characteristics over existing microtubule targeted drugs, it has a pKa value of about 7.8, exhibits oral bioavailability of 42.22 %, and has a half-life of 1.47 hours. This necessitates administering reasonably high doses of drugs (ED50), around 300 mg/kg body weight, to provoke anticancer responses (Madan et al., 2010).

Drug delivery systems based on nanoparticles have also been identified as a potential solution to these problems (Xiong et al., 2025). In particular, noscapine, when encapsulated within nanoparticles, possesses some advantages. Nanoparticles can enhance the pharmacokinetics and bioavailability of noscapine, and the increased

DOI: 10.1201/9781003559900-14

permeability and retention (EPR) effect supports targeting tumor tissue more effectively. Additionally, nanoparticles allow for the codelivery of several drugs, promoting synergistic combinations that have the potential to enhance efficacy while decreasing the chances of drug resistance (Singh et al., 2025; Shabnum et al., 2025). By exploiting the unique characteristics of noscapine in combination with nanoparticle technology, this approach overcomes most of the downsides of existing treatments for cancer. It fits with the developing precision medicine field.

14.2 NOSCAPINE: A NEW LIGHT TO THE ANTICANCER ARSENAL

Nature has provided humankind with numerous bounties, and one of the greatest is the treasure hoard of natural products with numerous medicinal or therapeutic uses. Noscapine phthalide isoquinoline alkaloid was first identified in the year 1817 from opium (Papaver Somniferum) by Professor Pierre-Jean Robiquet under the name of "Narcotine," and it was renamed "Noscapine" by A.M.A. council of drugs in the year 1958. Noscapine, which remained concealed for more than a century, became well known as an expectorant used for suppressing cough in the year 1960 and as an anti-tussive drug that has been widely utilized for decades (Rida et al., 2015). Noscapine, a potential anticancer drug, was identified by Harish Joshi's group members. Noscapine lowers the tension surrounding the kinetochore pairs and decreases microtubules bound to the kinetochores (Zhou and Giannakakou, 2005); now, in mitosis (Figure 14.1), until all the chromosomes have bound to the microtubules decorously along appropriate tension at chromosomes, anaphase is postponed by the SAC. Blind docking methods were employed to find possible docking positions, and a competitive binding strategy consisted of overlapping or nearly at the Colchicine binding site (Naik et al., 2011). No evident toxicity was detected in animals suffering from breast cancer, and it was also shown to reduce murine thymoma and human bladder tumors (Ye et al., 1998).

Unlike other alkaloids derived from opium, noscapine is not addictive or sedative, making it a promising candidate for use in the clinic. Apart from its antimitotic activity, noscapine exhibits anti-inflammatory, antithrombotic, and antioxidant activities, adding to its therapeutic value (Priyadarshani et al., 2024). Despite these benefits, the use of noscapine in the clinic has been limited by its low bioavailability and speedy removal from the body (Esnaashari and Amani, 2018). To address these limitations, current research aims to create sophisticated drug delivery systems, including nanoparticles, to enhance the pharmacokinetic profile of noscapine. These approaches increase bioavailability, facilitate targeted delivery to tumor locations, and minimize off-target toxicity.

14.3 VERSATILE ANTICANCEROUS AGENT IN COMBINATION AND IN PRECISION MEDICINE

With these developments, noscapine has great potential for combination therapies and for overcoming the challenges of existing cancer therapies. Synergizing noscapine

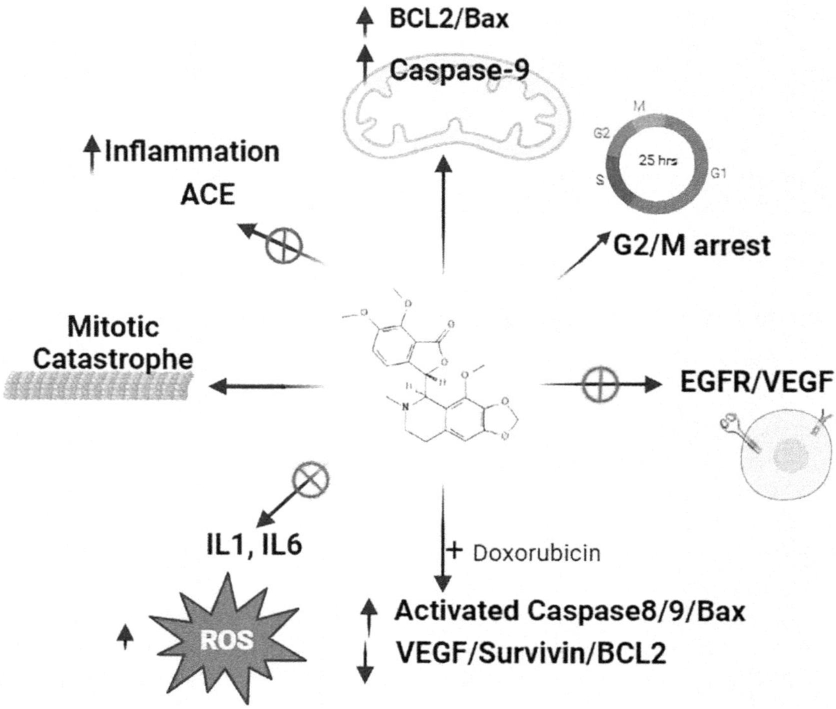

FIGURE 14.1 Mechanism action of noscapine with synergism with various proteins.

with other chemotherapeutic drugs or adjuvants forms an effective synergy that maximizes cancer cell destruction with reduced toxicity. Noscapine increases the efficacy of chemotherapeutic drugs such as doxorubicin, paclitaxel, and cisplatin, using lower doses while providing similar therapeutic effect. By causing mitotic arrest and apoptosis, it enhances cancer treatment efficiency with less toxicity (Debnath et al., 2025). It synergizes with other medications by acting through different mechanisms, e.g., destabilizing the mitotic spindle in addition to DNA intercalation, cell cycle arrest, and apoptosis or induction of ROS production, all enhancing cancer cell killing together. The major benefit of noscapine is its potential to resensitize resistant cancer cells by downregulating P-glycoprotein (Pgp) and anti-apoptotic proteins Bcl-2 and surviving (Han et al., 2022). Liposomal formulations of noscapine with doxorubicin or paclitaxel have successfully bypassed resistance in prostate cancer, while co-administration with cisplatin re-sensitizes resistant lung cancer. Further evidence points to noscapine enhancing immune checkpoint inhibitors (e.g., anti-PD-1/PD-L1), sensitizing tumors for immune system attack (Zughaier et al., 2010; Priyadarshani et al., 2024; Serpico et al., 2020). Noscapine's promise lies in its utility as part of personalized cancer therapy, especially when combined with nanotechnology. Its compatibility enables personalized drug formulations according to tumor molecular profiles, maximizing treatment efficacy.

In contrast to other microtubule-targeting drugs, noscapine does not depolymerize microtubules, minimizing neurotoxic side effects and making it more suitable for long-term therapy. Its effectiveness has been shown in diverse cancer models, such as breast cancer and glioblastoma, which emphasizes its potential in precision oncology. In addition to that, noscapine has proven effective in a broad spectrum of cancers through the penetration of hypoxic tumor microenvironments, inducing apoptosis of leukemia and lymphoma cells, and penetrating the blood-brain barrier for the treatment of glioblastoma (Patel, 2021).

14.4 NANOPARTICLE TECHNOLOGY: A GAME CHANGER IN DRUG DELIVERY

Nanoparticles are a revolutionary agent in cancer drug delivery, efficiently overcoming key issues like poor solubility, low bioavailability, and off-target effects. Nanoscale carriers can encapsulate hydrophobic drug such as noscapine, improving solubility, enhancing circulation, and ensuring even drug distribution (Basu et al., 2024). Through exploitation of the EPR (enhanced permeability and retention) effect and functionalization with ligand (e.g., antibodies, peptides, or small molecules) targeting particular cancer cell receptors, they reach targeted tumor accumulation without compromising healthy tissues (Wang et al., 2022). Nanoparticles provide sustained, controlled release of noscapine, sustaining drug levels at therapeutically relevant levels for extended periods, enhancing penetration into the tumor, and reducing the frequency of dosing (Yu et al., 2021). Also, noscapine-loaded nanoparticles with other chemotherapeutic drugs form synergistic drug combinations in a single delivery system, and thus it is safer to administer in cancer therapies.

14.4.1 Noscapine-Loaded Nanoparticles

Noscapine-loaded nanoparticles have been demonstrated in recent research to be effective in cancer therapy, and polymeric nanoparticles encapsulating noscapine have been found to increase cellular uptake, extend circulation time, and enhance tumor accumulation. These formulations enhance the therapeutic index of noscapine and make it a more effective anticancer agent. Nanotechnology has transformed drug delivery by encapsulating drugs, preserving them from degradation, and releasing them in a controlled manner at the target site. For noscapine, nanoparticle-based delivery systems have the potential to enhance solubility, bioavailability, and controlled release and maintain therapeutic concentrations at the tumor site for extended periods (Kumari et al., 2025). A number of nanoparticle types have been investigated for noscapine delivery, including liposomes, solid lipid nanoparticles, and polymeric nanoparticles. Preclinical studies have shown the enhanced therapeutic activity of nanoparticle-formulated noscapine compared to free noscapine with enhanced tumor suppression and decreased side effects.

14.4.2 Timeline of Development

The investigation of noscapine and nanoparticles commenced with the initial consideration of encapsulating noscapine in nanoparticles in order to overcome its pharmacokinetic drawbacks. Initial studies established the potential of nanocarrier systems such as PEGylated gelatin (Madan et al., 2011; Madan et al., 2013) and CREKA peptide-targeted lipid carriers (Chougule et al., 2010). Combination treatment has been demonstrated, with examples including noscapine with doxorubicin (Chougule et al., 2010) and gemcitabine (Chougule et al., 2011) in showing synergistic cytotoxicity, decreased tumor burden, and antiangiogenesis. Progress was also made on nanocarrier systems including PCL-NPs (polycaprolactone nanoparticles) for glioblastoma (Kumar, 2019), solid lipid nanoparticles for delivery to the brain (Madan et al., 2013), and targetable estrone-conjugated gelatin nanoparticles for ER+ breast cancer (Madan et al., 2014). Noscapine-drug combinations in cancers have been validated in apoptosis and antiangiogenic pathways by studies like TNBC (Doddapaneni et al., 2016) and NSCLC (Chougule et al., 2011). Computational methods and wider investigation have been used to enhance nanoparticle properties (Shalaby et al., 2014) and investigate biodegradable polymeric nanoparticles (Madan et al., 2013; Yadav et al., 2024). A recent development has been a renewed interest in custom combination therapies through the use of polymeric nanoparticles for synergistic Nos-Dox combinations in cancer, particularly breast cancer (Esnaashari et al., 2020).

14.4.3 Types of Nanocarriers for Noscapine Delivery

Various nanocarrier systems have been explored for encapsulating noscapine, including polymeric nanoparticles, lipid-based nanocarriers, dendrimers, micelles, metallic nanoparticles, and hybrid nanoparticles (Table 14.1). Of these, polylactic-co-glycolic acid (PLGA) (Kale et al., 2025) and chitosan nanoparticles confer biocompatibility, controlled release, and permeability (Aydin & Demir, 2025). Liposomes and solid lipid nanoparticles improve the stability and bioavailability of noscapine (Singh et al., 2025), whereas metal nanoparticles, upon functionalization with noscapine, show novel optical and therapeutic properties (Lin et al., 2025).

14.4.4 Combination Therapies with Noscapine and Nanoparticles

Noscapine's compatibility with other anticancer drugs provides the potential for synergistic combinations, adding to its effectiveness. The combination of noscapine and paclitaxel in a single nanoparticle formulation has demonstrated encouraging results in preclinical experiments, taking advantage of the microtubule-stabilizing and destabilizing activities of the two drugs to kill tumor cells more efficiently. Likewise, CREKA peptide-conjugated lipid nanocarriers co-encapsulated with noscapine and doxorubicin exhibited extensive tumor regression in ER− breast cancer cells (Chougule et al., 2010). PLGA nanoparticles with increased noscapine-doxorubicin cytotoxicity in 4T1 breast cancer cells resulted in improved tumor reduction in a mouse model. Noscapine-gemcitabine combinations exhibited extensive synergy in NSCLC models, resulting in 82.9% reduction in tumor volume in xenografted mice (Chougule et al., 2011). Noscapine pretreatment made TNBC MDA-MB-231 xenografts sensitive to docetaxel chemotherapy, decreasing tumor burden considerably. Stromal disruption

TABLE 14.1
Noscapine-Loaded Nanoparticles and Their Mechanism of Action

Noscapine-Loaded Nanoparticle	Mechanism	Reference
SLNs of NOS vs. Psoriasis	Get accumulated in dermis and release NOS that reduce inflammatory and proinflammatory cytokines, preventing APC maturation, reducing IL-17 producing T-cells	Rahmanian-Devin et al., 2023
NOS-loaded Lipid Nano capsules	Significant reduction of ROS level, inhibition of NO, inhibition of cancer cell proliferation, promotion of apoptosis via BCL2 gene and regulation of microtubule dynamics	Abdel-Wahab et al., 2024
NOS-aminoacid conjugate	Induces apoptosis in 4T1 mammary carcinoma tumor cells	Davarzani et al., 2023
HSA nanoparticles containing noscapine vs. SK-BR-3 cell line	Efficiently deliver maximum amounts of noscapine to target sites at the right rate and concentration for optimal therapeutic efficacy	Sebak et al., 2010
uPAR-targeted optical-MR imaging trackable nanoparticles of NOS vs. PC-3 cells	Inhibit PC-3 growth and maintain their T2 MRI contrast effect, so can be tracked by MRI	Abdalla et al., 2011
Estrone-conjugated noscapine-loaded gelatin nanoparticles (Nos-ES-GN) vs. MCF-7 cells	Follow 1st order kinetics and slow release, greater accumulation efficiency	Madan et al., 2014
9-Br-Nos-rapid release-nanostructured lipid particles vs. A549 cells	1.12-fold enhancement in half-life period in inhalation route while 3.75-fold in intravenous route administration	Jyoti et al., 2015
Mannosylated noscapine self-emulsifying solid dispersions (Mann-Nos_SESDs) vs. cisplatin-resistant NSCLC	Mann-Nos_SESDs showed saturating uptake across Caco-2 monolayers, controlled release, 40% bioavailability, and effectiveness in sensitizing H1650 SP cells to cisplatin in vitro and in lung tumor models	Andey et al., 2016
9-CH2OH Nos-Tel-SLN vs. H1299 cells	Has better MTT results than 9-CH2OH noscapine and may offer significant advantage in treatment of lung cancer	Singh et al., 2025
Nanonoscapine vs. LNCaP cells	Nanoparticle could potentially inhibit prostate cancer cells in G2/M and induce apoptosis by increased expression of GLI1 and BAX genes	Nazari et al., 2024

improved nanoparticle penetration in tumor tissue, increasing markers of apoptosis and decreasing survivin and Bcl-2, decreasing chemoresistance (Doddapaneni et al., 2016). Noscapine combinations with EGFR inhibitors (Esnaashari et al., 2020) or PI3K inhibitors (Rahmanian-Devin et al., 2021) may increase apoptosis in lung and breast cancers (Kumar, 2019). Noscapine nanoparticles co-loaded with siRNA against MDR genes have the ability to fight chemoresistance. In addition, new studies investigate the potential of combining noscapine nanoparticles with immune checkpoint inhibitors to enhance the immune responses against cancer.

14.4.5 Advantages of Nanotechnology with Noscapine in Cancer Therapy

Noscapine has garnered significant interest in cancer therapy due to its unique biological properties, non-narcotic nature, and minimal side effects compared to traditional chemotherapeutic agents. Its role in cancer treatment is being revolutionized with advancements in nanoparticle drug delivery systems and its integration into synergistic drug combinations. The encapsulation of noscapine within nanoparticles has revolutionized its delivery and efficacy in cancer therapy. Nanotechnology enhances noscapine's therapeutic index through:

I. Increased Bioavailability: Encapsulation with lipid-based vesicles improves water solubility and protects noscapine from premature degradation. PLGA (poly-lactic-co-glycolic acid) provides stable platforms that enhance pharmacokinetics of the drug.

II. Targeted Drug Delivery: Functionalized nanoparticles allow targeted delivery to cancer cells through ligands for specific receptors (e.g., folate receptors on ovarian and HER2 targets on breast cancer cells), minimizing off-target effects.

III. Controlled Release: Nanocarrier systems (PH or T-responsive carriers) provide sustained release of noscapine (85% release within 96 hours), maintaining therapeutic concentrations for extended periods and reducing dosing frequency.

IV. Enhanced Cellular Uptake: PEG (polyethylene glycol)-coated nanoparticles facilitate noscapine penetration into tumors, especially in solid tumors with poor vascularization via enhanced permeability and retention effects thus bioaccumulating in tumor cells.

V. Overcoming Multidrug Resistance (MDR): Noscapine-loaded nanoparticles have shown the ability to bypass drug efflux pumps, a key mechanism of multidrug resistance (MDR) in cancer cells. Gold conjugates and dual drug-loaded nanoparticles, combined with P-gp inhibitors, downregulate MDR-related proteins, resensitizing cancer cells to therapy and enhancing apoptosis in breast, prostate, and glioblastoma models. Preclinical studies in animal models have demonstrated its efficacy, leading to reduced tumor growth and improved survival rates.

14.5 TUMOR MICROENVIRONMENT (TME) AND TARGETING STRATEGIES

The tumor microenvironment (TME) plays a pivotal role in cancer growth, drug resistance, and the effectiveness of treatments. TME elements such as stromal cells, ECM, hypoxia, and aberrant vasculature pose barriers that hinder drug penetration and diminish therapeutic effects. Alterations aimed at specific TME features have improved the efficacy of noscapine by inhibiting tumor stroma, enhancing vascular permeability, and overcoming multidrug resistance. CREKA peptide-functionalized nanocarriers have increased noscapine nanoparticle deposition in tumor stroma (Chougule et al., 2010), whereas PEGylated lipid nanoparticles have increased stromal penetration and decreased ECM stiffness in in vivo cancer models (Priyadarshani et al., 2024; Serpico et al., 2020). Noscapine-loaded nanoparticles are capable of remodeling stromal architecture, relaxing ECM cross-linking, and enabling drugs to penetrate more effectively into deep tumor areas (Doddapaneni et al., 2016). Hypoxia-mediated resistance to chemotherapy is overcome by oxygen-releasing noscapine nanoparticles, which can be triggered by hypoxia-sensitive linkers. Phoenix-sensitive nanoparticles can realize tumor-selective drug release through regulation of burst release of noscapine at tumor sites (Rahmanian-Devin et al., 2021). PEGylated gelatin nanoparticles enhance noscapine bioavailability by shielding it from premature degradation and inducing release at the tumor site. Dual pH- and temperature-responsive systems have been investigated for synergistic cancer therapy, using co-loaded noscapine with other chemotherapeutic agents (Bardajee et al., 2020). Ligand-based targeting has been effective in increasing noscapine nanoparticle uptake by tumor cells and minimizing collateral damage to normal tissues.

Noscapine is not a P-glycoprotein efflux pump activator, and thus is an effective choice against MDR cancers. Co-loading of noscapine with P-gp inhibitors in liposomes has resulted in enhanced drug retention within cancer cells (Yu et al., 2021). MDR gene expression knockdown by siRNA-loaded noscapine nanoparticles proved to be more effective against chemo-resistant tumor models. Smart nanoparticle systems with active targeting ligands and TME-responsive triggers represent the future direction of noscapine-loaded nano-therapy. Hybrid lipid-polymer nanoparticles that offer controlled drug release and enhanced tumor specificity surpass one-material carriers.

14.6 PRECLINICAL AND CLINICAL ADVANCEMENTS

Various preclinical investigations have established the therapeutic potential of nanoparticle-enhanced noscapine, especially in combination regimens for cancer therapies. Loaded nanoparticles of noscapine demonstrated exceptional efficacy in animal models such as inhibited tumor growth and increased survival rates making them viable candidates for clinical translation. But the transition to clinical is complicated and must undergo strict assessment. In breast cancer, polymeric nanoparticles loaded with noscapine had inhibited tumor growth by 70% in xenograft models (Esnaashari et al., 2020). In glioblastoma, liposomal formulations of noscapine plus temozolomide greatly increased survival rates in animal models (Shahcheraghi et al., 2024). In prostate cancer, nanoparticles with dual drugs containing noscapine and docetaxel showed synergistic activity, especially in hormone-resistant cancers (Rabzia et al., 2028). Now early-phase clinical trials are ongoing. The positive findings from previous trials have generated optimism about the future of nanoparticle-enhanced noscapines.

14.7 CHALLENGES AND FUTURE DIRECTIONS

Nanoparticle-based drug delivery systems have been designed to overcome the majority of limitations, but some challenges still exist, such as formulation stability, drug loading efficiency, tumor-targeting specificity, and feasibility of large-scale manufacturing. Designing nanoparticles that selectively target cancer cells without harming healthy cells is also a challenging task. Long-term toxicity and safety are concerns, with nanoparticles containing metal or polymer potentially causing inflammatory responses and accumulating in the spleen and liver. Combination with other anti-cancer drugs may enhance these risks. Approvals for new formulations could take a long time, and commercial production could make manufacturing more expensive. New areas of research involve nanoparticle design and functionalization, optimization of combination therapy, genomics, personalized medicine, and siRNA-loaded nanoparticles. Long-term safety, efficacy, and side effects have to be determined by clinical trials. Long-term therapeutic effect relies on sustained release, but different nanoparticle systems are highly inconsistent in release. Prolonged release is facilitated by PEGylation, but drug burst release versus sustained exposure compromises has to be resolved.

Nanoparticles undergo plasma protein precipitation adsorption, fast reticulo-endothelial clearance, and enzymatic degradation, truncating their therapeutic lifespan in circulation. Active targeting using ligands improves breast and prostate cancer tumor selectivity but is extremely challenging to maximize to a balance between cellular uptake efficacy and ligand density. Noscapine nanoparticles showed limited penetration of the hypoxic tumor cores through thick ECM barriers. CREKA-functionalized nanoparticles improved tumor penetration by binding to ECM fibrin (Chougule et al., 2011), but more bioengineering solutions need to be engineered for deep tumor access. There is considerable heterogeneity in the degree of therapeutic gain, presumably due to differences in cancer cell lines used, nanoparticle formulations and routes of delivery, variability of animal models, and synergy combinations. Noscapine reached Phase I/II clinical trials in the form of a free drug (Singh et al., 2013), but noscapine nanoparticles are yet to be tested in human clinical trials.

In addition, regulatory hurdles, pharmacokinetic heterogeneity of nanoparticles, and lack of funds for the development of non-traditional chemotherapeutics have slowed down clinical translation. Reproducibility of manufacturing remains a major issue, and GMP compliance is required for noscapine nanoparticle manufacture. FDA and EMA regulations require stringent safety assessments of nanoparticle-based drugs, including long-term biocompatibility, immune response profiling, and toxicity. The lack of well-standardized regulatory routes for polymeric and lipid-based nanocarriers complicates approval schedules and results in delay in commercialization.

14.8 ADVANCES IN AI AND NANOPARTICLE OPTIMIZATION

AI-based research has transformed the production of polymeric, lipid, and hybrid nanoparticle systems to enhance reproducibility and therapeutic efficacy. Incorporating AI into noscapine nanoparticle optimization enables predictive modeling of physicochemical parameters, intelligent formulation tuning, and rapid identification of optimal delivery profiles (Kapoor et al., 2024). ANNs (Artificial Neural Networks) and ML (machine learning) algorithms have been applied to polymer composition, molecular weight, drug-polymer interaction, nanoprecipitation, and emulsification parameter optimization.

Machine learning algorithms combine quantum chemistry and molecular dynamics simulations to forecast stable noscapine-polymer mixtures for improved efficiency encapsulation. Automated platforms model different formulation conditions, minimizing trial-and-error experiments. Noscapine nano-systems need ideal particle size to enable tumor penetration with avoidance of premature clearance. ANN-based models can effectively forecast noscapine encapsulation efficiency and nanoparticle size by examining parameter inputs such as polymer-to-drug ratio, solvent evaporation rate, and surfactant concentration. Support vector regression (SVR) algorithms have precisely modeled noscapine release kinetics from PLGA and lipid-based nanoparticles, enhancing the efficiency of formulation. AI-discovered ligand functionalization strategies, including CREKA peptides, estrone, and uPAR ligands, enhance precision targeting. Machine learning programs model nanoparticle-cell interactions to optimize ligand density and affinity for maximum tumor uptake. Machine-learning models can predict other synergistic drug partners for optimal combinational therapy (Abd El-Hafeez et al., 2024). Scale-up production of noscapine-loaded nanoparticles in microfluidics-assisted AI-controlled systems makes it possible to achieve better batch-to-batch consistency in size, surface charge, and stability. Automated AI-based quality control systems test nanocarrier uniformity to ensure regulatory compliance for clinical use.

14.9 CLINICAL TRANSLATION AND FUTURE PERSPECTIVES

Future advancements must integrate AI-driven optimizations with real-time biofeedback platforms to engineer smart, adaptive noscapine delivery nanoplatforms. Large-scale scaling of AI-driven manufacturing technologies will be critical to facilitate clinical translation and commercialization of noscapine nanomedicine for broad applications in cancer treatment. Establishing common preclinical testing standards across research groups will help ensure that results are reliable and comparable, while greater funding and stronger collaborations will be essential to speed up early clinical testing of noscapine nanomedicine formulations. Noscapine-loaded nanoparticles are challenged by serious formulation, stability, targeting, and clinical issues, but several approaches (e.g., AI-optimized design, multifunctional carriers, and bioengineering techniques) can overcome these drawbacks. Clinical translation is the most pressing gap, and strong preclinical validation and phase I/II trials are thus crucial. Scalability and regulatory approval require concentrated investment to raise the chances of commercial success and worldwide availability. Formulation heterogeneity, inefficient drug loading, and unpredictable pharmacokinetics pose obstacles for clinical translation, though. To overcome these problems, phenotypic and pathway-based AI and ML strategies have been utilized to rationalize nanoparticle engineering, forecast drug release patterns, and maximize tumor-targeting efficiencies. There is excellent synergy between noscapine and DNA-targeting chemotherapies (doxorubicin, gemcitabine). Future interest resides in noscapine interactions with targeted drugs, immune checkpoint modulators, and gene therapy agents.

Future research should explore pH-responsive nanoparticles for tumor-selective drug delivery and co-delivery of complementary therapeutic agents. Clinical validation and PK/PD modeling are needed for practical use. Noscapine-containing nanoparticles have been shown to target tumor stroma, bypass hypoxia, and target MDR tumors using TME-directed strategies. CREKA peptide-conjugated nanocarriers and pH-responsive noscapine delivery systems hold potential but need clinical translation. Co-delivery systems that target multiple TME components in concert, incorporating antiangiogenic drugs, oxygen-releasing carriers, or immune-modulating molecules should be the focus of future research. Clinical trials must confirm the therapeutic efficacy and safety of TME-targeting noscapine nanocarriers, while ensuring reproducibility and scalability for real-world applications.

14.10 CONCLUSION

The combination of nanoparticle technology with noscapine and synergistic drug combinations is a revolutionary breakthrough in cancer therapeutics. With less association with traditional treatments, these provide the possibility of increased therapeutic efficacy with lower systemic toxicity and better patient outcomes, leading to the creation of more effective and possibly curative treatment modalities. In addition, the strategic co-administration of noscapine with traditional chemotherapeutics (paclitaxel and doxorubicin) not only enhances antitumor activity but also provides a potent means to circumvent drug resistance thereby opening the way for more efficient and targeted cancer therapies. In spite of promising developments, challenges remain, such as the scalable and reproducible manufacture of nanoparticles with uniform phytochemical characteristics, extensive toxicity profiling, and stringent clinical assessment through well designed trials. Further research and development in this area may result in more effective, targeted, and safer forms of therapy, ultimately enhancing patient outcomes and improving the quality of life for cancer patients across the globe.

REFERENCES

Abd El-Hafeez, T., Shams, M. Y., Elshaier, Y. A. M. M., Farghaly, H. M., & Hassanien, A. E. (2024). Harnessing machine learning to find synergistic combinations for FDA-approved cancer drugs. *Scientific Reports*, 14(1), 2428. https://doi.org/10.1038/s41598-024-52814-w

Abdalla, M. O., Karna, P., Sajja, H. K., Mao, H., Yates, C., Turner, T., & Aneja, R. (2011). Enhanced noscapine delivery using uPAR-targeted optical-MR imaging trackable nanoparticles for prostate cancer therapy. *Journal of Controlled Release*, 149(3), 314–322. https://doi.org/10.1016/j.jconrel.2010.10.030

Abdel-Wahab, N. D., Kabil, M. F., El-Sherbiny, I. M., Salama, M. F., El-Sayed, G., & El-Sherbini, E. S. (2024). Enhanced anticancer efficacy of noscapine-loaded lipid nanocapsules: In-vitro and in-vivo evaluation. *Egyptian Journal of Chemistry*, 67(8), 409–422. https://doi.org/10.21608/EJCHEM.2024.259976.9129

Andey, T., Patel, A., Marepally, S., Chougule, M., Spencer, S. D., Rishi, A. K., & Singh, M. (2016). Formulation, pharmacokinetic, and efficacy studies of mannosylated self-emulsifying solid dispersions of noscapine. *PloS One*, 11(1), e0146804. https://doi.org/10.1371/journal.pone.0146804

Aydın, R. S. T., & Demir, A. (2025). Artificial neural network prediction of TiO2-doped chitosan micro/nanoparticle size based on particle imaging measurements. *Colloid and Polymer Science*, 1–15.

Bardajee, G. R., Ghavami, S., & Hosseini, S. S. (2020). A review on pH and temperature responsive gels in drug delivery. *Journal of Chemical Reviews*, 2(2), 80–89.

Basu, B., Garala, K. K., Patel, R., Dutta, A., Ash, D., Prajapati, B., Singh, S., & Jha, S. K. (2024). Advanced targeted drug delivery of bioactive nanomaterials in the management of cancer. *Current Medicinal Chemistry*. https://doi.org/10.2174/0109298673283362231220115050

Chougule, M. B., Patel, A. R., Patlolla, R., & Sachdeva, M. (2010). Multifunctional CREKA peptide conjugated lipid nanocarriers of synergistically acting Noscapine and Doxorubicin for breast cancer therapy. *Cancer Research*, 70(8_Supplement), 5511–5511.

Chougule, M. B., Patel, A., Sachdeva, P., Jackson, T., & Singh, M. (2011). Enhanced anticancer activity of gemcitabine in combination with noscapine via antiangiogenic and apoptotic pathway against non-small cell lung cancer. *PloS One*, 6(11), e27394. https://doi.org/10.1371/journal.pone.0027394

Davarzani, Z., Salehi, P., Asghari, S. M., Bararjanian, M., Hamrahi Mohsen, A., & Dehghan Harati, H. (2023). Design, synthesis, and functional studies on noscapine and cotarnine amino acid derivatives as antitumor agents. *ACS Omega*, 8(48), 45502–45509. https://doi.org/10.1021/acsomega.3c05488

Debnath, P., Deb, V. K., Nath, S., Guha, A., & Adhikari, S. (2025). Structure-activity relationship of plant-derived bioactive compounds in anti-cancer therapy. In *Plant Derived Bioactive Compounds in Human Health and Disease* (pp. 66–94). CRC Press.

Doddapaneni, R., Patel, K., Chowdhury, N., & Singh, M. (2016). Noscapine chemosensitization enhances docetaxel anticancer activity and nanocarrier uptake in triple negative breast cancer. *Experimental Cell Research*, 346(1), 65–73. https://doi.org/10.1016/j.yexcr.2016.05.006

Esnaashari, S. S., & Amani, A. (2018). Optimization of noscapine-loaded mPEG-PLGA nanoparticles and release study: A response surface methodology approach. *Journal of Pharmaceutical Innovation*, 13, 237–246.

Esnaashari, S. S., Muhammadnejad, S., Amanpour, S., & Amani, A. (2020). A combinational approach towards treatment of breast cancer: An analysis of noscapine-loaded polymeric nanoparticles and doxorubicin. *AAPS PharmSciTech*, 21(5), 166. https://doi.org/10.1208/s12249-020-01710-3

Exertier, C., Antonelli, L., Fiorillo, A., Bernardini, R., Colotti, B., Ilari, A., & Colotti, G. (2024). Sorcin in cancer development and chemotherapeutic drug resistance. *Cancers*, 16(16), 2810. https://doi.org/10.3390/cancers16162810

Han, Z., Meng, L., Huang, X., Tan, J., Liu, W., Chen, W., Zou, Y., Cai, Y., Huang, S., Chen, A., Zhan, T., Huang, M., Chen, X., Tian, X., & Zhu, Q. (2022). Inhibition of p38 MAPK increases the sensitivity of 5-fluorouracil-resistant SW480 human colon cancer cells to noscapine. *Oncology Letters*, 23(2), 52. https://doi.org/10.3892/ol.2021.13170

Jyoti, K., Kaur, K., Pandey, R. S., Jain, U. K., Chandra, R., & Madan, J. (2015). Inhalable nanostructured lipid particles of 9-bromo-noscapine, a tubulin-binding cytotoxic agent: In vitro and in vivo studies. *Journal of Colloid and Interface Science*, 445, 219–230. https://doi.org/10.1016/j.jcis.2014.12.092

Kale, N., Edvall, C., Nandi, S., & Mallik, S. (2025). Nanocarrier mediated targeted delivery of natural products to prostate cancer. In *Discovery and Development of Anti-Prostate Cancerous Agents from Natural Products* (pp. 307–320). Elsevier.

Kapoor, D. U., Sharma, J. B., Gandhi, S. M., Prajapati, B. G., Thanawuth, K., Limmatvapirat, S., & Sriamornsak, P. (2024). AI-driven design and optimization of nanoparticle-based drug delivery systems. *Science, Engineering and Health Studies*, 24010003–24010003.

Ke, Y., Ye, K., Grossniklaus, H. E., Archer, D. R., Joshi, H. C., & Kapp, J. A. (2000). Noscapine inhibits tumor growth with little toxicity to normal tissues or inhibition of immune responses. *Cancer Immunology, Immunotherapy*, 49, 217–225. https://doi.org/10.1007/s002620000109

Kumar, S. S. (2019). Formulation and characterization of noscapine-loaded polycaprolactone nanoparticles. *Asian Journal of Pharmaceutics (AJP)*, 13(01).

Kumari, D., Jha, A., Mahato, P., Bhowmik, S., Sahu, J. K., Kumari, M., . . . & Dwivedi, M. (2025). Albumin nanoparticles as a versatile carrier in phytonanotechnology: Expanded scope for phytochemical delivery. In *Advances in Novel Phytopharmaceuticals* (pp. 213–229). CRC Press.

Landen, J. W., Lang, R., McMahon, S. J., Rusan, N. M., Yvon, A.-M., Adams, A. W., Sorcinelli, M. D., Campbell, R., Bonaccorsi, P., Ansel, J. C. (2002). Noscapine alters microtubule dynamics in living cells and inhibits the progression of melanoma. *Cancer Research*, 62(14), 4109–4114.

Lin, C. Y., Lin, Z. C., Chang, Y. T., Lin, T. J., & Fang, J. Y. (2025). Novel strategies in topical delivery for psoriasis treatment: Nanocarriers and energy-driven approaches. *Expert Opinion on Drug Delivery*, 1–17. Advance online publication. https://doi.org/10.1080/17425247.2025.2472968

Liu, Y., Yang, H., Fang, Y., Xing, Y., Pang, X., Li, Y., . . . & Liu, Y. (2023). Function and inhibition of Haspin kinase: Targeting multiple cancer therapies by antimitosis. *Journal of Pharmacy and Pharmacology*, 75(4), 445–465. https://doi.org/10.1093/jpp/rgac080

Madan, J., Dhiman, N., Parmar, V. K., Sardana, S., Bharatam, P. V., Aneja, R., Chandra, R., & Katyal, A. (2010). Inclusion complexes of noscapine in β-cyclodextrin offer better solubility and improved pharmacokinetics. *Cancer Chemotherapy and Pharmacology*, 65, 537–548. https://doi.org/10.1007/s00280-009-1060-3

Madan, J., Dhiman, N., Sardana, S., Aneja, R., Chandra, R., & Katyal, A. (2011). Long-circulating poly(ethylene glycol)-grafted gelatin nanoparticles customized for intracellular delivery of noscapine: preparation, in-vitro characterization, structure elucidation, pharmacokinetics, and cytotoxicity analyses. *Anti-Cancer Drugs*, 22(6), 543–555. https://doi.org/10.1097/CAD.0b013e32834159b8

Madan, J., Gundala, S. R., Kasetti, Y., Bharatam, P. V., Aneja, R., Katyal, A., & Jain, U. K. (2014). Enhanced noscapine delivery using estrogen-receptor-targeted nanoparticles for breast cancer therapy. *Anti-Cancer Drugs*, 25(6), 704–716. https://doi.org/10.1097/CAD.0000000000000098

Madan, J., Pandey, R. S., Jain, V., Katare, O. P., Chandra, R., & Katyal, A. (2013). Poly (ethylene)-glycol conjugated solid lipid nanoparticles of noscapine improve biological half-life, brain delivery and efficacy in glioblastoma cells. *Nanomedicine: Nanotechnology, Biology, and Medicine*, 9(4), 492–503. https://doi.org/10.1016/j.nano.2012.10.003

Naik, P. K., Santoshi, S., Rai, A., & Joshi, H. C. (2011). Molecular modelling and competition binding study of Br-noscapine and colchicine provide insight into noscapinoid-tubulin binding site. *Journal of Molecular Graphics & Modelling*, 29(7), 947–955. https://doi.org/10.1016/j.jmgm.2011.03.004

Nazari, M. H. D., Heidarian, R., Masoudnia, M., Dastjerdi, R. A., Talkhounche, P. G., & Tale-ahmad, S. (2024). Targeting GLI1 and BAX by nanonoscapine could impede prostate adenocarcinoma progression. *Scientific Reports*, 14(1), 18977. https://doi.org/10.1038/s41598-024-65968-4

Patel, N. (2021). *Drug delivery approaches for triple negative breast cancer and non-small cell lung cancer* (Doctoral dissertation, Florida Agricultural and Mechanical University).

Priyadarshani, A., Bhatia, R., & Shan, M. (2024). The Noscapine saga: Unravelling a valuable jewel from a poppy pod—past, present and future. *Psychoactives*, 3(1), 1–21.

Rabzia, A., Khazaei, M., Rashidi, Z., & Khazaei, M. R. (2017). Synergistic anticancer effect of paclitaxel and noscapine on human prostate cancer cell lines. *Iranian Journal of Pharmaceutical Research: IJPR*, 16(4), 1432.

Rahmanian-Devin, P., Askari, V. R., Sanei-Far, Z., Rahimi, V. B., Kamali, H., Jaafari, M. R., & Golmohammadzadeh, S. (2023). Preparation and characterization of solid lipid nanoparticles encapsulated noscapine and evaluation of its protective effects against imiquimod-induced psoriasis-like skin lesions. *Biomedicine & Pharmacotherapy*, 168, 115823. https://doi.org/10.1016/j.biopha.2023.115823

Rahmanian-Devin, P., Baradaran Rahimi, V., Jaafari, M. R., Golmohammadzadeh, S., Sanei-Far, Z., & Askari, V. R. (2021). [Retracted] Noscapine, an emerging medication for different diseases: A mechanistic review. *Evidence-Based Complementary and Alternative Medicine*, 2021(1), 8402517.

Rida, P. C., LiVecche, D., Ogden, A., Zhou, J., & Aneja, R. (2015). The noscapine chronicle: A pharmaco-historic biography of the opiate alkaloid family and its clinical applications. *Medicinal Research Reviews*, 35(5), 1072–1096. https://doi.org/10.1002/med.21357

Sebak, S., Mirzaei, M., Malhotra, M., Kulamarva, A., & Prakash, S. (2010). Human serum albumin nanoparticles as an efficient noscapine drug delivery system for potential use in breast cancer: Preparation and in vitro analysis. *International Journal of Nanomedicine*, 525–532. https://doi.org/10.2147/IJN.S10443

Serpico, A. F., Visconti, R., & Grieco, D. (2020). Exploiting immune-dependent effects of microtubule-targeting agents to improve efficacy and tolerability of cancer treatment. *Cell Death & Disease*, 11(5), 361. https://doi.org/10.1038/s41419-020-2567-0

Shabnum, S. S., Siranjeevi, R., Raj, C. K., Nivetha, P., & Benazir, K. (2025). A comprehensive review on recent progress in carbon nanotubes for biomedical application. *Environmental Quality Management*, 34(3), e70040.

Shahcheraghi, S. H., Alimardani, M., Lotfi, M., Lotfi, M., Uversky, V. N., Guetchueng, S. T., Palakurthi, S. S., Charbe, N. B., Hromić-Jahjefendić, A., Aljabali, A. A. A., Gadewar, M. M., Malik, S., Goyal, R., El-Tanani, M., Mishra, V., Mishra, Y., & Tambuwala, M. M. (2024). Advances in glioblastoma multiforme: Integrating therapy and pathology perspectives. *Pathology, Research and Practice*, 257, 155285. https://doi.org/10.1016/j.prp.2024.155285

Shalaby, K. S., Soliman, M. E., Casettari, L., Bonacucina, G., Cespi, M., Palmieri, G. F., Sammour, O. A., & El Shamy, A. A. (2014). Determination of factors controlling the particle size and entrapment efficiency of noscapine in PEG/PLA nanoparticles using artificial neural networks. *International Journal of Nanomedicine*, 9, 4953–4964. https://doi.org/10.2147/IJN.S68737

Shams, M., Abdallah, S., Alsadoun, L., Hamid, Y. H., Gasim, R., Hassan, A., & Hassan, A. (2023). Oncological horizons: The synergy of medical and surgical innovations in cancer treatment. *Cureus*, 15(11). https://doi.org/10.7759/cureus.49249

Sindhoor, S. M., Naveen, N. R., Rao, G. K., Gopan, G., Chopra, H., Park, M. N., Alshahrani, M. M., Jose, J., Emran, T. B., & Kim, B. (2022). A spotlight on alkaloid nanoformulations for the treatment of lung cancer. *Frontiers in Oncology*, 12, 994155. https://doi.org/10.3389/fonc.2022.994155

Singh, H., Singh, P., Kumari, K., Chandra, A., Dass, S. K., & Chandra, R. (2013). A review on noscapine, and its impact on heme metabolism. *Current Drug Metabolism*, 14(3), 351–360. https://doi.org/10.2174/1389200211314030010

Singh, S., Sewariya, S., Goel, T., Panchal, S., Singh, A., Kukreti, S., . . . & Chandra, R. (2025). Assessing dual drug 9-hydroxymethyl noscapine and telmisartan-loaded stearic acid nanoparticles against (H1299) non-small cell lung cancer and their mechanistic interaction with bovine serum albumin. *Materials Advances*. https://doi.org/10.1039/D4MA00958D

Wang, J., Zhou, T., Liu, Y., Chen, S., & Yu, Z. (2022). Application of nanoparticles in the treatment of lung cancer with emphasis on receptors. *Frontiers in Pharmacology*, 12, 781425. https://doi.org/10.3389/fphar.2021.781425

Xiong, Z., Yu, Y., He, L., & Chen, T. (2025). *Structure-driven medicinal applications of selenium nanoparticles: Challenges and opportunities in clinical translation*. Chem.

Yadav, K., Yadav, D., Kumar, S., Narra, K., El-Sherbiny, M., Al-Serwi, R. H., . . . & Mohamed, J. M. M. (2024). Natural biodegradable and polymeric nanoparticles for the delivery of noscapine for cancer treatment. *Biomass Conversion and Biorefinery*, 14(21), 27609–27621.

Ye, K., Ke, Y., Keshava, N., Shanks, J., Kapp, J. A., Tekmal, R. R., Petros, J., & Joshi, H. C. (1998). Opium alkaloid noscapine is an antitumor agent that arrests metaphase and induces apoptosis in dividing cells. *Proceedings of the National Academy of Sciences of the United States of America*, 95(4), 1601–1606. https://doi.org/10.1073/pnas.95.4.1601

Yu, X. Y., Jin, X., & Shou, Z. X. (2021). Surface-engineered smart nanocarrier-based inhalation formulations for targeted lung cancer chemotherapy: A review of current practices. *Drug Delivery*, 28(1), 1995–2010. https://doi.org/10.1080/10717544.2021.1981492

Zhou, J., & Giannakakou, P. (2005). Targeting microtubules for cancer chemotherapy. Current medicinal chemistry. *Anti-Cancer Agents*, 5(1), 65–71. https://doi.org/10.2174/1568011053352569

Zughaier, S., Karna, P., Stephens, D., & Aneja, R. (2010). Potent anti-inflammatory activity of novel microtubule-modulating brominated noscapine analogs. *PLoS One*, 5(2), e9165. https://doi.org/10.1371/journal.pone.0009165

15 Herbal Nanotechnology in Detection of Health Hazardous Molecules

Pranshul Sethi, Sonali Rastogi Neetu Panwar, Urvee Dave, Abhinav Verma, and Ayush Madan

15.1 INTRODUCTION TO HERBAL NANOTECHNOLOGY

Herbal nanotechnology merges herbal extracts' restorative properties with nanotechnology's advanced abilities to detect and treat illnesses. Through the use of nanoscale materials this technology delivers enhanced bioavailability while providing precise delivery mechanisms and advanced diagnostic tools. Herbal nanotechnology addresses traditional herbal therapy limitations such as poor solubility and bioavailability through the encapsulation of active compounds in nanometer-sized carrier particles (Panda & Mohapatra, 2024; Ai et al., 2024). Herbal nanotechnology includes applications in drug delivery systems as well as disease detection and environmental sample analysis (Setia et al., 2024).

Herbs have been used for centuries to treat ailments and feature as a fundamental component of traditional medicinal systems such as Ayurveda, Traditional Chinese Medicine (TCM), and Unani. Nanotechnology's integration with herbal treatments represents a recent development in medical science. Nanotechnology developments during the late 20th century entirely transformed the methods available for utilizing bioactive substances. Research showed that nanocapsules enhanced the curative benefits of herbal substances by promoting better absorption and precise delivery mechanisms (Mishra et al., 2022; Namdari et al., 2017). Curcumin derived from *Curcuma longa* has undergone extensive research due to its anti-inflammatory and antioxidant properties. The incorporation of substances into nanocapsules has provided improved solubility and biocompatibility for their use in detecting and neutralizing reactive oxygen species (ROS), which are dangerous to health (Na et al., 2021).

Herbal nanotechnology resolves multiple limitations that traditional diagnostic and therapeutic methods encounter. Nanotechnology provides solutions for early disease detection and targeted drug delivery while countering multidrug resistance and environmental applications which leads to its expanded use in diagnostic and therapeutic advancements (Chopra et al., 2023). The functionalization of gold nanoparticles with plant extracts creates herbal source-derived nanoparticles that show high sensitivity for biomolecule detection linked to health risks. For instance, biosensors based on nanotechnology have been fabricated for the identification of heavy metals, pesticides, and ROS in biological systems (Salek Maghsoudi et al., 2021).

DOI: 10.1201/9781003559900-15

Nanocarriers guarantee that herbal active ingredients are deposited directly at the site of concern with less systemic toxicity. For example, quercetin nanoliposomes have been used to target inflammatory conditions with accuracy and minimize oxidative stress (Mishra et al., 2022). Herbal nanoparticles increase the efficacy of classical therapies by evading resistance mechanisms as well. For example, silver nanoparticles synthesized using green chemistry with *Azadirachta indica* (neem) have been proven to overcome bacterial resistance (Namdari et al., 2017). Furthermore, herbal nanotechnology is used in environmental health monitoring through the detection of indirect toxicants that affect health (Segneanu et al., 2022). Phytosynthetically prepared nanoparticles are used more and more to biosense and remediate contaminants in water and soil.

15.1.1 Herbal Sources for Nanomaterials

15.1.1.1 Plants Used in Herbal Nanotechnology

Plants remain an invaluable source of nanomaterials because they contain an abundance of phytochemicals. Phytochemicals comprise alkaloids, flavonoids, terpenoids, and phenolic compounds that play a crucial role in reducing and stabilizing nanomaterials in their synthesis. For example, *Azadirachta indica* (neem) is widely researched for silver nanoparticle synthesis applications. Such nanoparticles show strong antimicrobial and anticancer activity (Wassihun, 2024). Another case is that of *Camellia sinensis* (green tea), whose polyphenols assist in synthesizing gold nanoparticles with excellent diagnostic and therapeutic properties (Nadaf et al., 2022). Another case is *Curcuma longa* (turmeric), whose curcumin acts as a catalyst for synthesizing particles with superior anti-inflammatory and antioxidant activity (Sindhu et al., 2024).

The versatility of plant-derived biologically active molecules makes it easy to synthesize particles using a simple method while providing an inherent healing property to the product nanomaterials. This combined property is important in solving intricate health issues and simplifying health-hazardous molecule detection.

15.1.1.2 Extraction and Preparation of Herbal Extracts

Extraction of phytoactive substances from plant-based sources is an important step in herbal nanotechnology. The purity and activity of phytochemicals employed in synthesizing nanoparticles depend on effective extraction processes. Maceration, Soxhlet extraction, and supercritical extraction are some commonly used extraction processes based on solubility and stability of target phytochemicals. Ethanol and methanol solvents are frequently utilized for the extraction of phenolic and flavonoid phytochemicals due to their high polarity and ability to dissolve a wide range of bioactive compounds (Manna et al., 2023).

After extraction, the herbal extracts are analyzed using advanced analytical techniques. High-Performance Liquid Chromatography (HPLC) is employed to quantify specific bioactive compounds, while Fourier-Transform Infrared Spectroscopy (FTIR) is used to evaluate their functional groups. These analytical methods ensure the quality and purity of the extracts, which are then utilized as reducing and capping agents in the synthesis of nanoparticles. This approach avoids the use of harmful chemicals, adheres to green chemistry principles, and enhances the safety profile of the produced nanoparticles (Bachheti & Bachheti, 2024).

15.1.1.3 Advantages of Herbal Sources

Natural sources provide considerable benefits compared to synthetic techniques for the creation of nanoparticles. To begin with, they are eco-friendly, making use of renewable plant materials and removing the necessity for harmful chemicals. This green approach diminishes the environmental impact of nanoparticle manufacturing, rendering it a sustainable option for extensive applications (Karnwal et al., 2024).

Furthermore, nanoparticles derived from plant extracts demonstrate superior biocompatibility and reduced toxicity compared to those produced through synthetic methods. As a result, they are exceptionally suitable for biomedical uses such as drug delivery, targeting, and imaging, as well as biosensing. For instance, quercetin-loaded nanoparticles sourced from plants have shown enhanced therapeutic efficacy with minimal adverse effects in clinical studies (Nadaf et al., 2022).

Moreover, the variety of phytochemicals present in herbs enables their modification with particular characteristics. This allows researchers to fabricate tailored nanoparticles with specific functionalities, such as directed delivery to malignant cells or neutralization of reactive oxygen species (ROS). The natural healing abilities of phytochemicals also amplify the efficacy of the nanoparticles by generating a synergistic effect that addresses multiple health issues simultaneously (Sindhu et al., 2024).

15.1.2 Methods of Synthesis of Herbal Nanoparticles

Nanotechnology has advanced considerably and has made a significant impact on the healthcare sector, with herbal nanoparticles taking the spotlight due to their capability to identify harmful health-related molecules. These particles are created through physical, chemical, and biological techniques. Each of these three methods offers distinct benefits and drawbacks, making them suitable for specific applications.

15.1.2.1 Physical Methods of Synthesis

Physical methods are top-down techniques that involve breaking down bulk materials into nanoparticles through mechanical or thermal means. Some examples include ball milling, evaporation-condensation, and laser ablation. For example, ball milling mechanically grinds materials into nanoparticles via high-energy impacts, enhancing uniformity in particle sizes and maintaining the stability of phytochemicals in herbal components (Yadav et al., 2023). Similarly, laser ablation employs a high-energy laser to vaporize materials into nanoparticles, which is ideal for creating herb-metal hybrid nanoparticles with consistent size distribution. Evaporation-condensation involves heating materials until they vaporize and then condensing that vapor into nanoparticles, effectively working with herbal extracts that contain volatile compounds.

Although more effective at producing particles with uniform sizes, physical processes have their own shortcomings. They demand high energy requirements, high-quality equipment, and might lack incorporation of biological activity of herbal material unless combined with an additional method of synthesis (Mandava, 2017).

15.1.2.2 Chemical Methods of Synthesis

Chemical synthesis is achieved through the reduction of metal salts into nanoparticles by using chemical reducing agents together with herbal extracts in most cases. The approach takes advantage of the inherent phytochemicals' bioactivity in flavonoids, alkaloids, and phenolics as stabilizing and reducing agents. For instance, herbal extracts can reduce metal salts like silver nitrate and gold chloride to produce nanoparticles with controlled shapes and sizes (Shah et al., 2015).

The use of solvents such as ethanol and water enhances the interaction between herbal compounds and metal ions, improving nanoparticle stability and reducing toxicity. Additionally, phytochemicals like tannins and saponins serve as stabilizers, extending the longevity of the nanoparticles. For example, multifunctional nanoparticles derived from Siberian ginseng have shown promise in diagnostic and therapeutic applications (Abbai et al., 2016). However, while chemical methods provide precise control over the morphology of nanoparticles, the potential use of toxic chemicals raises concerns regarding safety and environmental impact.

15.1.2.3 Biological Methods of Synthesis

Biological methods, also known as green synthesis, utilize plant extracts, enzymes, or microorganisms as eco-friendly reducing and stabilizing agents. This approach aligns with green chemistry principles by eliminating hazardous chemicals and reducing environmental pollution. For instance, silver nanoparticles exhibiting antimicrobial properties have been synthesized using plants like Swertia paniculata, leveraging their bioactive metabolites such as flavonoids and terpenoids (Ahluwalia et al., 2018).

Microbial synthesis broadens this method further with microorganisms including fungi and bacteria reducing metal ions to yield nanoparticles. Coupled with herbal extracts, microbial synthesis gives rise to improved structural and functional properties in the nanoparticles (Singh et al., 2016). Another viable technique is an enzymatic process through the use of enzymes from plant and microorganisms to catalyze metal ion reduction. These processes are specific and biocompatible as well as enable nanoparticle functionalization for application through targeted detection of health-hazardous molecules. Table 15.1 provides a summary and details about the various synthesis methods of herbal nanoparticles.

15.1.2.4 Advantages of These Methods in Biosensing

Herbal nanoparticles synthesized through these processes have various benefits in identifying toxic molecules. Sensitivity is achieved through their bioactive constituents making them capable of identifying trace levels of target molecules. Biological and chemical approaches more so fit into green chemistry practices as they minimize the environmental footprint during nanoparticle synthesis. These nanoparticles

TABLE 15.1
Comparison of Herbal Nanoparticle Synthesis Methods

S. No	Synthesis Method	Advantages	Limitations	Applications	References
1	**Physical Methods**	• Produces uniform nanoparticles with precise size and morphology • Eliminates the need for chemical reagents, reducing potential contamination	• Requires high energy input and expensive equipment • Does not incorporate bioactivity directly unless combined with biological agents	• Used for synthesizing hybrid nanoparticles for drug delivery • Suitable for biomedical imaging and catalysis	(Mishra et al., 2022; Segneanu et al., 2022)
2	**Chemical Methods**	• Allows for precise control over nanoparticle size and shape • Scalable for industrial production • Versatile for multiple types of applications	• Potential toxicity of reducing and stabilizing agents • Raises environmental safety concerns due to chemical waste	• Development of functionalized nanomaterials for diagnostics • Widely used in drug delivery and biosensors	(Nadaf et al., 2022; Sindhu et al., 2024)
3	**Biological Methods**	• Eco-friendly synthesis aligns with green chemistry principles • Produces biocompatible nanoparticles with intrinsic therapeutic properties • Sustainable and cost-effective for small-scale applications	• Variation in plant extract composition may lead to inconsistency • Scalability remains a challenge for large-scale production	• Effective for biosensing applications • Green synthesis of nanoparticles for environmental remediation • Biomedical applications like drug delivery	(Bachheti & Bachheti, 2024; Karnwal et al., 2024)

may be designed to be multifunctional and even incorporate sequential detection and elimination processes of toxic molecules (Abbai et al., 2016).

Through these synthesis processes, herbal nanotechnology is making advances in diagnostics to provide sustainable, biocompatible, and very effective solutions for health-hazards detection.

15.1.3 Mechanism of Detection Using Herbal Nanoparticles

Herbal nanoparticles lead cutting-edge detection technologies because they exhibit environmentally friendly synthesis, biocompatibility, and physicochemical properties

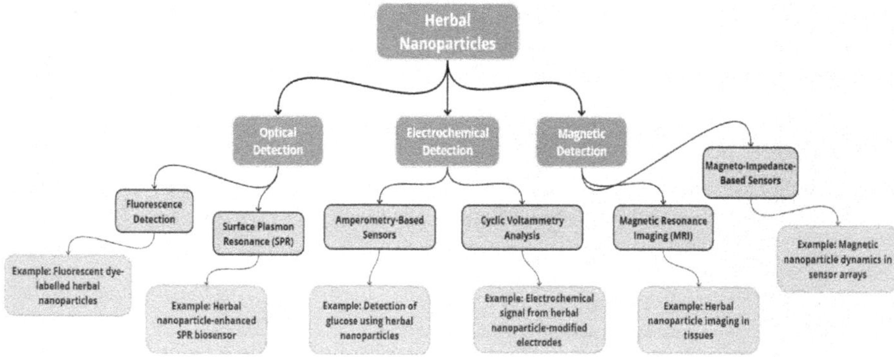

FIGURE 15.1 Mechanism of detection using herbal nanoparticles.

that set them apart. Herbal nanoparticles have facilitated groundbreaking progress in recognizing health-hazardous molecules with mechanisms like optical, electrochemical, and magnetic approaches. These mechanisms leverage particular properties tied to the use of nanoparticles and provide sensitivity, selectivity, and versatility for applications in various fields as shown in Figure 15.1.

15.1.3.1 Optical Detection

Optical detection techniques employ the interaction of light with nanoparticles to determine target molecules. One of the most outstanding mechanisms is based on localized surface plasmon resonance (LSPR), a phenomenon found in metal nanoparticles including gold and silver. When herbal nanoparticles interact with target analytes, they lead to a change in LSPR signals that is observable in real-time. Such a principle is used commonly in identifying hazardous molecules in biomedical and environmental applications (Zhang et al., 2022).

Fluorescence detection is another commonly used optical method. Nanoparticles made with plant extracts like Barbated Skullcap herb have been reported to show quenching in their fluorescence after they interact with certain molecules. Such quenching allows for sensitive detection of biomolecules and is an important tool in diagnostic fields (Wang et al., 2009).

In addition, colorimetric detection technologies have been on the increase because they are easy to use and visually interpretable. For example, herbal-derived Ce(IV) coordination polymer particles allow for hazard detection through visible color shifts. Such particles exhibit enzyme-like behavior by catalyzing reactions that yield specific colorimetric signals (Xu et al., 2024). Optical detection technologies are greatly preferred because they have a quick response time and are easy to use while they offer quantifying information.

15.1.3.2 Electrochemical Detection

Electrochemical detection technologies capitalize on the ability of nanoparticles to facilitate electron transfer and enhance electrochemical signals for heightened sensitivity in detection. Nanoparticles based on herbal sources play a vital role in modifying electrodes to improve the detection of harmful substances. Electrode modification enables the nanoparticles to interact with analytes, resulting in detectable changes in electrical current and potential.

Green-synthesized nanomaterials, including ZnO nanoparticles made from herbal extracts, have been employed to electrochemically detect biomolecules like silymarin. These nanomaterials enhance electron transfer rates, thereby increasing accuracy and sensitivity in detection (Sharma et al., 2018). Non-enzymatic electrochemical detection using green-synthesized Fe_2O_3-CuO nanocomposites is also utilized for identifying pollutants in industrial settings. Such methods are particularly advantageous in complex environments, such as industrial effluents, due to their specificity and robustness (Puthalapattu et al., 2023).

Electrochemistry allows for high sensitivity and detection at very low analyte concentrations, making it applicable in biomedical diagnostics and environmental monitoring. The integration of herbal nanoparticles into electrochemical sensors enhances their performance while adhering to eco-friendly production methods.

15.1.3.3 Magnetic Detection

Techniques that utilize magnetic detection leverage the magnetic characteristics of iron oxide-derived nanoparticles. These nanoparticles, based on iron oxide, are often modified with plant extracts to enhance their biosensing capabilities. Magnetic nanoparticles have extensive uses in areas such as biomedical imaging, drug delivery, and detection of harmful substances.

One of the primary uses of magnetic nanoparticles lies in the detection of analytes by observing alterations in their magnetic properties. This behavior is measured using advanced magnetometers capable of detecting minute quantities of hazardous substances (Qian & Kai, 2020). Additionally, magnetic molecularly imprinted nanoparticles (MMINPs) represent a novel tool for indirect electrochemical detection. Constructed from Fe_3O_4/SiO_2 cores, these particles exhibit a strong affinity for harmful molecules like matrine, an alkaloid derived from plants with toxic effects (Wei et al., 2024).

Incorporating magnetic features into plant-based nanomaterials enhances their utility in the separation and purification processes necessary for detecting hazardous substances within complex mixtures. Magnetic detection techniques are particularly valued for their ability to conduct rapid and efficient assessments across various fields.

Nanoparticles sourced from plants offer numerous distinct advantages in the detection of harmful molecules. Their synthesis aligns with green chemistry practices and minimizes reliance on toxic chemicals, thus reducing environmental impact. Furthermore, their biocompatibility makes them ideal candidates for biomedical applications. The multifunctional attributes of herbal nanoparticles enable the incorporation of optical, electrochemical, and magnetic qualities, facilitating the creation of versatile sensing platforms. These characteristics position herbal nanoparticles as a cornerstone for the advancement of contemporary detection technologies.

15.1.4 Application of Herbal Nanoparticles in the Detection of Hazardous Health Molecules

Sustainable and environmentally friendly herbal nanoparticles, known for their exceptional sensitivity, specificity, and biocompatibility, are increasingly being utilized to detect molecules that pose health risks due to their advantageous properties.

They are applicable in the detection of pathogens, toxins, pollutants, heavy metals, pesticides, and herbicides, thereby offering valuable solutions for environmental and public health safety.

15.1.4.1 Detection of Pathogens and Toxins

The use of herbal nanoparticles in identifying pathogens and toxins is transforming diagnostic techniques through swift and accurate detection. For instance, nanoparticles derived from medicinal plants have demonstrated remarkable effectiveness in recognizing microbial pathogens and fungal toxins. The molecular components of the pathogens interact with the herbal nanoparticles, producing measurable optical or electrochemical signals (Bahrulolum et al., 2021).

For example, zinc oxide nanoparticles sourced from herbs have been employed to detect rice pathogens, providing a more environmentally friendly alternative to traditional diagnostic methods (Ali et al., 2024). Furthermore, biosensors based on herbal nanoparticles can be utilized to detect mycotoxins in food products, thereby contributing to food safety (Balkrishna et al., 2024).

15.1.4.2 Detection of Pollutants

Herbal nanoparticles hold significant importance in detecting environmental pollutants such as pesticides, herbicides, and various chemical residues. Nanomaterials synthesized through green methods, like biogenic quantum dots and metal oxides, exhibit exceptional catalytic properties that enable the identification of trace pollutants in both water and soil (Rani et al., 2024).

Furthermore, biosensors based on herbal nanoparticles have been developed to detect water contaminants, such as halogenated hydrocarbons and other persistent organic pollutants. These sensors offer rapid detection and are vital for monitoring environmental health (Teschke & Xuan, 2022). The incorporation of nanoparticles derived from medicinal plant extracts has greatly improved the accuracy and effectiveness of pollutant detection systems.

15.1.4.3 Detection of Heavy Metals

Contamination from heavy metals poses a serious risk to both environmental and human health. Nanoparticles derived from herbs provide an eco-friendly method for detecting and capturing harmful metals, including lead, mercury, cadmium, and arsenic. Iron oxide nanoparticles made from herbal extracts have demonstrated high sensitivity in detecting heavy metals in samples of water and soil (Asiminicesei et al., 2024). Typically, these nanoparticles function as elements within electrochemical or optical sensors, where their interaction with heavy metals produces measurable signals. They also offer the added advantage of biocompatibility, reducing the likelihood of secondary environmental damage (Bahrulolum et al., 2021). Moreover, their ability to selectively bind to metal ions makes them a valuable resource for environmental remediation and monitoring.

15.1.4.4 Detection of Pesticides and Herbicides

Pesticides and herbicides, essential for agricultural productivity, also pose significant health risks as residues found in food and water. Herbal nanomaterials have proven to be highly effective in detecting these contaminating chemicals. Biogenic nanomaterials derived from herbal extracts have been shown to interact with pesticide molecules, leading to measurable changes in optical and electrochemical properties (Rad et al., 2022).

For example, biosensors based on herbal nanoparticles have been employed to monitor organophosphates and glyphosate in agricultural runoff, facilitating early intervention and reducing exposure risks (Alengebawy et al., 2021). Furthermore, the application of green-synthesized nanoparticles has been shown to improve the efficiency and cost-effectiveness of pesticide detection systems, as illustrated in Table 15.2.

15.1.5 Challenges and Future Directions in Herbal Nanotechnology

Herbal nanotechnology is a developing field that merges the advantages of natural products with nanotechnology. Despite its significant advancements, several challenges hinder its broader implementation. Tackling these challenges and planning for the future necessitates a balanced strategy to ensure effective integration into healthcare, agriculture, and environmental sectors.

15.1.5.1 Technical Challenges

Reproducibility and scalability in the synthesis of nanoparticles represent significant challenges for herbal nanotechnology. The inherent variations in the chemical makeup of plant extracts—stemming from differences in cultivation, harvesting, and extraction processes—complicate the standardization of nanoparticle production (Elkelish, 2024). Additionally, achieving stable and uniform nanoparticles with desired biochemical traits is technically demanding, particularly for applications that require precise particle size and surface characteristics (Kumar, 2023).

Another critical technical challenge lies in optimizing nanoparticle functionalization to improve treatment specificity and efficacy. While phytochemicals are inherently bioactive, their integration with advanced delivery systems or multifunctional nanocarriers remains a complex task (Ahmad et al., 2022).

15.1.5.2 Environmental and Safety Concerns

The increasing application of nanotechnology raises important environmental and health issues. The potential hazards posed by nanoparticles, especially those containing metal cores, lead to concerns regarding their effects on both humans and ecosystems. Evaluating the long-term repercussions of herbal-based nanoparticles is complicated by their diverse interactions within biological systems (Gottardo et al., 2021).

In addition, the regulatory frameworks surrounding herbal nanotechnology are still developing. The lack of uniform safety evaluations and toxicity testing standards obstructs the commercialization of herbal nanoparticles, particularly in the food, pharmaceutical, and cosmetic industries (Halamoda-Kenzaoui et al., 2022).

15.1.5.3 Innovation in Herbal Nanotechnology

To address these challenges, innovative strategies are essential. Green synthesis techniques, including biogenic and enzymatic methods, present more sustainable

TABLE 15.2
Uses of Herbal Nanoparticles in Detection

S. No	Health-Hazardous Molecule	Detection Mechanism	Example of Herbal Nanoparticles	Applications	References
1	**Pathogens**	**Optical detection**	Gold nanoparticles synthesized from *Azadirachta indica*	Gold nanoparticles exhibit strong localized surface plasmon resonance (LSPR), which can be used to detect changes in optical signals upon interaction with microbial pathogens. This method is rapid and highly sensitive, making it suitable for clinical diagnostics and disease surveillance.	(Nadaf et al., 2022; Bahrulolum et al., 2021)
2	**Toxins**	**Electro-chemical detection**	Silver nanoparticles derived from *Camellia sinensis*	Silver nanoparticles enhance electron transfer rates in biosensors, making them effective for detecting mycotoxins such as aflatoxins in food and agricultural products. This improves food safety and helps in monitoring toxin levels.	(Wassihun, 2024; Segneanu et al., 2022)
3	**Pollutants**	**Magnetic detection**	Magnetic iron oxide nanoparticles functionalized with neem	Magnetic nanoparticles are used in conjunction with magnetometers to detect and isolate pollutants such as hydrocarbons and persistent organic pollutants (POPs) in water and soil. This method is particularly valuable in environmental cleanup efforts and pollution monitoring.	(Asiminicesei et al., 2024; Rani et al., 2024)
4	**Heavy Metals**	**Electro-chemical detection**	Zinc oxide nanoparticles synthesized using herbal extracts	Electrochemical sensors leveraging zinc oxide nanoparticles provide high sensitivity for detecting heavy metals such as lead, mercury, and cadmium. These sensors are critical for monitoring water and soil quality and ensuring compliance with environmental regulations.	(Sharma et al., 2018; Asiminicesei et al., 2024)
5	**Pesticides and Herbi-cides**	**Optical and electro-chemical detection**	Quantum dots derived from herbal extracts	Quantum dots offer fluorescence-based detection of pesticides and herbicides like glyphosate and organophosphates. Electrochemical methods complement this by providing quantitative detection in agricultural runoff and food samples, aiding in ensuring food safety and environmental health.	(Rad et al., 2022; Balkrishna et al., 2024)

alternatives for producing nanoparticles (Dehkharghani & Ghahremanlou, 2023). These environmentally friendly practices not only reduce ecological damage but also enhance the biocompatibility and therapeutic efficacy of the nanoparticles.

Nanotechnology also facilitates the creation of novel delivery systems that utilize the bioactivity of herbal nanoparticles. Advanced drug delivery mechanisms, such as stimuli-responsive nanocarriers and hybrid nanoparticles, offer tailored treatment options with heightened bioavailability (Ahmad et al., 2022).

15.1.5.4 Integration with Other Technologies

The future of herbal nanotechnology is intertwined with its convergence with other technological fields. For example, merging herbal nanoparticles with IoT-enabled biosensors could revolutionize real-time monitoring and diagnostics within healthcare and agriculture (Acharya & Pal, 2020). Moreover, developments in artificial intelligence (AI) and machine learning (ML) are projected to enhance the design and optimization of herbal nanoparticles. By examining extensive data sets, AI can foresee interactions between herbal extracts and nanoparticles, resulting in more effective and targeted formulations (Wang et al., 2023).

Another promising avenue is the integration of herbal nanotechnology with agricultural nanotechnologies, such as nanofertilizers and nanosensors. These innovations can boost crop yields, enhance soil health, and lessen the environmental impact of agricultural practices (Pandey, 2018).

15.2 CONCLUSION

Herbal medicine utilizing nanoparticle technology is paving the way for a new chapter in nanomedicine. This review emphasizes the potential of herbal nanotechnology to create eco-friendly and biocompatible nanoparticles capable of detecting harmful agents such as pathogens, heavy metals, pesticides, and pollutants. The increased sensitivity, specificity, and therapeutic capabilities of these nanoparticles render them valuable for both medical and environmental applications.

Despite the significant potential in this field, it still confronts notable technical, environmental, and regulatory hurdles. Ongoing advancements in green synthesis and the integration of AI and IoT will play a crucial role in the evolution of this technology. Closing regulatory gaps and promoting interdisciplinary collaboration will enable herbal nanotechnology to achieve its full promise—providing safe and sustainable solutions for public health and environmental well-being.

REFERENCES

Abbai, R., R. Mathiyalagan, and J. Markus. "Green Synthesis of Multifunctional Silver and Gold Nanoparticles from the Oriental Herbal Adaptogen: Siberian Ginseng." *International Journal of Nanomedicine* 11 (2016): 2137–2145. Accessed December 30, 2024. https://www.tandfonline.com/doi/pdf/10.2147/IJN.S108549.

Acharya, A., and P. K. Pal. "Agriculture Nanotechnology: Translating Research Outcome to Field Applications by Influencing Environmental Sustainability." *Journal of Cleaner Production* 2020. Accessed December 30, 2024. https://www.sciencedirect.com/science/article/pii/S2452074820300264.

Ahluwalia, V., S. Elumalai, V. Kumar, and S. Kumar. "Nano Silver Particle Synthesis Using Swertia Paniculata Herbal Extract and Its Antimicrobial Activity." *Materials Science and Engineering: C* 33, no. 8 (2018): 1461–1467. Accessed December 30, 2024. https://www.sciencedirect.com/science/article/pii/S088240101731313X.

Ahmad, A., M. Imran, and N. Sharma. "Precision Nanotoxicology in Drug Development: Current Trends and Challenges in Safety and Toxicity Implications of Customized Multifunctional Nanocarriers for Drug Delivery." *Pharmaceutics* 14, no. 11 (2022): 2463. Accessed December 30, 2024. https://www.mdpi.com/1999-4923/14/11/2463/pdf.

Ai, S., Li, Y., Zheng, H., Zhang, M., Tao, J., and Liu, W. "Collision of Herbal Medicine and Nanotechnology: A Bibliometric Analysis of Herbal Nanoparticles from 2004 to 2023." *SpringerLink*, 2024. Accessed December 30, 2024. https://link.springer.com/.

Alengebawy, A., S. T. Abdelkhalek, S. R. Qureshi, and M. Q. Wang. "Heavy Metals and Pesticides Toxicity in Agricultural Soil and Plants: Ecological Risks and Human Health Implications." *Toxics* 9, no. 3 (2021): 42. Accessed December 30, 2024. https://www.mdpi.com/2305-6304/9/3/42/pdf.

Ali, M., A. Hussain, H. Manghwar, and A. Basit. "Pesticide, Herbicide and Nontechnology Application." In *Nanotechnology in Agriculture*. Springer, 2024. Accessed December 30, 2024. https://link.springer.com/chapter/10.1007/978-3-031-76000-6_7.

Asiminicesei, D. M., D. I. Fertu, and M. Gavrilescu. "Impact of Heavy Metal Pollution in the Environment on the Metabolic Profile of Medicinal Plants and Their Therapeutic Potential." *Plants* 13, no. 6 (2024): 913. Accessed December 30, 2024. https://www.mdpi.com/2223-7747/13/6/913/pdf.

Bachheti, R. K., and A. Bachheti. *Secondary Metabolites from Medicinal Plants: Nanoparticles Synthesis and Their Applications*. Accessed December 30, 2024. https://books.google.com/books?hl=en&lr=&id=oeC3EAAAQBAJ.

Bahrulolum, H., S. Nooraei, and N. Javanshir. "Green Synthesis of Metal Nanoparticles Using Microorganisms and Their Application in the Agrifood Sector." *Journal of Nanobiotechnology* 19 (2021): 110. Accessed December 30, 2024. https://link.springer.com/article/10.1186/s12951-021-00834-3.

Balkrishna, A., A. Kumari, A. Kumar, and V. Arya. "Biosensors for Detection of Pesticide Residue, Mycotoxins and Heavy Metals in Fruits and Vegetables: A Concise Review." *Microbiological Research* 2024. Accessed December 30, 2024. https://www.sciencedirect.com/science/article/pii/S0026265X24014048.

Chauhan, R. P. S., C. Gupta, and D. Prakash. "Methodological Advancements in Green Nanotechnology and Their Applications in Biological Synthesis of Herbal Nanoparticles." *ResearchGate*, 2012. Accessed December 30, 2024. https://www.researchgate.net/publication/233884927.

Chopra, H., Mohanta, Y. K., Rauta, P. R., Ahmed, R., Mahanta, S., Mishra, P. K., Panda, P., Rabaan, A. A., Alshehri, A. A., Othman, B., Alshahrani, M. A., Alqahtani, A. S., Al Basha, B. A., and K. Dhama. "An Insight into Advances in Developing Nanotechnology Based Therapeutics, Drug Delivery, Diagnostics and Vaccines: Multidimensional Applications in Tuberculosis Disease Management." *Pharmaceuticals (Basel, Switzerland)* 16, no. 4 (2023): 581. https://doi.org/10.3390/ph16040581.

Dehkharghani, F. M., and M. Ghahremanlou. "Future Energy and Therapeutic Perspectives of Green Nanotechnology: Recent Advances and Challenges." *Nanomedicine and Biotherapeutic Journal* 2023. Accessed December 30, 2024. https://www.nmb-journal.com/article_167151.html.

Elkelish, A. "New Plant Extracts toward Multidrug Resistance: The Convergence of Nanotechnology and Nanoscience." *SAS Journal of Medicine* 2024. Accessed December 30, 2024. https://sasj.journals.ekb.eg/article_396119.html.

Gottardo, S., A. Mech, J. Drbohlavová, and A. Małyska. "Towards Safe and Sustainable Innovation in Nanotechnology: State-of-Play for Smart Nanomaterials." *Nanotechnology*

2021. Accessed December 30, 2024. https://www.sciencedirect.com/science/article/pii/S2452074821000069.

Halamoda-Kenzaoui, B., R. Geertsma, and J. Pouw. "Future Perspectives for Advancing Regulatory Science of Nanotechnology-Enabled Health Products." *European Journal of Nanomedicine* 2022. Accessed December 30, 2024. https://link.springer.com/article/10.1007/s13346-022-01165-y.

Karnwal, A., A. Y. Jassim, A. A. Mohammed, and V. Sharma. "Nanotechnology for Healthcare: Plant-Derived Nanoparticles in Disease Treatment and Regenerative Medicine." *Pharmaceuticals* 17, no. 12 (2024): 1711. Accessed December 30, 2024. https://www.mdpi.com/1424-8247/17/12/1711.

Kumar, R. "Nanotechnology in Herbal Medicine: Challenges and Future Perspectives." *Journal of Herbal Science* 2023. Accessed December 30, 2024. https://www.sciencedirect.com/science/article/pii/B9780323995276000082.

Mandava, K. "Biological and Non-Biological Synthesis of Metallic Nanoparticles: Scope for Current Pharmaceutical Research." *ResearchGate*, 2017. Accessed December 30, 2024. https://www.researchgate.net/publication/320170333.

Manna, Indrani, Debalina Das, and Sejuty Mondal. "Use of Nanoparticles in the Delivery of Plant-Based Therapeutics." In *Advanced Nanomaterials for Drug Delivery*, edited by Maumita Bandyopadhyay, 55–78. Boca Raton, FL: CRC Press, 2023. Accessed December 30, 2024. https://www.taylorfrancis.com/chapters/edit/10.1201/9781003231721-3.

Mishra, Y., Amin, H. I. M., Mishra, V., and M. Vyas. "Application of Nanotechnology to Herbal Antioxidants as Improved Phytomedicine: An Expanding Horizon." *ScienceDirect*, 2022. Accessed December 30, 2024. https://www.sciencedirect.com/.

Na, Y., Woo, J., Choi, W. I., and D. Sung. "Novel Carboxylated Ferrocene Polymer Nanocapsule with High Reactive Oxygen Species Sensitivity and On-demand Drug Release for Effective Cancer Therapy." *Colloids and Surfaces. B, Biointerfaces* 200 (2021): 111566. https://doi.org/10.1016/j.colsurfb.2021.111566.

Nadaf, S. J., N. R. Jadhav, H. S. Naikwadi, and P. L. Savekar. "Green Synthesis of Gold and Silver Nanoparticles: Updates on Research, Patents, and Future Prospects." *Advances in Colloid and Interface Science* 15 (2022): 115–134. Accessed December 30, 2024. https://www.sciencedirect.com/science/article/pii/S235295202200038X.

Namdari, M., A. Eatemadi, and M. Soleimaninejad. "A Brief Review on the Application of Nanoparticle Enclosed Herbal Medicine for the Treatment of Infective Endocarditis." *ScienceDirect*, 2017. Accessed December 30, 2024. https://www.sciencedirect.com/.

Panda, P., and R. Mohapatra. "Herbal Nanoparticles: A Targeted Approach for Neurodegenerative Disorder Treatment." *Taylor & Francis*, 2024. Accessed December 30, 2024. https://www.taylorandfrancis.com/.

Pandey, G. "Challenges and Future Prospects of Agri-Nanotechnology for Sustainable Agriculture in India." *Journal of Cleaner Production* 2018. Accessed December 30, 2024. https://www.sciencedirect.com/science/article/pii/S2352186418301196.

Puthalapattu, R. P., Punyasamudram, S., Bathinapatla, A., Venkata, N. K. P., and Kanchi, S. "Electrochemical Non-Enzymatic Strategy with Green Synthesized Fe2O3CuO Nanocomposite for Detection of Amiprofos-Methyl Herbicide in Industrial Effluents and Soils." *Chemical Physics Impact*, 6 (2023): 100195. https://doi.org/10.1016/j.chphi.2023.100195

Qian, J., and G. Kai. "Application of Micro/Nanomaterials in Adsorption and Sensing of Active Ingredients in Traditional Chinese Medicine." *Journal of Pharmaceutical and Biomedical Analysis* 2020. Accessed December 30, 2024. https://www.sciencedirect.com/science/article/pii/S0731708520314345.

Rad, S. M., A. K. Ray, and S. Barghi. "Water Pollution and Agriculture Pesticide." *Clean Technologies* 4, no. 4 (2022): 615. Accessed December 30, 2024. https://www.mdpi.com/2571-8797/4/4/66/pdf.

Rani, M., A. Hosseini-Bandegharaei, and U. Shanker. "Pesticides Removal and Their Detection in Real Samples Using Green Synthesized Nanocomposites of Biogenic Quantum Dots and Metal Oxides: A Comprehensive Review." *Journal of Environmental Science* 2024. Accessed December 30, 2024. https://www.sciencedirect.com/science/article/pii/S0167732224019809.

Salek Maghsoudi, A., Hassani, S., Mirnia, K., and M. Abdollahi. "Recent Advances in Nanotechnology-Based Biosensors Development for Detection of Arsenic, Lead, Mercury, and Cadmium." *International Journal of Nanomedicine* 16 (2021): 803–832. https://doi.org/10.2147/IJN.S294417.

Setia, A., Vallamkonda, B., Challa, R. R., and A. K. Mehata. "Herbal Theranostics: Controlled, Targeted Delivery and Imaging of Herbal Molecules." *NCBI*, 2024. Accessed December 30, 2024. https://www.ncbi.nlm.nih.gov/.

Shah, M., D. Fawcett, S. Sharma, and S. K. Tripathy. "Green Synthesis of Metallic Nanoparticles via Biological Entities." *Materials* 8, no. 11 (2015): 7278–7308. Accessed December 30, 2024. https://www.mdpi.com/1996-1944/8/11/5377.

Sharma, D., M. I. Sabela, S. Kanchi, and K. Bisetty. "Green Synthesis, Characterization and Electrochemical Sensing of Silymarin by ZnO Nanoparticles: Experimental and DFT Studies." *Journal of Electroanalytical Chemistry* 2018. Accessed December 30, 2024. https://www.academia.edu/download/97449040/j.jelechem.2017.11.03920230118-1-1kdr4yo.pdf.

Sindhu, N., S. Kumar, and R. Kumar. "Nanointerventions for Detections of Viral Livestock Diseases." In *Nanotechnology in Wound Healing*, edited by C. S. John and K. A. Smith, 493–505. Singapore: Springer, 2024. Accessed December 30, 2024. https://link.springer.com/chapter/10.1007/978-981-16-1610-5_25.

Singh, P., Y. J. Kim, D. Zhang, and D. C. Yang. "Biological Synthesis of Nanoparticles from Plants and Microorganisms." *Trends in Biotechnology* 34, no. 7 (2016): 558–568. Accessed December 30, 2024. https://www.cell.com/trends/biotechnology/fulltext/S0167-7799(16)00040-8.

Socas-Rodríguez, B., and A. V. Herrera-Herrera. "Recent Applications of Magnetic Nanoparticles in Food Analysis." *Processes* 8, no. 9 (2020): 1140. Accessed December 30, 2024. https://www.mdpi.com/2227-9717/8/9/1140/pdf.

Teschke, R., and T. D. Xuan. "Heavy Metals, Halogenated Hydrocarbons, Phthalates, Glyphosate, Cordycepin, Alcohol, Drugs, and Herbs, Assessed for Liver Injury and Mechanistic Steps." *Frontiers in Bioscience* 27, no. 11 (2022): 314. Accessed December 30, 2024. https://www.imrpress.com/journal/FBL/27/11/10.31083/j.fbl2711314/htm.

Wang, H., Y. Chen, L. Wang, Q. Liu, and S. Yang. "Advancing Herbal Medicine: Enhancing Product Quality and Safety Through Robust Quality Control Practices." *Frontiers in Pharmacology* 2023. Accessed December 30, 2024. https://www.frontiersin.org/articles/10.3389/fphar.2023.1265178/full.

Wang, Y., X. He, K. Wang, X. Zhang, and W. Tan. "Barbated Skullcup Herb Extract-Mediated Biosynthesis of Gold Nanoparticles and Its Primary Application in Electrochemistry." *Colloids and Surfaces B: Biointerfaces* 2009. Accessed December 30, 2024. https://www.sciencedirect.com/science/article/pii/S0927776509001908.

Wassihun, Workayehu. "The Comparison of Silver Nanoparticle Optical Quality Synthesized Using Different Plant Extracts." Master's thesis, Addis Ababa University, 2024. Accessed December 30, 2024. http://ir.bdu.edu.et/handle/123456789/16168.

Wei, S., Y. Pan, J. Xu, J. Wang, and T. Li. "Magnetic Molecularly Imprinted Nanoparticles for Indirect Electrochemical Detection of Matrine." *Electrochimica Acta* 2024. Accessed December 30, 2024. https://www.sciencedirect.com/science/article/pii/S0013468624001944.

Xu, L., M. L. Luo, J. J. Dai, H. Zhu, P. Li, D. Wang, and F. Q. Yang. "Applications of Nanomaterials with Enzyme-Like Activity for the Detection of Phytochemicals and Hazardous

Substances in Plant Samples." *Chinese Medicine* 2024. Accessed December 30, 2024. https://link.springer.com/article/10.1186/s13020-024-01014-9.

Yadav, R. K., S. Jangeer, R. Parashar, and G. Sharma. "A Critical Review on Nanoparticles Synthesis: Physical, Chemical and Biological Perspectives." *ResearchGate* 2023. Accessed December 30, 2024. https://www.researchgate.net/publication/377443100.

Zhang, Y., D. Luo, S. K. Zhou, L. Yang, and W. F. Yao. "Analytical and Biomedical Applications of Nanomaterials in Chinese Herbal Medicines Research." *Journal of Analytical Chemistry* 2022. Accessed December 30, 2024. https://www.sciencedirect.com/science/article/pii/S016599362200173X.

Index

L

M

N

O

P

Q

R

S

T

For Product Safety Concerns and Information please contact our EU
representative GPSR@taylorandfrancis.com
Taylor & Francis Verlag GmbH, Kaufingerstraße 24, 80331 München, Germany

www.ingramcontent.com/pod-product-compliance
Ingram Content Group UK Ltd.
Pitfield, Milton Keynes, MK11 3LW, UK
UKHW022220190926
475702UK00003B/453
* 9 7 8 1 0 3 2 9 0 8 0 8 3 *